The Molecular Immunology of Neurological Diseases

The Molecular Immunology of Neurological Diseases

Edited by

SUNIL KUMAR, PHD
Associate Professor & Dean
Faculty of Biosciences
Institute of Biosciences and Technology
Shri Ramswaroop Memorial University
Barabanki, UP, India

ACADEMIC PRESS
An imprint of Elsevier

Academic Press is an imprint of Elsevier
125 London Wall, London EC2Y 5AS, United Kingdom
525 B Street, Suite 1650, San Diego, CA 92101, United States
50 Hampshire Street, 5th Floor, Cambridge, MA 02139, United States
The Boulevard, Langford Lane, Kidlington, Oxford OX5 1GB, United Kingdom

Notices

Knowledge and best practice in this field are constantly changing. As new research and experience broaden
our understanding, changes in research methods, professional practices, or medical treatment may become
necessary.

Practitioners and researchers must always rely on their own experience and knowledge in evaluating and
using any information, methods, compounds, or experiments described herein. In using such information or
methods they should be mindful of their own safety and the safety of others, including parties for whom
they have a professional responsibility.

To the fullest extent of the law, neither the Publisher nor the authors, contributors, or editors, assume
any liability for any injury and/or damage to persons or property as a matter of products liability, negligence or
otherwise, or from any use or operation of any methods, products, instructions, or ideas contained in
the material herein.

Library of Congress Cataloging-in-Publication Data
A catalog record for this book is available from the Library of Congress

British Library Cataloguing-in-Publication Data
A catalogue record for this book is available from the British Library

ISBN: 978-0-12-821974-4

For information on all Academic Press publications visit our website at
https://www.elsevier.com/books-and-journals

Publisher: Nikki Levy
Acquisitions Editor: Natalie Farra
Editorial Project Manager: Tracy I. Tufaga
Production Project Manager: Niranjan Bhaskaran
Cover Designer: Alan Studholme

Typeset by TNQ Technologies

Contents

Preface

Molecular Neuroimmunology comprises and integrates the fields of molecular neurology, molecular immunology, molecular virology, immunogenetics, and neuro-oncogenetics, each of which has seen considerable independent development in the past few decades. The common bond between them is the focus on the different immunological genes in neurological diseases. Although the advent of role of immunological genes has certainly taken the dread out of many neurological diseases, the threat of infection is still a fact of life: The roles of new viral pathogens are constantly being discovered.

The objective of this book of the molecular immunology of neurological diseases is to instill a broad-based knowledge of the etiologic organisms causing disease and the immunopathogenetic mechanisms, leading to clinically manifest infections causing neurological diseases into its users. This knowledge is a necessary prerequisite for the therapy and prevention of different neurological diseases. This book addresses primarily students of neuromedicine, immunology, and neurogenetics. Beyond this academic purpose, its usefulness extends to all medical professions and most particularly to physicians working in both clinical and private practice settings.

This book makes the vast and complex field of molecular neuroimmunogenetics more accessible by the use of four-color graphics and numerous illustrations with detailed explanatory legends. The many tables present knowledge in a cogent and useful form. Most chapters begin with a concise summary, and in-depth and supplementary knowledge are provided in boxes separating them from the main body of text.

This book has doubtlessly benefited from the extensive academic teaching and the profound research experience of its authors, all of whom are recognized authorities in their fields.

The editor would like to thank all colleagues whose contributions and guidance have been countless supports and who were so kind with design material. The editor is also grateful to the specialists at Elsevier and to the graphic design staff for their cooperation.

Dr. Sunil Kumar
Associate Professor and Dean
Faculty of Biosciences
Institute of Biosciences and Technology
Shri Ramswaroop Memorial University, Barabanki
(India)

Acknowledgments

There is a familiar aphorism that says that you never truly gain proficiency with a subject until you instruct it. We currently realize that you get familiar with a subject far and away superior when you expound on it. Setting up this book has furnished us with a magnificent chance to join our adoration for nervous system science and instructing and to impart our energy to understudies and specialists and clinical professionals all through the world. In any case, the undertaking has additionally been an overwhelming one on the grounds that such huge numbers of intriguing revelations have been made in the field of neuro-immunogenetics. The inquiry continually stood up to us: what immunogenetics information is most worth having? Addressing this inquiry required endeavoring to ace, however, much of the new material as could reasonably be expected and afterward choosing what to incorporate and, considerably harder, what to prohibit.

Be that as it may, we began from the scratch. We feel both blessed while writing this book. Lucky, in light of the fact that we had as our beginning stage the best molecular neuroimmunogenetics book at any point produced. To the degree that we have succeeded, we have done so as a result of the assistance of various contributors from around the globe.

Thanks go first and foremost to our authors of the different chanters of the book from all over the globe.

Not a word was written or an illustration constructed without the knowledge that bright, engaged readers would immediately detect vagueness or ambiguity. I especially thank the authors who have cheerfully contributed in making this book project successful.

I also thank our colleagues at Faculty of Biosciences, Institute of Biosciences and Technology, Shri Ramswaroop Memorial University, Barabanki (India) and Sanjay Gandhi Postgraduate Institute of Medical Sciences, Lucknow (India) who supported, advised, instructed me during this arduous task. I especially thank Prof. (Dr.) A. K Singh, The Vice Chancellor and Prof. (Dr.) Mukul Mishra, Director-Research and Consultancy of Shri Ramswaroop Memorial University for their continuous encouragement towards making this book mega successful.

I am also grateful to my colleagues throughout the world who served as reviewers for this book. Their thoughtful comments, suggestions, and encouragement have been of immense help to us in maintaining the excellence toward this book.

Dr. Sunil Kumar
Associate Professor and Dean
Faculty of Biosciences
Institute of Biosciences and Technology
Shri Ramswaroop Memorial University,
Barabanki (India)

List of Contributors

T.R. Anju, PhD
Assistant Professor
Department of Biotechnology
Newman College
Thodupuzha, Kerala, India

P.S. Baby Chakrapani, MSc, PhD
Director
Department of Biotechnology
Centre for Neuroscience
Cochin University of Science and Technology
Kochi, Kerala, India

Vijay R. Boggula, MSc, PhD
Research Associate
Department of Medical Genetics
Sanjay Gandhi Post Graduate Institute of
    Medical Sciences
Lucknow, Uttar Pradesh, India

Ayswaria Deepti, MSc, PhD
Research Associate
Department of Biotechnology
Centre for Neuroscience
Cochin University of Science and Technology
Kochi, Kerala, India

Chandrakanth Reddy Edamakanti, MSc, PhD
Research Assistant Professor
Davee Department of Neurology
Feinberg School of Medicine
Northwestern University
Chicago, IL, United States

Vivek Gaur, MSc
Junior Resident (VRDL)
Department of Microbiology
Baba Raghav Das Medical College
Gorakhpur, Uttar Pradesh, India

Devlina Ghosh, MSc, MBA
PhD Scholar
Amity Institute of Biotechnology
Amity University Uttar Pradesh
Lucknow Campus
Lucknow, Uttar Pradesh, India

Urmila Gupta, MSc
Microbiologist-Acute Encephalitis Cell
    (ICMR-New Delhi)
Department of Pediatrics
Baba Raghav Das Medical College
Gorakhpur, Uttar Pradesh, India

S. Jayanarayanan, PhD
Scientist
Athreya Research Foundation
Aluva, Kerala, India

Lakshmi Kesavan, MSc
Research Associate
Molecular Neurobiology Division
Rajiv Gandhi Centre for Biotechnology
Thiruvananthapuram, Kerala, India

Gayathri Krishna, MSc
Research Scholar
Virology Laboratory
Department of Biotechnology
Cochin University of Science and Technology
Kochi, Kerala, India

Alok Kumar, PhD
Associate Professor
Department of Molecular Medicine and Biotechnology
Sanjay Gandhi Postgraduate Institute of
    Medical Sciences
Lucknow, Uttar Pradesh, India

Vijay Kumar, PhD
Assistant Professor
Department of Biotechnology
Yeungnam University
Gyeongsan, Gyeongbuk, Republic of Korea

Anand Kumar Maurya, PhD
Assistant Professor
Department of Microbiology
All India Institute of Medical Sciences
Bhopal, Madhya Pradesh, India

Vishwa Mohan, BSc, MSc, PhD
Research Associate
Davee Department of Neurology
Feinberg School of Medicine
Northwestern University
Chicago, IL, United States

Mohind C. Mohan, MSc, PhD
Research Associate
Centre for Neuroscience
Department of Biotechnology
Cochin University of Science and Technology
Kochi, Kerala, India

Somnath Mukherjee, MSc, PhD
Research Associate
Bapu Nature Cure Hospital & Yogashram
New Delhi, India

Vinod Soman Pillai, MSc
Research Scholar
Virology Laboratory
Department of Biotechnology
Cochin University of Science and Technology
Kochi, Kerala, India

Divisha Rao, MSc
Post-graduate Fellow
Department of Molecular Medicine and Biotechnology
Sanjay Gandhi Postgraduate Institute of
  Medical Sciences
Lucknow, Uttar Pradesh, India

Vyom Sharma, MSc, PhD
Scientist
Charles River Laboratories
Skokie, IL, United States

Gajendra Singh, MS
PhD Fellow
Department of Molecular Medicine and Biotechnology
Sanjay Gandhi Postgraduate Institute of Medical
  Sciences
Lucknow, Uttar Pradesh, India

Aditi Singh, PhD
Associate Professor
Amity Institute of Biotechnology
Amity University Uttar Pradesh
Lucknow Campus
Lucknow, Uttar Pradesh, India

Amresh Kumar Singh, MD
Assistant Professor and Head
Department of Microbiology
Baba Raghav Das Medical College
Gorakhpur, Uttar Pradesh, India

Neeraj Sinha, PhD
Professor
Centre of Biomedical Research
SGPGIMS-Campus
Lucknow, Uttar Pradesh, India

Niraj Kumar Srivastava, MSc, PhD
Biochemistry Consultant
School of Life Sciences (SOS)
Jawaharlal Nehru University (IGNOU)
New Delhi, India

Kumari Swati, MSc
Research Associate
Department of Biotechnology
Yeungnam University
Gyeongsan, Gyeongbuk, Republic of Korea

Gyanesh M. Tripathi, MSc, PhD
Senior Scientific Officer
Department of Molecular Medicine
Vivekanand Polyclinic and Institute of Medical
  Sciences
Lucknow, Uttar Pradesh, India

Swati Tripathi, MSc
Research Fellow
Department of Microbiology
Integral University
Lucknow, Uttar Pradesh, India

Mohanan Valiya Veettil, PhD
Senior Principal Scientist (F)
Institute of Advanced Virology (IAV)
Bio 360 Life Sciences Park, Thonnakkal
Thiruvananthapuram, Kerala, India

Ramakant Yadav, MD, DM
Professor & Head
Department of Neurology
UP (Uttar Pradesh) University of Medical Sciences
Etawah, Uttar Pradesh, India

CHAPTER 1

Molecular Basis of Neurological Disorders

GAJENDRA SINGH, MS • DIVISHA RAO, MSC • ALOK KUMAR, PHD

1.1 INTRODUCTION

Neurology is the branch of medicine dealing with the nervous system. It is now discernible that the emergence of neurological disorders is a priority health problem worldwide, which has been reflected in the studies of Global Burden of Diseases-published by the World Health Organization (WHO) and various other groups (Menken et al., 2000). Alzheimer's disease (AD), Parkinson's disease (PD), dementia and epilepsy are among the most common neurological disorders and hit hundreds of millions of people globally. More than 47 million people get affected by dementia alone globally. Neurological disorders and mental health related issues, that depend on the extension of life expectancy and the aging factor of the general population is significantly increased in both developed and developing countries (Janca and Prilipko, 1997).

The global burden of illness patterns has also changed, due to the epidemiological transitions. Enhancement in the maternal and child health and newly cognized nervous system disorders are some of the factors that have contributed to the change in the illness pattern among people. Therefore, there is only 1.4% of death rate in neurologic and psychiatric disorders, however they account for 28% of all the years of disabled life (Menken et al., 2000). India and other developing countries are going through a phase called epidemiological transition with an expanded burden of noncommunicable diseases (NCDs) (Gourie-Devi, 2014).

It is observed, that the most critical concern is the lack of effective therapies in diseases of central nervous system which is devastating for an individual. The emergence of molecular genetics approach to mapping techniques and identification of diseased gene took place during the 1990s, which basically attributed to the foundation for an enormous advancement in our understanding of the pathogenicity of various neurological disorders. Several scientific studies in this area focussed on inherited disorders, which significantly affect health of human population. Given this criterion, apprehending the reason for neuron degeneration in different clinic-pathological entities has been an important aspect, with evident importance for the development of the therapy (La Spada and Ranum, 2010). A neurological gene map, describing the exact position and location of the chromosome for these diseases, such as AD, HD, Charcot−Marie−Tooth syndrome, neurofibromatosis, myotonic dystrophy, and Duchenne−Becker muscular dystrophy (DMD) are determined by linkage analysis (Rosenberg, 1993).

It is observed that changes in protein synthesis are immediately followed by functional changes in the brain. Their detection is one of the most important aspects for understanding the physiological regulation. Detection of protein levels has routinely been done by immunoprecipitation or two-dimensional gel analysis. The level of mRNA encoding a protein has been assessed and gene expressions analyzed by hybridization using radiolabeled DNA probes to the isolated mRNA or by translation of mRNA in vitro in a cell-free system. The whole-organ analysis helps to detect changes in a relatively rare mRNA, which probably is less likely to be present except for the organs which are predominantly composed of a cell type in which mRNA is present (Griffin and Morrison, 1985). Enormous amounts of communicational and computational capacities need to be performed to carry out diverse functions simultaneously. In the late 1950s, it became noticeable that both chemical and electrical signalling mechanisms are used to operate the nervous system, whereas according to the current understandings, the chemical (neurotransmitters) transmission through synapses is a key point in neuron to neuron signalling. Several amounts of various neurotransmitter candidates have been reported other than the "classical"

The Molecular Immunology of Neurological Diseases. https://doi.org/10.1016/B978-0-12-821974-4.00014-5

(amines, amino acids, acetylcholine [ACh]) ones. Radically unique types of transmitters have been discovered such as nitric oxide and protons (Francis, 2005). The disorders or substances alter the production, release, breakdown, or re-uptake of neurotransmitters, change their numbers or the affinity of receptors, which may cause psychiatric or neurologic symptoms.

Along with this, there are also various other external stimuli that lead to neurological disorders. For instance, acute and tenacious viral infection is initiated in the periphery, often starting at epithelial and endothelial surfaces of cell. A tissue-specific antiviral response is generated at the site of infection as an antiviral response including both autonomous response and paracrine signalling to surround uninfected cells and protect them from secreted cytokines. Though, at chronic phase, the infection gets cleared by the activity of infection-specific antibodies and T cells due to adaptive immune response. A poor evolutionary path is carried out by viruses to invade the host nervous system because of the possibly damaging and deadly nature of these infections. Zoonotic type of infection occurs in the CNS, which has no apparent advantage for the host or the pathogen. Zoonotic viral infections are frequently less pathogenic in their natural hosts but can be thoroughly virulent and microinvasive in their non-natural hosts. The lethality of the zoonotic virus infections may result from the elicitation of the cytokine storm after the primary infection took place. Apart from this CNS infections, there are some other human adaptive viruses that gain entry into the CNS, which might be the result of the declined defense mechanism of host that failed to restrict peripheral infections. In the following chapter, we will discuss molecular basis of neurological disorders including neuroviral infectious diseases and will provide overview on the use of advance tools in these disorders (Koyuncu et al., 2013).

1.2 MOLECULAR GENETICS

Over the past few years, neurology has seen a crucial transmogrification especially in the cognizance of neurological disorders where the inheritance is either recessive, autosomal dominant, or X-linked. Gene markers of neurological disorders have been identified now because of the formation of alliances between the Departments of Molecular Genetics and Neurology. Prime authorities have been selected to anatomize the present molecular basis of neurological disorders such as AD, HD, mitochondrial encephalomyopathies, prion disease, Charcot–Marie–Tooth syndrome,

neurofibromatosis, myotonic dystrophy, DMD, Gaucher disease, skeletal muscle sodium-channel diseases, and certain potential areas of treatments, as well as gene therapy. Therefore, for delineating the positional chromosome location for each and every disease, a neurological gene map has been included by linkage analysis of the disease (Rosenberg, 1993).

1.2.1 Genetic Neurological Disorders

Genetic neurological disorders are not common in individuals, but if gathered together, they attribute to a notable bundle of disability in most probably younger groups of age than they are affected by various other neurological diseases. Typically, in a neurological clinic, inherited neurological diseases are certainly not the most common conditions that are beheld. Resolving the problem of etiopathogenesis by the techniques used in molecular biology is not only scientifically enthralling but also providing traces of information to the basic mechanism that is involved in neurological disorders. The steps involved in it are quite straightforward: (1) localization of chromosome and its linkages; (2) its mapping; (3) identification of the mutated gene(s); (4) identifying the function of the gene; and (5) its wise therapy. Throughout the genome, there are fragments of DNA that are well recognized for their precise location on the chromosome (e.g., Duffy blood group, chromosome 1, and MHC on chromosome 6). In the autosomes, numerous other markers have also been recognized (Rosenberg, 1993).

1.2.2 Linkage Analysis

In the analysis of the family linkage, each member in the family is analysed to determine the presence of the diseases of interest and their position (RFLP; restriction fragment length polymorphisms). If the locus of the disease and the marker is close enough on the same chromosome (i.e., they are linked together), then at the time of meiosis, an independent assortment of both the genes will be rare and the offspring will get both the traits. In its analysis, the very first step involved is to determine the linkages between the recognized chromosome markers and the family that has inherited the disease. When the LOD score (LOD being the logarithm of probability ration of known marker cosegregating with the putative disease marker) is greater than 3 (or odds of 1000:1), such linkages are said to be found. After the chromosome has once been confined either to its short (p) or long (q) arm, further minute analysis of DNA can be undertaken in these segments. RFLP technique is then used to cleave DNA at known sites by restriction endonucleases. Here the linkage is carried out at a single chromosome level, as those RFLP-applied fragments of DNA proceed in the exact relation to each other in the chromosome exchange, and therefore, the chromosome "walks" in between the known marker points in the anticipation of finding a marker, which is closely related or linked to the family's inheritance. Thus, the RFLP approach of linkage analysis assists the development of flanked markers (recognized markers on both sides of abnormal genes). The majority of neurological disorders have now progressed to this extent. For eg; Charcot–Marie–Tooth disease, which is a hereditary motor sensory neurological disease, is agnized with the two chromosome markers, 1 (HMSN 1b) and 17 (HMSN 1a) (Cumming, 1992).

As was the case for DMD, isolation of the gene, responsible for the disorder, was done and then cloned and sequenced, and it was concluded that it encoded for 400-kDa protein, i.e., dystrophin. This strategy is referred to as "**reverse genetics**" (first, the gene was recognized and then was the gene product), which is the path that is to be followed with great pace to decode the molecular basis of neurological disorders. Expeditious progress is expected by this approach for identifying the gene products; furthermore, the gene products by the technique of reverse genetics have already been identified for certain diseases such as retinitis pigmentosa, retinoblastoma, chronic granulomatous disease, cystic fibrosis, neurofibromatosis-1, and DMD (MacMillan and Harper, 1994).

1.3 GENE EXPRESSION ANALYSIS FOR NEUROLOGICAL DISEASES

The changes in protein synthesis accompany the functional or the maturational changes in the brain. For the discernment of physiological regulations and how it is disrupted by a disease, it is important to detect such changes. The procedure of making messenger RNA that certainly will be translated into proteins, called the expression of genes, is one of the important aspects of particular interest. In a variety of tissues, comprising the brain tissue as well, the gene expression has been quantitated by hybridization of radiolabeled DNA probes to separate out mRNA, or it can also be quantitated by the in vitro translation of mRNA in a system that is cell-free (Griffin and Morrison, 1985). For biomedical research, the studies related to gene expression are an essential factor because first, the cell specification is dictated by the pattern of gene expression and, second, the complements of genes that are expressed by any cell are subject to change and to regulate

extensively. Genes expressed in the brain are unusually heterogeneous. Certain mRNAs are expressed in only certain brain regions, often within a determined neuron or glial cells (Carter et al., 2010). mRNAs being localized frequently to vertebrate's axons and for axon path-finding or branching their local translation are required in the course of their development and maintenance, repair, and neurodegeneration in their postdevelopmental period. Therefore, to ensure the recruitment of mRNAs and their function in a whole animal, the procedure that enables the mRNAs visualization in situ should all be ideally combined with the transcriptome analyses. The novel in situ hybridization (ISH) technologies have been developed which detect the RNAs at a single-molecule level, which is specifically beneficial for the analysis of subcellular localization of mRNA, since mRNAs that are localized are found at lower levels (Baleriola et al., 2015).

To scrutinize the RNA expression in a tissue, differs by short or single nucleotide sequences at a single-cell level, has been bordered by the specificity and susceptibility of the technique of ISH. It is a critical capability to distinguish the divergent expression of RNA variants in tissue because the splicing and editing of mRNA is altered, in addition to it coding single nucleotide polymorphism (SNP), have both been correlated with several neurological and psychiatric disorders (Erben et al., 2018).

1.3.1 In situ Hybridization

The brain, which is an anatomically complicated organ, has various functional roles accompanying location and connectivity (e.g., neural circuits); here the mRNA molecule is expressed, and we need to test and visualize as to "where" (i.e., in vivo cellular location) the mRNA molecule is expressed to know its major functional implications. Microarrays and next-generation sequencing are used to examine global gene expression by various researchers. Therefore, there is an augmented need of this wide genome expression data to complement with the experimental techniques addressing anatomical entanglements.

In Situ Hybridization (ISH) is the best method to achieve the desired result for analyzing genetic expression in brain, tissue sections, or cell culture samples, being a practical and a convenient tool developed for visualizing the structural expression pattern of mRNA. The procedure involves the hybridization of a tagged sequence of DNA/RNA to its complementary mRNA, i.e., ISH involves the nucleotide sequences being tagged or labeled with a molecule that has been detected (i.e., a probe) (Carter et al., 2010). A homologous, tagged sequence of DNA will then hybridize to its complementary sequences of mRNA to be detected in a consistent number under an appropriately strict conditions in the tissue or cell culture samples (Griffin and Morrison, 1985). Radiolabeled hybrids accommodating complementary mRNA can then be detected in a cell by using standard autoradiography techniques, i.e., the tag or a label allows or enables the subsequent detection and visualization of the precise location of the mRNA that is of our interest and RNase pretreatment, and nucleic acid opposite-strand sequences are used as a standard control for these methods (Carter et al., 2010).

Cultured neurons are the ones, on which ample work on mRNA recruitment and local translation in a mature axon has been performed. This basically is correct for transcriptome analyses, as the specialized culture methods are available allowing axon isolation from somatodendritic compartments (Baleriola et al., 2015). To recognize mRNA profiles in isolated axons, these high-throughput transcriptome analyses have turned out to be useful in *in-vitro* studies but have certain limitations for *in-vivo* studies, as axons are never found in isolation in a whole tissue but are found merged with the glial cells, neuronal cell bodies, and other types of cells. Therefore, for the confirmation of the subcellular localization of mRNAs, imaging techniques have to be combined with such analyses. Hence, for the identification and visualization of particular RNA sequences, in tissue and a cell, RNA ISH technique is used (Baleriola et al., 2015).

Conventionally, ISH that uses radioactively labeled probes and fluorescently labeled RNA complementary probes has imparted enough sensitivity to better analyze various transcripts at a cellular level, but to identify cells expressing unusual splice variants, it failed unambiguously. Recent improvements and evolutions in ISH, which uses various nonradio isotopic oligonucleotide pairs of probes that target a single transcript incorporated with chemical signal amplification helps codetection of infrequent transcripts "multiplexing" (Erben et al., 2018).

1.4 GENE REGULATION ANALYSIS IN NEUROLOGICAL DISEASES

In the neurology, neurodegenerative disorders are one of the leading health challenges in modern medical history because the molecular pathogenesis of these disorders is not yet clear. Some of the major factors that basically contributes to these neurodegenerative diseases include pathogenic mutations, amyloid precursor proteins, superoxide dismutase, and DNA- and

RNA-binding proteins. RNA metabolism is the rudimentary procedure by which RNA is generated, transported, regulated, stored, and translated. Recent studies have shown that mutations in RNA-binding proteins are a key cause of several human neuronal-based diseases. Measurement of RNA regulation levels in different brain regions and cell types would define the neuropathology at the molecular level (Shyu et al., 2008).

Lack of available animal models for most neurological diseases restricts the analysis of neuropathological human brain disorders. The ability to quantitate alterations in specific RNA level in the brain would better define the neuropathology with respect to the brain region and to specific transcriptional or posttranscriptional changes. Recent literature has revealed the role of RNA with specific emphasis on roles of RNA translation, microRNAs in neurodegeneration, RNA-mediated toxicity, repeat expansion diseases, RNA metabolism and molecular pathogenesis in neurological diseases (such as amyotrophic lateral sclerosis, frontotemporal dementia and spinal muscular atrophy (SMA) (Giuditta et al., 2012).

Messenger RNA (mRNA) is a single-stranded RNA molecule that is complementary to one of the DNA strands of a gene. mRNA is synthesized in the nucleus using the nucleotide sequence of DNA as a template. The mRNA is an RNA version of the gene that leaves the cell nucleus and moves to the cytoplasm where proteins are made. The mRNA directly involved in the synthesis of proteins occurs in the cytoplasm. During protein synthesis, an organelle called a ribosome moves along the mRNA, reads its base sequence, and uses the genetic code to translate each codon, into its corresponding amino acids (Thelen and Kye, 2020). According to recent progressive years, many studies revealed that many neurological illnesses arise because of the RNA-based genetic regulation: pre-mRNA splicing, mRNA translational regulation, and non-splicing RNA modifications (Faustino and Cooper, 2003).

1.4.1 Pre-mRNA splicing and human disease
Pre-mRNA splicing is an essential step in gene expression. It is mostly a cotranscriptional process during which the noncoding introns are excised from pre-mRNA molecules and the flanking exons are spliced together, resulting in translation-competent mature mRNA molecules (Patel and Steitz, 2003). Pre-mRNA is processed in the nucleus in conjunction with RNAs and proteins collectively called the spliceosome. In most metazoan organisms, pre-mRNA splicing is carried out by two separate spliceosomes that function in

parallel but specialize in different intron types. The first one is the major (U2-dependent) spliceosome that removes the bulk of the introns in addition to the nearly invariant GT-AG sequences and relatively divergent consensus sequences in 5 and 3' terminals. It is collectively approximately 99.5% of all introns that are called major or U2-type introns. On the other hand, the minor (U12-dependent) spliceosome that removes only a small subset of introns that contain conserved 5' splice sites and branch point sequences (BPS). These minor introns are found in approximately 700−800 genes in humans and represent around 0.5% of all human introns. Many studies suggest that the vast majority of human genes contain introns and that most pre-mRNAs undergo alternative splicing (Jutzi et al., 2018). Alternative splicing is the joining of different splice sites (5' and 3'), allowing individual genes to express multiple mRNAs that encode proteins with diverse and even antagonistic functions. Therefore, the disruption of normal splicing patterns can cause or modify human disease. Some diseases caused by a mutation in pre-mRNA splicing are discussed in the below sections.

1.4.2 Alzheimer's Disease

AD is caused by the abnormal buildup of proteins in and around the brain cells. One of the proteins involved is called amyloid, deposits of which form plaques around brain cells. The other protein is called tau protein. It is characterized by aggregation of the microtubule-associated protein tau into neuronal cytoplasmic inclusions that are also related to the cause of several other neuropathological conditions characterized by progressive dementia including Pick's disease, frontotemporal dementia, and Parkinsonism all linked to chromosome 17 (FTDP-17) (Medeiros et al., 2011). FTDP-17 is an autosomal dominant disorder caused by mutations in the *MAPT* gene that encodes tau. As we know that tau is required for microtubule assembly and its function, it plays a major role in microtubule-dependent transport in axons. So mutation in *MAPT gene* alters the biochemical properties of the protein that may not be bound to microtubules, which is proposed to be a subject to hyperphosphorylation and aggregation (Buée et al., 2000).

1.4.3 Spinal Muscular Atrophy

SMA is the most common motor neuron disease. The pathological hallmarks of the degeneration of motor neurons in the anterior horn of the spinal cord and brain stem are associated with muscle atrophy. The disease is caused by decreased levels of the SMN protein due to loss or mutation of the *SMN1* gene (Cooper et al., 2009).

1.4.4 mRNA Translation

Regulation of mRNA translation is essential for the survival and the function of neural cells. Reductions in translation initiation due to mutations in the translational machinery or improper activation of the incorporate stress response may contribute to pathogenesis in a subset of neurodegenerative disorders (Beckelman et al., 2016). Dysfunction of central components of the elongation machinery such as the tRNAs and their associated enzymes that cause translational regulation results in neurodegeneration. Eukaryotic initiation factor 2 (*eIF2*) is a eukaryotic initiation factor that is responsible for the translation initiation. eIF2 is an essential factor for protein synthesis (Bordone et al., 2019). Dysregulation of eIF2 is associated with many neurodegenerative diseases. For example, leukoencephalopathy is a disorder that is associated with the loss of astrocytes, oligodendrocytes, and axons in the CNS white matter. In this disorder, mutations in any subunit of eIF2B are responsible for the leukoencephalopathy (Vanbatalo and Soinila, 1998).

1.5 MOLECULAR BASIS OF CHEMICAL NEUROTRANSMISSION

The brain is a tremendously active organ with extreme demand for energy because of an excessive number of neurons present and so there is higher need to sustain a delicate balance among the metabolism of energy, neurotransmission, and plasticity. If there occurs a disturbance in the energetic balance or the control of the quality of mitochondria or the disturbance in the metabolic interaction of glia and neuron, this may further lead to malfunctioning of the brain circuit or even more critical disorders of the CNS (Lodish et al., 2000a,b).

According to current studies on neuron to neuron signalling, chemical transmission by means of synapses is far more important. The small organic molecules that are synthesized by the series of enzymatic reactions are said to be the "classical neurotransmitters" (amines, amino acids, and ACh); a number of criteria hold good for considering a substance, a neurotransmitter. However, from the past few decades, the substances that do not meet the criteria to be neurotransmitters have also been discovered to be neurotransmitters. This has exerted a lot of effects onto the neuronal signalling, varying from activation of impulse propagation to the precise transition of impulse traffic (Bachiller et al., 2018). The human brain utilizes almost 20% of oxygen and almost 25% of glucose, although it contributes to only 2% of the body weight. Therefore, the CNS utilizes 2%−8% energy of the basal metabolism. The total glucose consumption by the brain is the definite function of the number of neurons and so the huge amount of energy use in humans may be elucidated by its immense number of neurons (Lodish et al., 2000a,b).

At various chemical synapses, several small molecules manufactured in the cytosol of axon terminals work as neurotransmitters. The synaptic vesicles, which are uniform-sized organelles of 40–50 nm diameters, accumulate the "classic" neurotransmitters. The classic neurotransmitters are amino acids or their derivatives, with the exception of ACh. Various molecules serve as neurotransmitters, e.g., nucleotides such as ATP and the corresponding nucleosides. Generally, one type of classic neurotransmitters is only generated by each neuron. Then after the neurotransmitters exocytose to the synaptic cleft from the synaptic vesicles, on the plasma membrane of postsynaptic cells, the neurotransmitters adhere to specific receptors, inducing a difference in the ion's permeability (Kandimalla and Reddy, 2017).

Microglia are basically the cells that originate from the yolk sac, and during embryogenesis migrate to the CNS. Microglia serve as a supportive role as well as protective function as an immune cell within the CNS. Being determined to the anatomical region, microglia of the entire cell population of the human brain account for 0.5%–16.6%, being 5%–12% in the mouse brain. Microglial regulation of phenotype is based on their interaction with the molecules that the surrounding cells release such as neurons, astrocytes, microglial cells, and so on, which are recognized by membrane-bound pattern recognition receptors (PRRs). Conventionally, classic (M1) and alternative (M2) are the two different and opposite states in which macrophages and microglial activations have been classified. The proinflammatory state is considered by M1 phenotype; here, cytokines such as tumor necrosis factor-α (TNF-α), interleukin 1β (IL-1β), interleukin 12 (IL-12), ROS, and nitrogen reactive species (NRS) are produced and released by microglial cells. On the contrary, M2 microglia phenotype is considered to be an anti-inflammatory state, comprehending the release and production of trophic factors such as tumor growth factor-β (TGF-β) and brain-derived neurotrophic factor (BDNF). But currently, this classification is unresponsive to diverse microglial phenotypes found in brain. Rather, a vast assemblage of activation profile and phenotypes has newly been described, particularly in correspondence with neurodegenerative diseases (Francis, 2005).

In the course of human brain development, aging accompanies a decline in energy metabolism. The progression in metabolic deficiency is suggested by the decline in energy metabolism, ensuing in age-associated cognitive reduction and general disturbances in brain functioning. Aging is reviewed as a chief risk factor for several neurodegenerative diseases and chronic altered inflammation incorporating modification in microglial morphology, phenotype, and activity. In consideration of diminished energy metabolism followed by aging, neurodegeneration is the distinguishing factor. Calling attention to aging is a destabilizing factor for "healthy" brain energetics, which makes it more vulnerable to neurodegenerative diseases (Lodish et al., 2000a,b).

1.5.1 Alzheimer's Disease

AD is the most frequent form of dementia, caused by neurodegeneration; its pathology and underlying roots are not clearly understood (Francis, 2005). AD pathogenicity has got two hypotheses. On one hand, the amyloid hypothesis suggests that the brain atrophy followed by nerve cell death is caused by the occupancy of extracellular amyloid-beta (Aβ) plaques and tangles of intracellular neurofibrillary. On the contrary, on the other side, the biochemical, histological, and clinical features of sporadic AD are unified by the mitochondrial cascade hypothesis. AD is linked to the deposition of Aβ plaques extracellularly, because of the increased production of and/or lack of removal of peptides of Aβ procured from amyloid precursor protein (APP) cleavage and by intraneuronal aggregation of hyperphosphorylated tau protein abnormally. This is associated with cognitive impairments and neurodegenerative disorders (Lodish et al., 2000a,b).

The endogenous chemicals that enable neurotransmission are the neurotransmitters. The signals are transmitted across the synapse and neuromuscular junctions through neurotransmitters. Neurotransmitters are generally stored in the synaptic vesicles and are usually liberated with an appropriate signal into the synapse. The neurotransmitters are synthesized endogenously from amino acids. ACh is synthesized from serine; dopamine is synthesized from L-phenyl alanine/L-tyrosine; GABA is synthesized from decarboxylation; histaminergic is synthesized from L-histidine; serotonin is synthesized from L-tryptophan; and NMDA is synthesized from aspartic acid and arginine. These neurotransmitters are significant and play a crucial role in AD pathogenesis (except dopamine). The neurotransmitters that are released get diffused on to the synaptic cleft and on the postsynaptic neuronal membrane; neurotransmitters bind to an appropriate receptor for potential synaptic transmission. On the postsynaptic membrane, there are two neurotransmitter receptors present, i.e., metabotropic and ionotropic. Secondary messenger-acting membrane receptors are metabotropic, whereas ionotropic receptors are ligand-gated channels. In respect to this, metabotropic receptors such as muscarinic AChRs, histamine receptors for histamine, $GABA_B$ receptors for GABA, adenosine receptors for adenosine, serotonin receptors (G protein–coupled receptors for serotonin), and ionotropic neurotransmitter receptors such as nicotinic AChRs, $GABA_{A\&A\text{-}\rho}$ receptors also for GABA, and additionally 5-HT, are all included in acetylcholine receptors (AChRs) (Lloyd et al., 1975).

AD does have pieces of evidence for the existence of cholinergic and glutamatergic involvements in its etiopathogensis. In patients with AD, the levels of acetylcholine (ACh) (which is accountable for memory processing and learning) is found reduced, in both concentration and function. This shortfall and other presynaptic cholinergic shortfalls, together with cholinergic neurons loss and decreased acetylcholinesterase activity, accentuate the cholinergic hypothesis of AD. The cognitive deterioration in Alzheimer's patients due to the overactivation of N-methyl-D-aspartate (NMDA) receptors causing neuronal damage because of glutamate is the other hypothesis named glutamatergic hypothesis. The prolonged low-level activation of NMDA receptors may be a consequence of deficiency in reuptake of glutamate by astroglial cells in the synaptic cleft. The choline acetyltransferase (enzyme responsible for catalyzing the synthesis of ACh from acetyl coenzyme A and choline) activity is reduced to 35%–50% of the normal levels in AD patients.

Various input and output pathways between the hippocampus and neocortex enable the capacity of thinking and remembering, and all such pathways dependent on signalling are effectuated by the neurotransmitter glutamate. Neuropathologic studies have shown that the glutamate reuptake levels in the frontal and temporal cortices in patients with AD are reduced, perhaps due to the modifications caused by oxidation of glutamate transporter 1 molecule. In correspondence with this hypothesis, inadequate expulsion of free glutamate from the synapse accounts for abnormally high level of glutamate under relaxed conditions (Werner and Coveñas, 2013).

1.5.2 Parkinson's Disease

The effect of PD is found to be somewhat different among individuals, but in substantia nigra pars compacta, the degeneration of dopamine neurons occur due to the oxidative stress, which is marked as the primary neurological marker for it. The basal ganglia are presupposed in various different behavioral functions, but one of its most notable functions is in movements. Therefore, for the initiation of movements, the dopamine in this pathway interacts with the subcortical brain areas, and the patients with PD lacks dopamine, which causes them to have rigidity or akinesia. However, the focus is on the neurotransmitter, i.e., dopamine, there happens to be some downstream effects on the other neurotransmitters as well, such as GABAergic neurons, glutamate, and muscarinic cholinergic neurons (Kondziella, 2017). In the perikarya and processes of dopaminergic neurons, alpha-synuclein could be found having Lewy bodies. After 1-dopa or dopamine agonist treatment in PD, there happens to be an improvement in Parkinsonian motor symptoms; however, dyskinesia occurred.

In PD, imbalance in a dopaminergic–cholinergic neurotransmitter occurs between DA deficiency through D2 receptors and an excess muscarinic cholinergic neuron through M4 receptors. Antagonistic interaction is exerted between dopaminergic and muscarinic cholinergic neurons, which facilitates through presynaptic GABAergic neuron by way of GABA receptors and glutamatergic neurons via NMDA receptors. Nonetheless, dopaminergic neurons are partly activated by nicotinic cholinergic (nAch) neurons. NAch neurons in the putamen activate dopaminergic neurons via β2 receptors. 40% of the patients with PD evolve dementia; this could be the result of loss of cholinergic neurons in the nucleus basalis of Meynert, from where cerebral cortex inputs are originated.

In the globus pallidus internus and externus and in the striatum, GABA exerts a presynaptic inhibitory function through $GABA_A$ and $GABA_B$ receptors. GABA hypofunction enhances dopamine deficiency while having AMPA/NMDA glutamatergic neurons and $GABA_A$ receptors antagonistic interactions (Moon et al., 2018).

1.5.3 Epileptic Seizures

Epileptic seizures are basically caused by outrageous and hypersynchronous cortical glutamatergic activity, which is treated typically by benzodiazepines such as lorazepam or diazepam, which enhances GABAergic inhibition. It needs to be ensured that seizures must be halted as soon as possible because internalization of postsynaptic GABA$_A$ receptors increasingly make benzodiazepine inefficient, which probably causes refractory status epilepticus and encourages epileptogenesis (Bugiani et al., 2010).

1.5.4 Depression and Related Disorders

For the treatment of migraine and cluster headache, triptans stimulating 5-HT$_{1B/1D}$ receptors are used (Bugiani et al., 2010). Dyskinesia is associated with the 5-HT hyperactivity in the caudate nucleus, and in the striatum, the 1-dopa-treated 6-hydroxy-dopamine lesion rats were shown in the studies. Apparently 5-HT hyperactivity compensates dopamine loss. It has been shown that dopamine and serotonin interact in the putamen across glutaminergic neurons through NMDA receptors and 5-HT$_{2A}$ receptors, which helps 5-HT to exert function (Moon et al., 2018). Despite the substantial efforts put in understanding basic neurochemical research, the essential knowledge that has proven relevant is lesser in amount. Nevertheless, every passing day, a huge number of patients with disorders such as PD, AD, epilepsy, and myasthenia are benefited from this knowledge with regard to improved survival and quality of life (Bugiani et al., 2010).

1.6 MOLECULAR GENETICS AND NEUROMUSCULAR DEGENERATIVE DISORDERS

Muscular dystrophy is a large, heterogeneous group of inherited diseases that result in progressive weakness and degeneration of muscles over the time. This damage and weakness of muscles are occurred due to the lack of a protein called dystrophin. Dystrophin is necessary for the normal functioning of muscle in the human body. The absence of this protein can cause problems in walking, swallowing, and cardiac complications including heart failure and irregular heart rhythms and affect the muscles coordination (Emery, 2002). Muscular dystrophy was first described in the early 1800s; since then, many types and causes of it have been identified. Now there have been more than 30 types of muscular dystrophies characterized, which vary in symptoms and severity of the disease (Kanagawa and Toda, 2006). There are some of the most common types of this disease which are - Duchenne muscular dystrophy (DMD), Becker muscular dystrophy (BMD), congenital muscular dystrophy, myotonic dystrophy, facioscapulohumeral dystrophy (FSHD), limb-girdle muscular dystrophy, oculopharyngeal muscular dystrophy (OPMD), and Emery–Dreifuss muscular dystrophies (Kanagawa and Toda, 2006).

Clinically, at any point from birth till middle age, these diseases present with muscle ineffectiveness, showing significant abnormality and defect due to frequently high serum creatine kinase levels (Murray et al., 1982). The molecular genetic mapping techniques, their development, and advancements have disclosed that an amount of clinically similar problems are associated with a variety of marked single-gene disorders. At the muscle pathology level, the characterization of muscular dystrophies are by necrotic and regenerating fibers, a fiber size the variation that goes on elevating, splitting of fibers, and myonuclei being centrally located. The necrosis and substitution of muscles with fatty and fibrous tissues is a consequence of successive degeneration and regeneration of muscle fibers (Lovering et al., 2005).

A collaborative endeavor of clinicians and scientists was made to portray and discover cytogenetic abnormalities associated with DMD and BMD, which helped to determine and identify a large mRNA transcript that is expressed in the striated muscle. DMD and BMD have similar signs and symptoms and are caused by different mutations in the same gene (Bonilla et al., 1988). The two disease conditions differ in their severity, age of onset, and rate of progression. In boys with DMD, muscle weakness appears in early childhood and worsens rapidly, while children may have delayed motor skills, such as sitting, standing, and walking. In BMD, the signs and symptoms are usually milder and more varied. In most cases, muscle weakness becomes apparent later in childhood or in adolescence and worsens at a much slower rate (Okubo et al., 2016).

DMD (also known as Meryon's disease) is the commonest form of neuromuscular disorders affecting 1 in 3500 to 5000 newborn worldwide that primarily affect boys due to an X-linked mutation in the DMD gene that encodes the 427-kDa cytoskeletal protein called dystrophin (Hoffman et al., 1987). The finding of the dystrophin gene was the first step forward in comprehending the molecular basis of muscular dystrophy (Monaco et al., 1986). Dystrophin is missing in DMD patients or reduced in amount in BMD patients. When dystrophin is solubilized from the sarcolemmal fraction, it is associated with a large oligomeric complex of sarcolemmal proteins and glycoproteins called dystrophin-associated glycoprotein complex (DGC). The proteins in this complex also designate the components as dystrophin-associated proteins (DAPs). It is a large protein that shows major structural role to stabilize muscle fibers by forming a link between the internal cytoskeleton and a dystrophin-associated protein complex (DAPC) in the plasma membrane that is connected to the extracellular matrix. The amino terminus of dystrophin binds to F-actin, and the carboxyl terminus binds to the DAPC at the sarcolemma. The DAPC includes the dystroglycans, sarcoglycans, integrins, and caveolin, and mutations in any of these components cause autosomally inherited muscular dystrophies. Genetic mutations resulting in the absence of dystrophin causes DMD, and those resulting in insufficient or abnormal dystrophin cause BMD (Ehmsen et al., 2002).

The dystrophin gene contains 79 exons that encode a 14-kb mRNA and produce a 427-kDa membrane protein and around 2.6 million base pairs of DNA. The most common mutation in the X chromosome in the Xp21 region found responsible for DMD and BMD. Such deletion accounts for 60%—70% of all DMD cases and 80%—85% BMD cases. Approximately 60% of dystrophin mutations are large insertions or deletions that lead to frameshift errors downstream, whereas approximately 40% are found point mutations or small frameshift rearrangements (Nowak and Davies, 2004).

1.7 IMMUNOMOLECULAR BIOLOGY OF NEUROVIRAL INFECTION AND OTHER CHRONIC NEUROLOGICAL DISEASES

A virus is the smallest type of parasite that contains a single nucleic acid (RNA or DNA) core surrounded by a protein coat. Sometimes, enzymes are required to initiate viral replication within the host cells because viruses have the natural ability to deliver genetic material to cells. Viruses sometimes invade the body and infect various organs that cause everything from mild disturbances to serious problems such as infections of the brain, which leads to transient or permanent neurologic or psychiatric problems. Moreover, some recent studies evident the hypothesis of the role of neurotropic viruses from the Herpesviridae family, especially Human herpesvirus 1 (HHV-1), Cytomegalovirus (CMV), and Human herpesvirus 2 (HHV-2), in AD and post-encephalitic Parkinsonism after certain viral infections such as H5N1, coxsackie virus, Japanese encephalitis B., St. Louis viral encephalopathy, and HIV. In most of these neurological diseases, the cause of neuronl cell death is still not known however which neurons are involved is much clear. For example, PD is caused by a loss of substantia nigra dopamine neurons and narcolepsy is caused by loss of hypothalamic hypocretin neurons (Lodish et al., 2000a,b).

More commonly, viruses enter the central nervous system (CNS; the brain and spinal cord) either by crossing the blood—brain barrier (BBB) or via the peripheral nervous system (PNS), nerve tissue outside of the CNS systemic infections. Some studies suggest that the infection of neurons and neighboring glial cells and the accompanying increase in cytokine proinflammatory mediators can trigger cell dysfunction and increase neuronal death via several mechanisms, including necrosis, apoptosis, and autophagy. Virus preferentially infects neurons and can cause severe, and sometimes fatal, brain inflammation. When foreign substances entered into the CNS, they activate microglia via pattern recognition receptors (PRRs) found on the cell surface as well as on endosomal membranes. These receptors include families of scavenger receptors and Toll-like receptors (TLRs) such as (TLR2, TLR5) typically identify and bind pathogen-associated molecular patterns (PAMPs) such as bacterial- and viral-derived carbohydrates, nucleic acids, and lipoproteins. Upon activation, microglia release proinflammatory molecules. When a proinflammatory pathway is activated, microglia contribute to oxidative stress in the microenvironment through the release of cytokines and reactive oxygen species that adversely impact adjacent neurons and cause neuronal cell death (Béraud and Maguire-Zeiss, 2012).

There are some diverse mechanisms by which immune responses contribute to neurodegeneration (Krishnamurthy and Laskowitz, 2016). Chronic diseases that found occur due to viral infection are described below.

1.7.1 Alzheimer's Disease

AD is a neurodegenerative disorder that is characterized by amyloid-β (Aβ) plaques, neurofibrillary tangles, loss of synapses, neuronal cell death, and chronic neuroinflammation. Depending on the stages of the disease, the Aβ uptake or inflammatory responses directly or indirectly activate specific TLRs and coreceptors that depend on the disease stage. At early stage, fibrillar Aβ has the capability to directly interact with TLR2, TLR4, and CD14 to induce microglial Aβ phagocytosis and TLR3-mediated signal enhances neuronal Aβ autophagy. Similarly, TLR7, TLR8, and TLR9 can enhance microglial Aβ uptake, but over time they contribute to increases in neuronal apoptosis, neuroinflammatory responses, and neuroinflammation. In normal conditions, active microglia and astrocytes exert a neuroprotection effect through Aβ phagocytosis, whereas as the disease progresses, they fail in Aβ clearance and exert detrimental effects, including neuroinflammation and neurodegeneration (Gambuzza et al., 2014; Elsa Gambuzza et al., 2015).

1.7.2 Parkinson's Disease

PD is a progressive nervous system disorder that affects body movement. It is characterized by motor symptoms such as tremor, rigidity, bradykinesia, and gait imbalance, as well as nonmotor symptoms such as sleep disturbances and depression. Pathologically, PD is defined by the selective degeneration of dopamine-producing neurons in the substantia nigra pars compacta. As we know neuronal cell death is responsible for the accumulation of α-synuclein in substantia nigra pars compacta. The accumulation of α-synuclein in PD and how this relates to inflammatory pathways is not fully clear, but the TLR level, especially TLR2 signalling associated with major pathway mediating inflammation, is correlated with the accumulation of pathological α-synuclein in PD (Dzamko et al., 2017).

1.8 MITOCHONDRIAL BIOLOGY AND NEUROLOGICAL DISEASES

The brain is the most energy-consuming part of the body, and it has a complex neurological network. These neurological functions are directly related with precise mitochondrial energy metabolism; however, dysfunction in such energy metabolism leads to various neurological diseases, including ALS, AD, HD, parkinsonism, and stroke (Morotz et al., 2012). Such mitochondrial dysfunctions are due to disruption of oxidative phosphorylation (OXPHOS) or integral mitochondrial function, which is the result of nuclear and mitochondrial DNA (mtDNA) mutations (Zhunina et al., 2020).

In neurological disease, various gene mutations occur in both circular 16,569 bp mtDNA and the nuclear DNA. Each mitochondrion has numerous circular DNAs and depends on the energy metabolism of the cell type; every circular mt-DNA encodes 37 genes including 22 tRNAs and 2 rRNAs, which are required for 13 encoded respiratory chain subunits (Stenton and Prokisch, 2020). The nuclear DNA encodes more than 1000 mitochondrially localizing proteins, translated in the cytoplasm, and transports into the mitochondria with the help of protein import machinery (Zhou et al., 2020). Several mitochondrial gene mutations are recognized as responsible for neurological diseases. Mutations in α-synuclein, parkin, PINK1, LRRK2, DJ-1, and OMI/HTRA2 genes are involved with mitochondrial dysfunction. Decreased activity of OXPHOS proteins of respiratory complex I, III, IV, and V have been relatively occurred by mutations (Gonçalves et al., 2020).

In 2015, first mitochondrial gene mutation *CHCHD2*, a homolog of CHCHD10 and encoded by nuclear DNA (nDNA), was identified in PD and thereafter has been reported in PD, ALS, FTD, and AD. CHCHD2 is a member of a family of proteins containing CHCH (coiled coil—helix—coiled coil—helix) domain, characterized by a pair of cysteines separated by nine amino acids, named CX9C motifs. Evidences supporting a direct role of CHCHD2 in PINK1/Parkin-mediated mitophagy is lacking. CHCHD2 and CHCHD10 form homodimer along with CHCHD2—CHCHD10 heterodimer to maintain Mitochondrial Contact Site and Cristae Organising System (MICOS). It is a protein complex located in the mitochondrial inner membrane (IM). The mammalian MICOS is composed of key components Mitofilin/Mic60 and MINOS1/Mic10, with other components including CHCHD3 and CHCHD6. CHCHD2 mutation which causes the impairment of CHCHD2—CHCHD10 complex, leading to destabilized MICOS and OXPHOS complexes. In CHCHD2, various disease-related heterozygous mutations, missense, and nonsense mutations have been identified in different neurological diseases, such as Pro2Leu in sporadic PD, ET (essential tremor); Ser5Arg, Ser85Arg, Ala32Thr in AD and FTD (frontotemporal dementia); Arg8His, Ala32Thr, Pro34Leu, The61Ile, Ala71Pro, Ala79Ser, Ile80Val, Gln120X, Arg145Gln, in PD; Val66Met in MLA (Zhou et al., 2020).

Neurological diseases are associated with different types of mtDNA genetic mutations, such as SNP, which is 79.2% of total mutation rate and CNVs (copy number variations) of 2.6%; other genetic mutations are reported as deletion of 15.1%, insertion of 1.6%, and remaining for over replication and rearrangement of 1.6%. In most of mtDNA, variants are reported with the all 13 mitochondrial proteins coding regions (Cruz et al., 2019).

Various 148 SNP point mutations are also found associated with mtDNA protein-coding regions and four mutations with noncoding regions. The first mutation is reported as m.476A>G, which is associated with ASD, ID, and microcephaly (Avdjieva-Tzavella et al., 2012) whereas m.114C>T and m.16300A>G are associated with bipolar disorder (Kageyama et al., 2018) and fourth mutation, m.8271A>G, in the mitochondrial noncoding locus MT-NC7. These noncoding regions mutations are associated with ND-related symptoms. According to mtDNA genetic analysis, it has been reported that NARP, dementia, RTT, learning disability, ID, and psychiatric disorders have low number of SNP mutations. In NARP, mtDNA gene *MT-ATP6* shows only three mutations, m.8993T>G, m.8993T>C, and m.8989G>C, whereas dementia has only one m.5549G>A in *MT-TW* gene *locus*. In Rett syndrome mutation associated with one SNP variation in the *MECP2* gene, the m.2835C>T in the *MT-RNR2* gene *locus*. The m.1659T>C and m.10044A>G SNP mutations are found in the MT-TV transcript and the MT-TG transcript locus respectively, in particular learning disability associated with pathogenic conditions, including hemiplegia and a movement disorder (Valenti et al., 2014).

In intellectual disorder, the mutation m.14709T>C in the mitochondrial transcript MT-TE is associated with deafness and ataxia. In schizophrenia, it has non-synonymous mutations which are related to the MT-ND6, MT-ATP6, MT-CYB, and MT-ND mtDNA genes. In bipolar disorder, the m.114C>T and the m.16300A>G, associated with the D-Loop region and other three mutations m.3666G>A, m.4564G>A, and m.15784T>C are found in the MT-ND1, MT-ND2, and MT-CYB protein-coding genes, respectively (Fries et al., 2017).

1.9 CONCLUSION

The chief obstacle in the successful diagnosis and treatment of the neurological disorder is the access to the central nervous system (CNS). Due to the presence of blood brain barrier and blood-cerebrospinal fluid barrier, which is the anatomical and biochemical dynamic barriers present in the CNS, offers great restrictions in the delivery of drugs to the central nervous system. However, to have a major impact in the neurological research, advances in nano-medicines are made which help in delivery of durg molecules in closed CNS system. Moreover, due to major advancement in other molecular tools, significant contribution in diagnosis and treatment of neurological disorders have also been made in recent years. For instance, therapeutic projects for the triplet repeat disease, the utilisation of nucleic acids to lower the expression of the target gene in a non-specific and allele-specific manner, is being carried out in recent studies. Furthermore, advancement of neurological tools of brain stimulation i.e., DBS (Deep Brain Stimulation) device has also developed as a new therapeutic aid (it customarily has a quadripolar electrode inserted into the brain of the patient). DBS has been used in the patients with PD in there later stages, where patients receiving very limited effective pharmacological treatments. These strategies are likely

to have a broad impact for neurological disease and can be corrected through this approach.

The recognition of the familiarity of neurological disorders and accessible strategies of its prevention and treatment, are more important to considerably reduce the diseases impact. The advance development of molecular tools to understand the molecular basis of neurological diseases is needed further, especially for the generation of effective therapies. It is anticipated that this knowledge will allow a more precise diagnosis of many other neurological disorders into clinical practice.

REFERENCES

Avdjieva-Tzavella, D., Mihailova, S., Lukanov, C., Naumova, E., Simeonov, E., Tincheva, R., Toncheva, D., 2012. Mitochondrial DNA mutations in two Bulgarian children with autistic spectrum disorders. Balkan J. Med. Gen. 15 (2), 47–53. https://doi.org/10.2478/bjmg-2013-0006.

Bachiller, S., Jiménez-Ferrer, I., Paulus, A., Yang, Y., Swanberg, M., Deierborg, T., Boza-Serrano, A., 2018. Microglia in neurological diseases: a road map to brain-disease dependent-inflammatory response. Front. Cell. Neurosci. 12, 488.

Baleriola, J., Jean, Y., Troy, C., Hengst, U., 2015. Detection of axonally localized mRNAs in brain sections using high-resolution in situ hybridization. JoVE 100, e52799.

Beckelman, B.C., Zhou, X., Keene, C.D., Ma, T., 2016. Impaired eukaryotic elongation factor 1a expression in alzheimer's disease. Neurodegener. Dis. 16 (1–2), 39–43.

Béraud, D., Maguire-Zeiss, K.A., 2012. Misfolded α-synuclein and toll-like receptors: therapeutic targets for Parkinson's disease. Park. Relat. Disord. 18, S17–S20.

Bonilla, E., Samitt, C.E., Miranda, A.F., Hays, A.P., Salviati, G., DiMauro, S., Kunkel, L.M., Hoffman, E.P., Rowland, L.P., 1988. Duchenne muscular dystrophy: deficiency of dystrophin at the muscle cell surface. Cell 54 (4), 447–452.

Bordone, M.P., Salman, M.M., Titus, H.E., Amini, E., Andersen, J.V., Chakraborti, B., Diuba, A.V., Dubouskaya, T.G., Ehrke, E., Espindola de Freitas, A., Braga de Freitas, G., 2019. The energetic brain—A review from students to students. J. Neurochem. 151 (2), 139–165.

Buée, L., Bussière, T., Buée-Scherrer, V., Delacourte, A., Hof, P.R., 2000. Tau protein isoforms, phosphorylation and role in neurodegenerative disorders. Brain Res. Rev. 33 (1), 95–130.

Bugiani, M., Boor, I., Powers, J.M., Scheper, G.C., van der Knaap, M.S., 2010. Leukoencephalopathy with vanishing white matter: a review. J. Neuropathol. Exp. Neurol. 69 (10), 987–996.

Carter, B.S., Fletcher, J.S., Thompson, R.C., 2010. Analysis of messenger RNA expression by in situ hybridization using RNA probes synthesized via in vitro transcription. Methods 52 (4), 322–331.

Cooper, T.A., Wan, L., Dreyfuss, G., 2009. RNA and disease. Cell 136 (4), 777–793.

Cruz, A.C.P., Ferrasa, A., Muotri, A.R., Herai, R.H., 2019. Frequency and association of mitochondrial genetic variants with neurological disorders. Mitochondrion 46, 345−360.

Cumming, W.J., 1992. Molecular biology of neurological diseases. Postgrad. Med. 68 (798), 237.

Dzamko, N., Gysbers, A., Perera, G., Bahar, A., Shankar, A., Gao, J., Fu, Y., Halliday, G.M., 2017. Toll-like receptor 2 is increased in neurons in Parkinson's disease brain and may contribute to alpha-synuclein pathology. Acta Neuropathol. 133 (2), 303−319.

Ehmsen, J., Poon, E., Davies, K., 2002. The dystrophin-associated protein complex. J. Cell Sci. 115 (14), 2801−2803.

Elsa Gambuzza, M., Maria Salmeri, F., Soraci, L., Soraci, G., Sofo, V., Marino, S., Bramanti, P., 2015. The role of toll-like receptors in chronic fatigue syndrome/myalgic encephalomyelitis: a new promising therapeutic approach? CNS Neurol. Disord. - Drug Targets 14 (7), 903−914.

Emery, A.E., 2002. The muscular dystrophies. Lancet 359 (9307), 687−695.

Erben, L., He, M.X., Laeremans, A., Park, E., Buonanno, A., 2018. A novel ultrasensitive in situ hybridization approach to detect short sequences and splice variants with cellular resolution. Mol. Neurobiol. 55 (7), 6169−6181.

Faustino, N.A., Cooper, T.A., 2003. Pre-mRNA splicing and human disease. Gene Dev. 17 (4), 419−437.

Francis, P.T., 2005. The interplay of neurotransmitters in Alzheimer's disease. CNS Spectr. 10 (S18), 6−9.

Fries, G.R., Bauer, I.E., Scaini, G., Wu, M.J., Kazimi, I.F., Valvassori, S.S., Zunta-Soares, G., Walss-Bass, C., Soares, J.C., Quevedo, J., 2017. Accelerated epigenetic aging and mitochondrial DNA copy number in bipolar disorder. Transl. Psychiatry 7 (12), 1−10.

Gambuzza, M.E., Sofo, V., Salmeri, F.M., Soraci, L., Marino, S., Bramanti, P., 2014. Toll-like receptors in Alzheimer's disease: a therapeutic perspective. CNS Neurol. Disord. - Drug Targets 13 (9), 1542−1558.

Giuditta, A., Zomzely-Neurath, C., Kaplan, B.B. (Eds.), 2012. Role of RNA and DNA in Brain Function: A Molecular Biological Approach, vol. 3. Springer Science & Business Media.

Gonçalves, A.M., Pereira-Santos, A.R., Esteves, A.R., Cardoso, S.M., Empadinhas, N., 2020. The mitochondrial ribosome: a World of opportunities for mitochondrial dysfunction toward Parkinson's disease. Antioxidants Redox Signal. 1−51. https://doi.org/10.1089/ars.2019.7997.

Gourie-Devi, M., 2014. Epidemiology of neurological disorders in India: review of background, prevalence and incidence of epilepsy, stroke, Parkinson's disease and tremors. Neurol. India 62 (4), 588.

Griffin, W.S.T., Morrison, M.R., 1985. In situ hybridization—visualization and quantitation of genetic expression in mammalian brain. Peptides 6, 89−96.

Hoffman, E.P., Brown Jr., R.H., Kunkel, L.M., 1987. Dystrophin: the protein product of the Duchenne muscular dystrophy locus. Cell 51 (6), 919−928.

Janca, A., Prilipko, L., 1997. The World Health Organization's global initiative on neurology and public health. J. Neurol. Sci. 145 (1), 1−2.

Jutzi, D., Akinyi, M.V., Mechtersheimer, J., Frilander, M.J., Ruepp, M.D., 2018. The emerging role of minor intron splicing in neurological disorders. Cell Stress 2 (3), 40.

Kageyama, Y., Kasahara, T., Masaki, K., Sakai, S., Deguchi, Y., Tani, M., Kuroda, K., Hattori, K., Yoshida, S., Goto, Y., Kinoshita, T., Inoue, K., Kato, T., 2018. The relationship between circulating mitochondrial DNA and inflammatory cytokines in patients with major depression. J. Affect. Dis. 223, 15−20. https://doi.org/10.1016/j.jad.2017.06.001.

Kanagawa, M., Toda, T., 2006. The genetic and molecular basis of muscular dystrophy: roles of cell−matrix linkage in the pathogenesis. J. Hum. Genet. 51 (11), 915−926.

Kandimalla, R., Reddy, P.H., 2017. Therapeutics of neurotransmitters in Alzheimer's disease. J. Alzheim. Dis. 57 (4), 1049−1069.

Kondziella, D., 2017. The top 5 neurotransmitters from a clinical neurologist's perspective. Neurochem. Res. 42 (6), 1767−1771.

Koyuncu, O.O., Hogue, I.B., Enquist, L.W., 2013. Virus infections in the nervous system. Cell Host Microbe 13 (4), 379−393.

Krishnamurthy, K., Laskowitz, D.T., 2016. Cellular and molecular mechanisms of secondary neuronal injury following traumatic brain injury. In: Translational Research in Traumatic Brain Injury. CRC Press/Taylor and Francis Group.

La Spada, A., Ranum, L.P., 2010. Molecular Genetic Advances in Neurological Disease: Special Review Issue.

Lloyd, K.G., Davidson, L., Hornykiewicz, O., 1975. The neurochemistry of Parkinson's disease: effect of L-dopa therapy. J. Pharmacol. Exp. Therapeut. 195 (3), 453−464.

Lodish, H., Berk, A., Zipursky, S.L., Matsudaira, P., Baltimore, D., Darnell, J., 2000a. Neurotransmitters, synapses, and impulse transmission. In: Molecular Cell Biology, fourth ed. WH Freeman.

Lodish, H., Berk, A., Zipursky, S.L., Matsudaira, P., Baltimore, D., Darnell, J., 2000b. Viruses: structure, function, and uses. In: Molecular Cell Biology, fourth ed. WH Freeman.

Lovering, R.M., Porter, N.C., Bloch, R.J., 2005. The muscular dystrophies: from genes to therapies. Phys. Ther. 85 (12), 1372−1388.

MacMillan, J.C., Harper, P.S., 1994. Clinical genetics in neurological disease. J. Neurol. Neurosurg. Psychiatr. 57 (1), 7.

Medeiros, R., Baglietto-Vargas, D., LaFerla, F.M., 2011. The role of tau in Alzheimer's disease and related disorders. CNS Neurosci. Ther. 17 (5), 514−524.

Menken, M., Munsat, T.L., Toole, J.F., 2000. The global burden of disease study: implications for neurology. Arch. Neurol. 57 (3), 418−420.

Monaco, A.P., Neve, R.L., Colletti-Feener, C., Bertelson, C.J., Kurnit, D.M., Kunkel, L.M., 1986. Isolation of candidate cDNAs for portions of the Duchenne muscular dystrophy gene. Nature 323 (6089), 646−650.

Moon, S.L., Sonenberg, N., Parker, R., 2018. Neuronal regulation of eIF2α function in health and neurological disorders. Trends Mol. Med. 24 (6), 575—589.

Morotz, G.M., De Vos, K.J., Vagnoni, A., Ackerley, S., Shaw, C.E., Miller, C.C., 2012. Amyotrophic lateral sclerosis-associated mutant VAPBP56S perturbs calcium homeostasis to disrupt axonal transport of mitochondria. Hum. Mol. Genet. 21 (9), 1979—1988.

Murray, J.M., Davies, K.E., Harper, P.S., Meredith, L., Mueller, C.R., Williamson, R., 1982. Linkage relationship of a cloned DNA sequence on the short arm of the X chromosome to Duchenne muscular dystrophy. Nature 300 (5887), 69—71.

Nowak, K.J., Davies, K.E., 2004. Duchenne muscular dystrophy and dystrophin: pathogenesis and opportunities for treatment: third in molecular medicine review series. EMBO Rep. 5 (9), 872—876.

Okubo, M., Minami, N., Goto, K., Goto, Y., Noguchi, S., Mitsuhashi, S., Nishino, I., 2016. Genetic diagnosis of Duchenne/Becker muscular dystrophy using next-generation sequencing: validation analysis of DMD mutations. J. Hum. Genet. 61 (6), 483—489.

Patel, A.A., Steitz, J.A., 2003. Splicing double: insights from the second spliceosome. Nat. Rev. Mol. Cell Biol. 4 (12), 960—970.

Rosenberg, R.N., 1993. An introduction to the molecular genetics of neurological disease: recent advances. Arch. Neurol. 50 (11), 1123—1128.

Shyu, A.B., Wilkinson, M.F., Van Hoof, A., 2008. Messenger RNA regulation: to translate or to degrade. EMBO J. 27 (3), 471—481.

Stenton, S.L., Prokisch, H., 2020. Genetics of mitochondrial diseases: identifying mutations to help diagnosis. EBioMedicine 56, 102784.

Thelen, M.P., Kye, M.J., 2020. The role of RNA binding proteins for local mRNA translation: implications in neurological disorders. Front. Mol. Biosci. 6, 161.

Valenti, D., de Bari, L., De Filippis, B., Henrion-Caude, A., Vacca, R.A., 2014. Mitochondrial dysfunction as a central actor in intellectual disability-related diseases: an overview of Down syndrome, autism, Fragile X and Rett syndrome. Neurosci. Biobehav. Rev. 46, 202—217.

Vanbatalo, S., Soinila, S., 1998. The concept of chemical neurotransmission—variations on the theme. Ann. Med. 30 (2), 151—158.

Werner, F.M., Coveñas, R., 2013. Classical neurotransmitters and neuropeptides involved in major depression in a multi-neurotransmitter system: a focus on antidepressant drugs. Curr. Med. Chem. 20 (38), 4853—4858.

Zhou, W., Ma, D., Tan, E.K., 2020. Mitochondrial CHCHD2 and CHCHD10: roles in neurological diseases and therapeutic implications. Neuroscientist 26 (2), 170—184.

Zhunina, O.A., Yabbarov, N.G., Grechko, A.V., Yet, S.F., Sobenin, I.A., Orekhov, A.N., 2020. Neurodegenerative diseases associated with mitochondrial DNA mutations. Curr. Pharmaceut. Des. 26 (1), 103—109.

CHAPTER 2

Immunological Genes Expression in the Aged Brain

DEVLINA GHOSH, MSC, MBA • ADITI SINGH, PHD • NEERAJ SINHA, PHD

2.1 INTRODUCTION

Aging is an inevitable part of life cycle; every organism, big or small, undergoes this phenomenon. Several prominent changes take place during this time-associated process; it could be biological, chemical, biochemical, and genetic. In human beings, aging could have several dynamic aspects, which include but may not restrict to physical, psychological, cognitive, and neurological changes.

It is a natural and gradual process, with increased entropy and degradation in cells, tissues, and organs with decreased functionality, which may lead to the development of age-associated neurodegenerative diseases (Harman, 2006). In 1961, American anatomist Leonard Hayflick demonstrated that normal human cells divide around 40–60 times before senescence starts (popularly known as *Hayflick limit*) and thus for the first time interpreted aging at the cellular level. Cells experience senescence when the doubling rate slows down before stopping altogether (Shay and Wright, 2000). The human brain may be considered the most complex organ in the body. This jelly-like mass of tissues, weighing approximately 1.4 kg, processes and controls all our thoughts, memory, and feelings. This organ is composed of an astounding 100 billion nerve cells known as neurons. Neurons are distinct due to their branch-like projections called axons and dendrites that collect and transmit the electrochemical signals. Every neuron can make connections with hundreds or thousands of other neurons by very minute junctions called synapses; this connectivity among huge number of cells is what makes the brain so complex. The pattern and strength of these connections are ever changing and so no two brains are the same. Apart from neurons, there are other types of cells in brain called glial cells; surprisingly, they outnumber neurons by almost 10 times. Initially thought to be as only support cells, these amplify neural signals and also play a significant role in maintaining tissue homeostasis. There are three types of glia cells in mature central nervous system (CNS): microglia cells, astrocytes, and oligodendrocytes (von Bartheld et al., 2016).

To understand the concept of *neuroimmunology*, we should discuss the immune cells in brain which are microglia cells. In CNS, these cells represent the special population of macrophages that are the first responders in case of any pathogen invasion, tissue injury, or any other changes. For several years, the role of microglia was not clear. However, now we know that these cells mediate primary immune responses in the CNS by working as macrophages, clearing up all cell debris and dead neurons and other unwanted matter from nervous tissue by the method of phagocytosis (cell eating). Thus, they play a pivotal role in maintaining CNS homeostasis. Microglia cells never rest and actively scan the microenvironment with their long projected processes; they quickly become activated when encountered with even slight perturbations as they are extremely sensitive in nature (Wolf et al., 2017).

Inflammation may trigger microglia activation in the CNS, which may lead to neurological degenerative diseases such as Parkinson's disease (PD), Alzheimer's disease (AD), or infectious diseases such as Creutzfeldt–Jakob disease. Various research studies suggest that microglia can hinder the development of disorders in the brain that are caused by infectious elements called prions by removing prion-damaged cells (Ulland et al., 2017). The quantity and function of microglia are strongly regulated by the local microenvironment under physiological conditions, as they constantly interact with the surrounding cells. Microglia have the capability to modify into different states in response to any insult, for instance, changing its morphology by retracting the processes, increased phagocytic activity, and antigen presentation (Sierra et al., 2013). They also start secreting proinflammatory factors such as cytokines and chemokines (Kettenmann et al., 2011). Interestingly, in different regions of the brain where microglia are in proximal communication with neurons, astrocytes, and oligodendrocytes, the local environment differs in different pathologies of diseases. Therefore, it can be said that the phenotype of microglia is basically dependent on the disease condition and may be controlled by biochemical and cellular composition, neuronal subpopulations and circuitries, and neurotransmitters among other factors. There is one more dominant factor that is strictly related to microglial cell activation and that is aging, which has been widely studied in terms of microglial cell response (Mosher and Wyss-Coray, 2014). Aging is a key factor associated with alterations in gene expression as well as in the formation of dystrophic microglia, which have been related to abnormal cytoplasmic configuration, decreased ramification, and disjointed processes (Streit et al., 2004). All these alterations related to aging might have a role in the development of neurodegenerative disorders.

Conventionally, microglia upon activation have been distinguished in two different and contrary states: the classic (M1) and alternative (M2). The M1 phenotype is a proinflammatory state, as this microglial cell generates and discharges tumor necrosis factor-α (TNF-α), interleukin-1β (IL-1β), IL-12, reactive oxygen species (ROS), and nitric oxide (NO). On the contrary, the M2 is classified as an antiinflammatory state that produces and secretes trophic factors, such as tumor growth factor-β (TGF-β) and brain-derived neurotrophic factor (BDNF) (Tang and Le, 2016). It has been demonstrated that microglial cells play a precise role in neurodegenerative disorders based on apolipoprotein E (APOE) and activating receptor expressed on myeloid cells 2 (TREM2) (Krasemann et al., 2017).

Researchers have established clear expressions of CD11b, CD40, CD45, CD80, CD86, F4/80, TREM-2b, CXCR3, and CCR9 throughout striatum, cerebral cortex, hippocampus, spinal cord, and cerebellum (de Haas et al., 2008).

The blood—brain barrier (BBB) is yet another characteristic that makes brain extraordinary. In 19th century, scientists discovered that when a dye is injected in the bloodstream, it spreads to all other organs except brain and spinal cord; only when it is injected directly into the CSF, it tints the brain and spinal cord. This emphasized that BBB provides an ingenious protective layer composed of special tightly bound cells that make a semipermeable doorway that keeps all toxins, pathogens, and other damaging substances away, thereby allowing only essential nutrients and oxygen to pass through in the bloodstream (Obermeier et al., 2016).

Cellular aging has many signature characteristics; however, when we talk about brain, things are quite different.

The phenomenon of aging is almost inevitable and impacts every individual differently. Throughout our life span, brain modifies more than any other organs. In the third week of gestation when the brain begins to develop till old age, its complex structures and functions are altering, and networks and pathways form links and unlink. As we know that brain controls our thought process, planning, remembering, organizing, making decisions, learning, and many more (Langa and Levine, 2014), cognitive abilities enable people to do their everyday work in a proper way. However as they age, some people might experience changes in their abilities. For instance, difficulty in multitasking, forgetting names as they undergo attention deficit. Being said that, aging may also bring positive changes in some people; research has shown that people may learn new things, and they can even construct new memories (Venegas and Heneka, 2017).

The impact of aging on brain and cognitive abilities is extensive and has multiple causes. Aging has its influence on molecules, cells, vasculature, overall morphology, and cognition. As we age, our brains reduce in size, particularly the area of frontal cortex. As our vasculature undergoes aging and our blood pressure goes up, the likelihood of suffering from stroke and ischemia increases and our white matter may develop lesions (Asha Devi and Satpati, 2017). Retaining memory may also be a downhill process. Genetics, neurotransmitters, hormones, learning, and knowledge all have key roles to play in brain aging. But, like a cloud with silver lining, higher levels of education or occupational accomplishment may work as a shielding factor. Also, other factors such as healthy food, reduced alcohol intake, and regular physical exercise might have a positive role to play. Biological aging may be loosely bound to chronological aging, and possibly the former may be slowed down, thereby even reducing the possibility of developing neurodegenerative disorders.

A research study at the Albert Einstein College of Medicine in New York, United States, underlines the significance of "good genes." Under this Longevity Genes Project, more than 500 individuals called superagers (over 90 years of age) are observed to identify particular genes that support healthy aging.

2.2 AGE-ASSOCIATED CHANGES IN BRAIN

As we have discussed so far, aging is a complex process and is yet to be characterized completely. There are several unknown aspects that researchers are trying to understand. In this section of the chapter, we will discuss various factors that may contribute to the process (Fig. 2.1).

2.2.1 Oxidative Stress

Several research studies so far have demonstrated that oxidative stress (OS) may be a link between normal brain aging and neurodegenerative disease condition. We know that free radicals with unpaired electrons are highly reactive/unstable and do harm at the cellular level. Buildup of ROS and free radicals may add to the decay of brain functioning with time. In healthy condition, our body has its own defense system against free radical damage, in the form of antioxidant enzymes such as glutathione (GSH), superoxide dismutase (SOD), catalase, glutathione peroxidase, and so on. The brain uses around 20% of the cardiac output to meet its energy demand, and that is why, it is more prone to superoxide buildup. Also, it has poor antioxidant defense when compared with other body organs. Brain has a comparatively greater concentration of PUFA (polyunsaturated fatty acids) (da Silva et al., 2008), and OS is considered a prime factor in normal aging in brain along with decay in neurological function. Neurons and glial cells undergo increased OS during normal aging and neurological disorders (Raefsky and Mattson, 2017)

A study done on early-stage PD patients reveals that increased OS is a prominent trait during early stages of disease progression, preceding significant neuron loss (Coune et al., 2012). This suggests that uninhibited ROS generation may be a possible contributory factor in dopaminergic neuron death, rather than being a secondary response to progressive neurodegeneration. It is

evident that ROS buildup is a factor in various detrimental molecular pathways during early stage of PD, before initiation of neuron death. Unwarranted accrual of ROS may also participate in apoptosis (intrinsic and extrinsic), cytoplasmic cell death, and autophagic cell death (Morris et al., 2018).

Interestingly, to neutralize the ROS and reactive nitrogen species (RNS) buildup, Nrf2 (nuclear factor erythroid 2–related factor) transcription factor turns on. Activation of Nrf2 pathway upregulates the expression of several endogenous antioxidants. Aged animals demonstrate reduced Nrf2 activation and expression compared with young, along with increase in expression of inflammatory genes (Gounder et al., 2012). For this reason, Nrf2 may be recommended as a prospective therapeutic target in neuroinflammatory disorder and OS-related condition (Salim, 2017).

M1 activation phenotype of microglia is predominantly characterized with reduced antioxidant production and ROS buildup along with proinflammatory responses. Increased levels of ROS on the other hand lead to M1 microglial activation, suggesting that a vicious cycle of prooxidant microglial activation may be present within the aging substantia nigra pars compacta (SNc) region of brain. Significantly, age-associated microglial activation takes place preferentially in the susceptible vSNc and may add to the selective susceptibility of nigral subregion to OS insult in PD (Trist et al., 2019).

Thus, understanding the detailed role of oxidative stress in terms of age-associated diseases may, therefore, expose new objectives for therapeutic intervention and preclinical diagnosis.

2.2.2 Shortening of Telomeres

Telomeres are the end regions of chromosome that keep shortening with every cell division, when they are lost to a critical level, cells stop dividing due to deleterious loss of genetic material. Telomere length is often considered a biomarker for biological age in contrast to chronological age. Interestingly, researchers have recently found a correlation between telomere length with initiation of age-associated disorders and structural differences in different regions of brain; so for two people with the same chronological age, the one with shorter telomere is at greater risk of developing neurological disorders. OS adds on to telomere shrinking that occurs in leukocytes and other mitotic cells during aging and chronic disease states. Prominently, the 5′-TTAGGG-3′ repeats in telomere sequence are vulnerable to oxidative damage (8-oxodG) and -stimulated DNA breaks (Puhlmann et al., 2019).

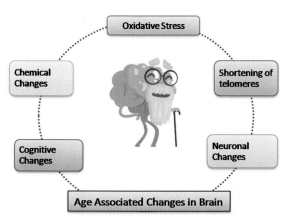

FIG. 2.1 Depicts different age related changes in brain.

As we mentioned before, when cells enter senescence phase, they cease to divide; instead, they grow in size, generate proinflammatory cytokines, and express proteins p21 and p16Ink4a. Senescence may be the outcome of glial cells and neural progenitor cells during aging as signified by high expression of p16Ink4a related with decreased amount of propagating progenitors in the subventricular zone (Regulski, 2017). Studies have shown that human neural progenitor cells in laboratory conditions show a restricted number of population doublings and then experience aging. Scientific research has demonstrated that shortening of telomeres may be slowed down or even reinstated at appropriate conditions (Puhlmann et al., 2019).

Future studies involving selective removal of senescent cells from the brain in aged animal models and age-related neurodegenerative disorders may explain whether cell senescence is a true characteristic of brain aging.

2.2.3 Neuronal Changes

Neurons are very specialized cells in the nervous system, communicating information in the form of electric signals, thus receiving sensory input and sending messages to other nerve cells, gland, or muscle cells. These are the fundamental units of the brain, and thus, their interaction defines our brain. They have a cell body (soma, where the nucleus is), dendrites (branches where neuron receives signal from other cells), and axon (the output structure that generates and sends action potential).

At individual level, changes in neuron may result in reduced size and cortical thinning of aging brain. As neurons shrink, they withdraw their dendrites while the fatty myelin sheath that wraps around axons degrades. Learning and memory may also get hindered, as a number of connections, or synapses, connecting brain cells plummet (Petralia et al., 2014).

When compared to chemical changes, which will be discussed in the next section, synaptic alterations have greater impact on cognitive decline. Scientists have observed shrinkage in dendrites with increasing age especially in prefrontal cortex and hippocampus region; the proliferating branches become less complex; and dendritic spines, the small protrusions that receive chemical signals, also experience a significant decrease in the number. In a study on rhesus monkeys, researchers found that aging greatly impacts a specific group of spines more profoundly, called the *thin* spines. These tiny, slender protrusions are extremely plastic structures and can elongate and retract faster than larger *mushroom* spines. This led to the speculation that *thin*

spines might have a key role in working memory, which needs high amount of synaptic plasticity. Therefore, decrease in *thin* dendritic spines could weaken neuronal talk and lead to cognitive decline (Patterson, 2015).

Neurogenesis, i.e., the development of new neurons also reduces with age. Research suggests that this phenomenon is rare in adult brain or it is so minuscule that goes undetected. And even if in adult brain, new neurons appear, scientists have not determined if they could incorporate into forming strong brain networks or affect cognition. However, it has been found that regular physical exercise can improve neurogenesis in mice and in turn improve cognitive function (Ma et al., 2017).

Another aspect of neuronal functioning is the glucose metabolism, which in turn controls the brain glucose uptake. GLUT1 (glucose transporter) is mainly responsible for transporting brain glucose across endothelium into astrocytes and then is transmitted into neurons, generally via GLUT3 and GLUT4, contributing to the glycolytic pathway. Reduced glucose consumption and a decrease in neuronal GLUTs have been reported in aged rat brains. During aging, changes in mitochondria energy-transducing capability, as well as glucose accessibility, take place, along with disruption in neuronal glucose usage, increased oxidant production, and decline in electron transport chain activity (Yin et al., 2016).

During maturation, the vast majority of the biophysical highlights of neurons stay flawless. It has been discovered that in hippocampus and prefrontal cortex area, there are no distinctions in young and matured neurons as far as resting membrane potential, ascent time, span of activity potential, and membrane time constant are considered. In any case, there is critical improvement in Ca^{2+} conductance, which no doubt adds to age-related changes in pliancy (long-term potentiation [LTP] and long-term depression [LTD]) (Burke and Barnes, 2006).

2.2.4 Chemical Changes

Neurons in brain do the cross-talk by using signaling molecules known as neurotransmitters. These are special chemical messengers that send signals through a chemical synapse from one neuron to another or other cells of muscle/gland and thus communicate information throughout our brain and body. Dopamine (DA), serotonin (5-HT), gamma-aminobutyric acid (GABA), and glutamate (Glu) are some of the major neurotransmitters involved in synaptic transmission. Neurotransmitters may be excitatory, inhibitory, neuromodulators, or neurohormones. But as we age, the level of chemical messengers may decrease in brain.

Various researchers have revealed that mature brain integrates less dopamine, and there are less dopamine receptors. Significant age-associated decline is reported in dopamine receptors, D2 and D3 in different brain regions, particularly in anterior cingulate cortex, lateral temporal cortex, hippocampus, medial temporal cortex, and amygdala. Loss in dopamine signaling may impair cognitive flexibility as exemplified in PD (Kaasinen et al., 2000). One investigation found that people older than 60 years with slight cognitive impairment had less serotonin in their brains, and thus scientists pondered, controlling serotonin levels may help avert and treat cognitive decline.

Scientific evidence suggests that deficit in serotonergic neurotransmission may give rise to age-related depression in elderly due to changes in 5-hydroxytryptamine (5-HT) neurons. It has also been suggested that disorder in cholinergic and serotonergic function can contribute to cognitive decline in AD, with serotonergic dysfunction most likely contributing for a major portion of the behavioral side of the disease. Postmortem examination of brain tissue to establish the role of serotonergic dysfunction in depression, suggests changes in 5-HT metabolite and receptor concentrations; the neuroendocrine reaction to serotonergic challenge and antidepressants acts by altering the 5-HT neurotransmission (Meltzer, 1998). The amount of GABA, a major inhibitory neurotransmitter in the human brain, declines in frontal and parietal regions due to aging (Porges et al., 2017). Thus, chemical changes may have a profound effect, as aging progresses.

2.2.4 Cognitive Changes

Changes in cognitive abilities are evident as aging progresses, such as difficulty in processing new information and recalling names and numbers that may take longer. It has been found that autobiographical recollection of happenings and gathered knowledge of learned facts and information take a downfall with age; however, procedural memory such as knowing how to drive a car or tie a bow remains mostly intact. The decline in working such as remembering numbers and passwords may start as early as age 30. Fluid intelligence, for instance, processing rate and problem-solving, too decreases with age. Multitasking such as holding a conversation while driving may also become more challenging with the increase in age (Langa and Levine, 2014). However, not everything turns gloomy with age; in fact some cognitive aptitudes improve in middle age.

The Seattle Longitudinal Study, which followed the psychological capacities of thousands of subjects in the course of recent years, indicated that individuals really improved on trial of verbal capacities, spatial thinking, math, and theoretical thinking in middle age than they did when they were young. Neuroscientists are exploring new insights about our brains that suggest that this dynamic organ remains pretty "plastic" as we age, implicating neural connections are rerouted to adapt to new challenges and jobs (Asha Devi and Satpati, 2017; Patterson, 2015).

Our minds experience various changes during the maturing procedure. Notwithstanding, researchers are learning each day how leading a healthy and active life can defer or limit the negative results of these progressions. In the next section of the chapter, the immune cells present in our CNS, mainly brain, and how their role is still an enigma for scientists are discussed.

2.3 MICROGLIA: THE IMMUNE CELLS IN BRAIN

Unlike other body organs, the brain is conventionally considered as immunologically privileged, primarily because the BBB does not allow immune cells to reach brain. In 1940s, Peter Medawar did an interesting experiment in which he transplanted heterologous tissues into CNS; surprisingly, they did not go through immunological rejection. This led to the notion that brain and immune system exist as two separate entities, isolated from one another (Spierings and Fleischhauer, 2019). However, as scientists gathered more facts on how immune system has a prominent role in brain aging and neurodegenerative disorders, it became more evident that the two systems are certainly not isolated from each other; rather, there is a cross-talk between them. For instance, epidemiological studies showed that prolonged usage of antiinflammatory medicines decreases the risk of AD and PD significantly. This emphasizes the function of neuroinflammation in age-associated neurodegeneration (Sudhakar and Richardson, 2019; Tang and Le, 2016).

The immune system in CNS is unique in the sense that peripheral immune system is off limits to the entire region of the brain parenchyma; instead microglia cells (resident macrophages of the brain) mediate the parenchymal immunity. On the contrary, the meningeal borders of the CNS do witness the circulation of peripheral immune cells, and many lymphocytes inhabit mostly in the meninges and choroid plexus region. Significantly, this isolation is by all accounts basic for appropriate working of the CNS, as the presence of adaptive immunity inside the brain parenchyma is reminiscent of chronic brain infection or autoimmunity. In any case, a few lines of proof exhibit the significance of meningeal immune cells and their inferred cytokines in brain work from worms to warm blooded creatures. The most predominant immune cells in the cerebrum are microglia, which establish 80% immune cells in brain. Other immune cells perceived in the brain incorporate myeloid cells, monocytes/macrophages, dendritic cells, T cells, B cells, and natural killer (NK) cells (Rankin and Artis, 2018). As mentioned before, lymphocytes (including T cells, B cells, and NK cells), which are recognized as a CD45hi population, are sparse in the CNS (Pösel et al., 2016). And therefore, in this chapter, our prime focus will be on microglia cells that are discussed in detail in the next section.

2.3.1 About Microglia Cells

Microglia cells have been in the center of debate, uncertainty, and curiosity since decades due to extraordinary diversity in structure as well as function. The foundation of microglial tale can be traced back to the efforts of Franz Nissl in the late 19th century while he first recognized the glial cell population, named as rod cells in neurodegenerative brain. Providing their description, he even observed their migratory and phagocytic capabilities. In 1932, Del Rio Hortega for the first time distinguished microglia cells in brain and noted that any change in CNS microenvironment triggers ramified microglia cells to transform into ameboid, migratory, and phagocytic variant in case of immune challenges (de Castro, 2019).

Microglia cells play a key role in this cross-talk, though in normal physiological conditions these cells are in resting phase and constantly surveying the microenvironment and thus maintain the tissue homeostasis; in case of any injury or infection, microglia promptly become reactive and respond in a variety of ways. Therefore, they are aptly called the first line of defense in brain cells. Such interesting functional profile necessitates sensory tools such as receptors, ion channels that account for the appearance, abnormal concentration, or changed features of soluble and insoluble factors as they can point out danger and harm. It also requires a range of releasable messengers (cytokines, chemokines, and other trophic factors) to communicate with other cells especially in emergency conditions. Recently, several studies have tried to decipher the diverse molecular and functional features of reactive response of microglia in pathogenesis ranging from axonal injury, traumatic injury, ischemia, tumors, and neurodegenerative diseases to autoimmune disorders.

Microglia spread everywhere throughout the brain in assorted form and shift among species. In people,

microglia represent up to 16% of the CNS cell populace, and this is reliant on brain regions. In people, microglia are available at a higher density in white matter than gray matter, whereas in rodents, 5%—12% of the CNS cell populace is microglia, with higher thickness in the gray matter (Mittelbronn et al., 2001).

2.3.2 Developmental Origin of Microglia in Brain

Microglia are very distinct from peripheral macrophages and other tissue-resident macrophages (such as Kupffer cells and alveolar macrophages) phenotypically as well as developmentally. It has been established that microglia arise from yolk-sac fetal macrophages, while other tissue macrophages originate from precursors generated slightly later in development. Though hematopoietic cells do not contribute to microglia homeostasis in normal development and adulthood; however, peripheral hematopoietic cells may add to the microglia pool in the brain in pathological conditions. For instance, in case of chronic stress and irradiation of the brain that breaches the BBB, peripheral hematopoietic cells can enter the neural tissue and become component of the microglia/macrophage pool in the parenchyma. Every one of these realities recommends that microglia have an extraordinary formative starting point and tissue condition that drives their particular development. When they arrive at a last differentiated state in brain, microglia replication is limited. In any case, several researchers report that microglia may multiply after traumatic injury. The impact of aging on microglial replication is controversial with the consensus that the number of microglia does not increase with age (Mildner et al., 2007; Wohleb et al., 2013).

One more interesting feature worth mentioning here is that, depending on specific anatomical organization or activation profile, microglia possess different morphological features such as lysosome content, electrophysiological activities (i.e., hyperpolarized resting potentials and differential membrane capacitance), membrane composition, and gene transcriptome profile. For example, in the basal ganglia, the expressions of genes related to classical microglial profile such as branch dynamics and cytoskeletal modulation, inflammatory signaling, immune function, and homeostasis have a propensity to be preserved across different sections. Interestingly, the expressions of genes engaged with mitochondrial functioning, cell digestion, oxidative flagging, ROS homeostasis, and lysosomal work are differentially communicated in various basal ganglia locales (De Biase et al., 2017).

2.3.3 Microglia Quickly Adapt to Changes in Central Nervous System Microenvironment

Microglia are involved in the immune defense of the brain and also participate in the maintenance of homeostasis condition. These cells alter their phenotype when CNS homeostasis is compromised in a state described as microglia activation. In this condition, microglia may discharge several cytotoxic mediators such as ROS, proinflammatory cytokines, adenosine triphosphate (ATP), and arachidonic acid (AA), which are involved in numerous CNS disorders (Maher et al., 2005).

In developed brain, microglia for the most part keep up homeostasis and exist in a resting state portrayed by ramified morphology with exceptionally motile, profoundly dynamic, long cellular processes more than one cell breadth long. These cell forms experience ceaseless patterns of expansion, withdrawal, and de novo formation to persistently check their condition for disturbances in homeostasis (Morris et al., 2018). While resting adult microglia express very low level of many activation markers, they secrete neurotrophic factors that include BDNF, insulin-like growth factor-1 (IGF-1), TGF-β, and nerve growth factor (NGF). However, resting microglia express high levels of microRNA-124 and have decreased expressions of CD46, major histocompatibility complex-II (MHC-II), and CD11b, which contribute to cell maintenance and survival (Orihuela et al., 2016). Subsequently, microglia are in resting state during typical physiological conditions. During this state, they are continually reviewing and examining the microenvironment to identify any modification in homeostasis.

In response to any trauma/pathogen invasion, microglia have the capability to alter into different functional forms, transforming their propagation and morphology. Microglia cells represent a rather heterogeneous community with subpopulations being distinct by their functional ability. Two types of microglia activation modes have been studied: classical M1 microglia and alternative M2 phenotype. M1 phenotype produces an excess of inflammatory mediators, including ROS, superoxide anion matrix metalloproteinase (MMP)-9, proinflammatory cytokines, and chemokines. On the contrary, M2 phenotype releases protective cytokines such as IL-10, transforming growth factor b, IL-4, and IL-13, which play significant roles in tissue repair and curing lesion (Loane & Kumar, 2016).

2.4 EXPRESSION OF IMMUNOLOGICAL GENES IN AGING BRAIN

As we have discussed so far, aging may lead to subtle changes and impairment in many different aspects of brain functioning. Gene expression profile in terms of aging is indicative of inflammatory response, oxidative stress, and decreased neurotrophic support in neocortex and cerebellum. At the transcriptional level, brain aging in mice displays parallels with human neurodegenerative disorders (Lee et al., 2000). The transmembrane protein TMEM119 has been identified as a specific marker of microglia and thus has enabled global gene expression profiling of highly pure preparations of microglia feasible (Bennett et al., 2016).

As we know so far, regulation of microglial phenotype generally relies on their communication with molecules released by nearby cells (neurons, microglial cells, astrocytes, etc.) in the microenvironment via membrane-bound pattern recognition receptors (PRRs). These PRRs can be distinguished on the basis of their affinity for molecules related to pathogens (pathogen-associated molecular patterns, PAMPs) or cellular injury (danger-associated molecular patterns, DAMPs) (Venegas and Heneka, 2017). However, apart from PRRs, microglial cells themselves play a key role in regulation, as these cells are also loaded with a range of receptors to detect various different molecules such as hormones and neurotransmitters (Kettenmann et al., 2011). It is thought that based on signals received, microglia can polarize to two contrasting states either M1 or M2 state, where M1 is proinflammatory and M2 is phagocytic and supports in tissue regeneration (Ransohoff, 2016). However, sequencing of microglial transcriptomes in resting (surveillance) and reactive states has found that their reactive signature is very diverse and complex to characterize (Hirbec et al., 2018).

Conventionally, the proinflammatory phenotype of microglia, which is the M1 state, can generate and discharge ROS, NRS, and cytokines such as TNF-α, IL-1β and IL-12. On the contrary, the M2 phenotype, the antiinflammatory form, produces and releases trophic factors, such as TGF-β and BDNF (Tang and Le, 2016). Detailed studies involving transcriptional differences in microglial cells have been recently emphasized to be associated with disease progression. For instance, it has been established that microglial cells play a precise role in neurodegenerative disorders based on apolipoprotein E (APOE) and activating receptor found on myeloid cells 2 (TREM2). APOE intonation of microglial cells activation via TREM2 control contributes to the neuronal loss in an acute model of neurodegeneration (Krasemann et al., 2017). Thus, APOE-TREM2-reliant microglial regulation is important to control homeostatic microglia. It is very significant to understand that reestablishment of homeostatic microglial function stops the neuronal damage in AD model (Krasemann et al., 2017).

Regulation of the microglial metabolism has also been linked to TREM2 by supporting cellular energetic and biosynthetic metabolism (Ulland et al., 2017), which is very important to preserve the high microglial activity required in the pathology, to deal with the surplus of amyloid pathology and plaque deposition. Significance of the microenvironment in defining the microglial profile has further been demonstrated when after irradiation, brain-engrafting macrophages have diverse phenotype compared with resident microglial cells, thus highlighting unique phenotype of microglial cells in the brain (Cronk et al., 2018).

de Haas et al. studied the expression profile of microglial cells and established that throughout areas of striatum, hippocampus, spinal cord, cerebellum, and cerebral cortex, there is a clear expression of CD11b, CD40, CD45, CD80, CD86, TREM-2b, F4/80, CXCR3, and CCR9, whereas for either major histocompatibility complex II (MHCII) or CCR7, no expression was found (de Haas et al., 2008).

Aging is considered the principal danger factor for numerous neurodegenerative disorders and is connected with chronic altered inflammation involving alterations in microglial morphology, phenotype, and activity. In this regard, it has been demonstrated that region-dependent transcripts are differentially controlled during senescence under physiological conditions. Scientists have also suggested an immunophenotypic distinction of brain microglia, in which cortical and striatal microglia are alike while hippocampal microglia represent an intermediate profile between pro- and antiinflammatory states (Hanamsagar and Bilbo, 2017).

It is interesting to know that most neurological conditions take place in a region-defined way and show discrepant regulatory method of gene expression. Also in trauma-inflicted lesions, the spinal white matter displayed increased microgliosis than spinal gray matter. So it may be speculated that myelin composition is a factor that manipulates the microglial activity. As a matter of fact, increased inflammatory response was established in white matter compared with gray matter in both brain and spinal cord (Batchelor et al., 2008).

Researchers have found that inflammation inside CNS escalates with age. Increased oxidative stress and lipid peroxidation are considered hallmarks of brain aging, so it might be considered that buildup of free radical damage over time directs the way to amplified inflammation in brain. Consistently, many microarray experiments show that there is, in general, rise in inflammatory and prooxidant genes, while there is a decrease in development and antioxidant genes in the brain of older mice compared with adults (Godbout et al., 2005). Experimental evidence suggests that protein levels of several inflammatory cytokines, including IL-1β and IL-6, in the brain of aged rodents increase significantly, along with decrease in several antiinflammatory cytokines including IL-10 and IL-4 (Maher et al., 2005). This showcases that there is an age related change in the inflammatory profile of the brain. Also there is amplified expression of inflammatory markers such as MHC II and complement receptor 3 (CD11b) in the aged brain of humans; almost 25% of microglia from aged mice were MHC II positive, compared with only 2% of microglia from adult mice (Frank et al., 2006). In consistence with the increased inflammatory profile of microglia, it is also evident that there is presence of activated morphology in the brain of aged mice. When microglia are stained against ionized calcium-binding adaptor protein-1 (Iba1), a protein expressed on the surface of microglia suggests that microglia from healthy brains in aged animals have shorter and less branched dendritic processes than young ones.

A large section of microglia research is carried out on mouse model, while relatively less is done in human model. Though only around 30% of microglia genes found in humans are enriched in mouse microglia, there is a core set of significant microglia genes whose expression is conserved between mice and human (Thion et al., 2018). Learning the functional effect of genes that are either differentially expressed or not expressed at all in one or the other species might be challenging, particularly when the divergence is closely related with neurodevelopmental disorders. However, some prominent similarities validate the use of mouse models to learn microglial role all through development. Also it has been found that microglia inhabit the human brain over a similar timeline to that in rodents (Konishi et al., 2017; Zhang et al., 2016).

2.5 THERAPEUTIC ASPECTS IN NEURODEGENERATIVE DISORDERS

The major threat for several brain disorders is aging; a significant number influence brain structure and function. Due to AD, PD, and other types of dementia, anomalous proteins aggregate and give rise to plaques and tangles that cause harm to brain cells. Apart from that, other factors such as medications, loss in vision, hearing deficit, insomnia, and depression may hinder with normal brain function; as a result, cognitive ability declines.

We know that the nervous system and immune system work together directly to control the body's immune responses, plus inflammatory reaction. In accordance with that, the term neuroimmune system refers collectively to the immune system and other components of the nervous system that regulate immune responses. This area also encompasses hormones and other signaling molecules that help communicate between the immune and nervous systems. It is widely known that innate immune system, a part of the neuroimmune system, is a complex network of cells and signaling molecules that are present from birth and form the first line of defense system in the body including inflammatory reactions.

We have discussed earlier in this chapter that microglia are very sensitive cells and can easily become activated; during the initial stages of activation, there are secretion of signaling molecules and subtle morphological changes along with increased production of molecules involved in immune responses (major histocompatibility complex [MHC]) as well as of Toll-like receptors (Graeber, 2010). Activation of microglia and astrocytes also enhances proinflammatory agents, including TNF-α, which alters neurotransmission, including signal transmission regulated by the excitatory neurotransmitter glutamate.

Thanks to the extensive research on aging, scientists can intervene in the process at the molecular and cellular level. Over the past decades, researchers have unfolded molecular mechanisms involved in the process of aging that should enable them to mediate in and slow it down.

Although neurodegenerative diseases have been on focus, there is lack of therapies that can rectify the underlying processes. It has been noticed that as the disease advances, the effectiveness of pharmacological treatments is decreased, despite increase in medication dosage. As a result, the benefit to side effect ratio decreases. Intracerebral drug delivery, especially gene therapy, is a hopeful approach for tackling deficits in medical administration. By modifying the expression of specific proteins, gene therapy may pave way for neuroprotection, neurorestoration, and finally, correction of the underlying pathogenic mechanism.

One of the drugs, prescribed to organ transplant patients to suppress immune response, is called rapamycin, which basically targets and inhibits mTOR and may promote phagocytosis, which is accountable for clearing up senescent cells and debris. This drug may prove advantageous in AD and PD during which there is plaque buildup in brain tissues. Another very popular drug for diabetes, metformin, may also improve autophagy and reduce DNA damage by participating in mTOR inhibition. It has been shown to lower the occurrence of heart ailments and cancer, with reduced cognitive impairment and longer life span (Inyang et al., 2019)

In another front, gene therapy is gaining potential as a potent therapeutic strategy for a variety of neurodegenerative disorders, including AD, PD, and HD. Gene therapy can be generally defined as the transfer of specific genetic material to particular target cells in a patient so as to prevent or improve the disease outcome. Vectors are the carriers to deliver therapeutic genetic material, which may be viral, but several nonviral techniques are also being used. The technique usually implies introducing altered genes into the body of a patient instead of just the products of cells with altered genes. Genetic engineering that involves genetic alteration of living cells, as well as implantation of genetically engineered cells into the living body, is also a form of gene therapy. Neurosurgeons refer to gene therapy as "cellular and molecular" neurosurgery (Åkerud et al., 2001; Coune et al., 2012).

With advancement in technology, now several viral and nonviral vectors can successfully be transferred into the diseased tissues of human and animals. These transgenes are capable of expressing therapeutic proteins, antibodies, Cas9/gRNA for gene editing, micro-RNAs, and small interfering RNA (siRNA). For neurodegenerative disorders, the most commonly applied vector is one of the adeno-associated viruses (AAVs) (Wang et al., 2019). Additionally, greater amount of capsids can be used in different species to successfully target multiple cells or tissues in the CNS, together with neurons, oligodendrocytes, and astrocytes (Xiang et al., 2020). Several preclinical studies have established that delivery of gene could be done via numerous routes including but not restricted to subpial, intrathecal, intracerebroventricular, intravitreal, intraparenchymal, and subretinal injection and can attain adequate extent of gene in diseased tissues (Karumuthil-Melethil et al., 2016; Morabito et al., 2017). With advancement in AAVs and nonviral delivery systems, gene therapies have promising results in terms of wider transgene expression and therapeutic safety. Significantly, tremendous functional outcomes have been reported in experimental models of various neurodegenerative disorders, including PD, AD, HD, and aromatic-L-amino-acid decarboxylase (AADC) deficiency (Axelsen and Woldbye, 2018).

AD, described by aggregation of extracellular β-amyloid plaques, intracellular neurofibrillary tangles, and neurodegeneration with indicative symptoms including dynamic psychological decrease and cognitive decline with synaptic loss, is a typical issue in aging. Early determination is not possible, and the broadly affirmed treatment uses cholinesterase inhibitors to postpone the beginning of cognitive decline without altering the disease progression. Proposed treatments with gene therapy approaches expect to lessen amyloid plaques and tangles or acquaint neurotrophic factors, which decrease the fatality of synapses. Changes in the SORL1 gene that may underlie pathology in irregular instances of AD could offer further direct roads to controlling the hereditary reasons for the most widely recognized type of AD (Rogaeva et al., 2007).

In **PD**, two different ways of treatment have been used so far in AAV2-mediated gene therapy clinical trials. First is based on improving clinical symptoms rather than modifying the disease succession called the symptomatic. In another strategy, gene transfer of glutamic acid decarboxylase (GAD), which catalyzes production of GABA from glutamate or AADC, a rate-limiting enzyme for dopamine production that converts levodopa to dopamine, is being done. In neurorestorative therapies, the focus is on restoring cellular functions impacted by the disease process. It basically involves in vivo gene relocation of a neurotrophic factor, either glial-derived neurotropic factor (GDNF) or neurturin (Sudhakar and Richardson, 2019).

HD is an autosomal dominant disorder that is caused by presence of extreme amount of trinucleotide CAG repeats on exon 1 of chromosome 4, giving rise to polyglutaminated huntingtin (mHtt) protein (MacDonald, 1993). Elevated levels of mHtt aggregation inside striatal medium spiny neurons and pyramidal cortical neurons are accountable for HD pathological phenotype. Patients in their 40s are often challenged with motor, cognitive, and psychiatric dysfunction. Motor symptoms may differ as the disease progresses. In the beginning, patients show choreiform movements with gait impairment, which leads to bradykinesia and rigidity. In HD, cognitive alteration includes reduction in attention and mental flexibility, whereas psychiatric manifestations include impulsivity, depression, and apathy (Ross and Tabrizi, 2011).

Gene therapy is a well-known treatment elective for HD, as it offers direct focusing of dysfunctional nuclei to diminish levels of mHtt. Preclinical gene therapy procedures for HD have progressed and now are engaged around utilizing RNA interference (RNAi) to limit pathologic degrees of mHtt. RNAi uses little RNA particles like short interfering RNA, short hairpin RNA, and micro-RNA to target mRNA molecules and destroy them by the RNA-induced silencing complex (Hutvagner and Simard, 2008).

In spite of all the attractive features, gene therapy is quite complex and may be challenging avenue for researchers.

2.6 CHALLENGES IN GENE THERAPY

There are numerous preclinical and clinical investigations progressing depending on gene therapy methodologies for forestalling or treating a wide scope of neurodegenerative sicknesses. One of the significant concerns is the well-being issue that is perhaps the greatest obstacle to its effective clinical application. Gene therapy may cause serious toxicity in the target cells and tissues due to either overexpression of transgene in focused tissues or expression in off-target cells. This toxicity may have several damaging effect including debilitated ambulation, ataxia, harmed dorsal root ganglia, and raised transaminases (Hinderer et al., 2018). Another important factor to be considered is host response that may affect the extent and safety of gene therapy treatment. Patients with adaptive immune responses sometimes generate corresponding neutralizing antibodies, which hinder the vectors from getting to their respective cells (Vandamme et al., 2017).

There are cases where insertional mutagenesis and genotoxicity have posed a challenge when defined transgenes are injected with high-dose vectors (Chandler et al., 2017). Keeping the aforementioned challenges in mind, gene therapy−based treatment strategy for neurodegenerative disorders should be strictly scrutinized. Safety measures should be of prime focus to reap the benefits of such therapy.

With advancement in science and technology, the average life span of human beings is increasing worldwide; understanding the brain mechanisms especially the neuroimmunology aspect plays a crucial role in healthy living.

REFERENCES

Åkerud, P., Canals, J.M., Snyder, E.Y., Arenas, E., 2001. Neuroprotection through delivery of glial cell line-derived neurotrophic factor by neural stem cells in a mouse model of Parkinson's disease. J. Neurosci. 21 (20), 8108−8118. https://doi.org/10.1523/JNEUROSCI.21-20-08108.2001.

Asha Devi, S., Satpati, A., 2017. Oxidative stress and the brain: an insight into cognitive ageing. In: Topics in Biomedical Gerontology. Springer, Singapore, pp. 123−140. https://doi.org/10.1007/978-981-10-2155-8_8.

Axelsen, T.M., Woldbye, D.P.D., 2018. Gene therapy for Parkinson's disease, an update. J. Parkinsons Dis. 8 (2), 195−215. https://doi.org/10.3233/JPD-181331.

Batchelor, P.E., Tan, S., Wills, T.E., Porritt, M.J., Howells, D.W., 2008. Comparison of inflammation in the brain and spinal cord following mechanical injury. J. Neurotrauma 25 (10), 1217−1225. https://doi.org/10.1089/neu.2007.0308.

Bennett, M.L., Bennett, F.C., Liddelow, S.A., Ajami, B., Zamanian, J.L., Fernhoff, N.B., Mulinyawe, S.B., Bohlen, C.J., Adil, A., Tucker, A., Weissman, I.L., Chang, E.F., Li, G., Grant, G.A., Hayden Gephart, M.G., Barres, B.A., 2016. New tools for studying microglia in the mouse and human CNS. Proc. Natl. Acad. Sci. USA 113 (12), E1738−E1746. https://doi.org/10.1073/pnas.1525528113.

Burke, S.N., Barnes, C.A., 2006. Neural plasticity in the ageing brain. Nat. Rev. Neurosci. 7 (1), 30−40. https://doi.org/10.1038/nrn1809.

Chandler, R.J., Sands, M.S., Venditti, C.P., 2017. Recombinant adeno-associated viral integration and genotoxicity: insights from animal models. Hum. Gene Ther. 28 (4), 314−322. https://doi.org/10.1089/hum.2017.009.

Coune, P.G., Schneider, B.L., Aebischer, P., 2012. Parkinson's disease: gene therapies. Cold Harb. Prespect. Med. 2 (4), a009431. https://doi.org/10.1101/cshperspect.a009431.

Cronk, J.C., Filiano, A.J., Louveau, A., Marin, I., Marsh, R., Ji, E., Goldman, D.H., Smirnov, I., Geraci, N., Acton, S., Overall, C.C., Kipnis, J., 2018. Peripherally derived macrophages can engraft the brain independent of irradiation and maintain an identity distinct from microglia. J. Exp. Med. 215 (6), 1627−1647. https://doi.org/10.1084/jem.20180247.

da Silva, T.M., Munhoz, R.P., Alvarez, C., Naliwaiko, K., Kiss, Á., Andreatini, R., Ferraz, A.C., 2008. Depression in Parkinson's disease: a double-blind, randomized, placebo-controlled pilot study of omega-3 fatty-acid supplementation. J. Affect. Disord. 111 (2−3), 351−359. https://doi.org/10.1016/j.jad.2008.03.008.

De Biase, L.M., Schuebel, K.E., Fusfeld, Z.H., Jair, K., Hawes, I.A., Cimbro, R., Zhang, H.-Y., Liu, Q.-R., Shen, H., Xi, Z.-X., Goldman, D., Bonci, A., 2017. Local cues establish and maintain region-specific phenotypes of basal ganglia microglia. Neuron 95 (2), 341−356.e6. https://doi.org/10.1016/j.neuron.2017.06.020.

de Castro, F., 2019. Cajal and the Spanish neurological school: neuroscience would have been a different story without them. Front. Cell. Neurosci. 13 https://doi.org/10.3389/fncel.2019.00187.

de Haas, A.H., Boddeke, H.W.G.M., Biber, K., 2008. Region-specific expression of immunoregulatory proteins on microglia in the healthy CNS. Glia 56 (8), 888−894. https://doi.org/10.1002/glia.20663.

Frank, M.G., Barrientos, R.M., Biedenkapp, J.C., Rudy, J.W., Watkins, L.R., Maier, S.F., 2006. mRNA up-regulation of MHC II and pivotal pro-inflammatory genes in normal brain ageing. Neurobiol. Ageing 27 (5), 717−722. https://doi.org/10.1016/j.neurobiolageing.2005.03.013.

Godbout, J.P., Chen, J., Abraham, J., Richwine, A.F., Berg, B.M., Kelley, K.W., Johnson, R.W., 2005. Exaggerated neuroinflammation and sickness behavior in aged mice following activation of the peripheral innate immune system. FASEB. J. 19 (10), 1329−1331. https://doi.org/10.1096/fj.05-3776fje.

Gounder, S.S., Kannan, S., Devadoss, D., Miller, C.J., Whitehead, K.S., Odelberg, S.J., Firpo, M.A., Paine, R., Hoidal, J.R., Abel, E.D., Rajasekaran, N.S., 2012. Impaired transcriptional activity of Nrf2 in age-related myocardial oxidative stress is reversible by moderate exercise training. PloS One 7 (9), e45697. https://doi.org/10.1371/journal.pone.0045697.

Graeber, M.B., 2010. Changing face of microglia. Science 330 (6005), 783−788. https://doi.org/10.1126/science.1190929.

Hanamsagar, R., Bilbo, S.D., 2017. Environment matters: microglia function and dysfunction in a changing world. Curr. Opin. Neurobiol. 47, 146−155. https://doi.org/10.1016/j.conb.2017.10.007.

Harman, D., 2006. Ageing: overview. Ann. N. Y. Acad. Sci. 928 (1), 1−21. https://doi.org/10.1111/j.1749-6632.2001.tb05631.x.

Hinderer, C., Katz, N., Buza, E.L., Dyer, C., Goode, T., Bell, P., Richman, L.K., Wilson, J.M., 2018. Severe toxicity in nonhuman primates and piglets following high-dose intravenous administration of an adeno-associated virus vector expressing human SMN. Hum. Gene Ther. 29 (3), 285−298. https://doi.org/10.1089/hum.2018.015.

Hirbec, H., Marmai, C., Duroux-Richard, I., Roubert, C., Esclangon, A., Croze, S., Lachuer, J., Peyroutou, R., Rassendren, F., 2018. The microglial reaction signature revealed by RNAseq from individual mice. Glia 66 (5), 971−986. https://doi.org/10.1002/glia.23295.

Hutvagner, G., Simard, M.J., 2008. Argonaute proteins: key players in RNA silencing. Nat. Rev. Mol. Cell Biol. 9 (1), 22−32. https://doi.org/10.1038/nrm2321.

Inyang, K.E., Szabo-Pardi, T., Wentworth, E., McDougal, T.A., Dussor, G., Burton, M.D., Price, T.J., 2019. The antidiabetic drug metformin prevents and reverses neuropathic pain and spinal cord microglial activation in male but not female mice. Pharmacol. Res. 139, 1−16. https://doi.org/10.1016/j.phrs.2018.10.027.

Kaasinen, V., Vilkman, H., Hietala, J., Någren, K., Helenius, H., Olsson, H., Farde, L., Rinne, J., 2000. Age-related dopamine D2/D3 receptor loss in extrastriatal regions of the human brain. Neurobiol. Ageing 21 (5), 683−688. https://doi.org/10.1016/s0197-4580(00)00149-4.

Karumuthil-Melethil, S., Marshall, M.S., Heindel, C., Jakubauskas, B., Bongarzone, E.R., Gray, S.J., 2016. Intrathecal administration of AAV/GALC vectors in 10-11-day-old twitcher mice improves survival and is enhanced by bone marrow transplant. J. Neurosci. Res. 94 (11), 1138−1151. https://doi.org/10.1002/jnr.23882.

Kettenmann, H., Hanisch, U.-K., Noda, M., Verkhratsky, A., 2011. Physiology of microglia. Physiol. Rev. 91 (2), 461−553. https://doi.org/10.1152/physrev.00011.2010.

Konishi, H., Kobayashi, M., Kunisawa, T., Imai, K., Sayo, A., Malissen, B., Crocker, P.R., Sato, K., Kiyama, H., 2017. Siglec-H is a microglia-specific marker that discriminates microglia from CNS-associated macrophages and CNS-infiltrating monocytes. Glia 65 (12), 1927−1943. https://doi.org/10.1002/glia.23204.

Krasemann, S., Madore, C., Cialic, R., Baufeld, C., Calcagno, N., El Fatimy, R., Beckers, L., O'Loughlin, E., Xu, Y., Fanek, Z., Greco, D.J., Smith, S.T., Tweet, G., Humulock, Z., Zrzavy, T.,

Conde-Sanroman, P., Gacias, M., Weng, Z., Chen, H., Butovsky, O., 2017. The TREM2-APOE pathway drives the transcriptional phenotype of dysfunctional microglia in neurodegenerative diseases. Immunity 47 (3), 566–581.e9. https://doi.org/10.1016/j.immuni.2017.08.008.

Langa, K.M., Levine, D.A., 2014. The diagnosis and management of mild cognitive impairment. J. Am. Med. Assoc. 312 (23), 2551. https://doi.org/10.1001/jama.2014.13806.

Lee, C.-K., Weindruch, R., Prolla, T.A., 2000. Gene-expression profile of the ageing brain in mice. Nat. Genet. 25 (3), 294–297. https://doi.org/10.1038/77046.

Loane, D.J., Kumar, A., 2016. Microglia in the TBI brain: the good, the bad, and the dysregulated. Exp. Neurol. 275, 316–327. https://doi.org/10.1016/j.expneurol.2015.08.018.

Ma, C.-L., Ma, X.-T., Wang, J.-J., Liu, H., Chen, Y.-F., Yang, Y., 2017. Physical exercise induces hippocampal neurogenesis and prevents cognitive decline. Behav. Brain Res. 317, 332–339. https://doi.org/10.1016/j.bbr.2016.09.067.

MacDonald, M., 1993. A novel gene containing a trinucleotide repeat that is expanded and unstable on Huntington's disease chromosomes. Cell 72 (6), 971–983. https://doi.org/10.1016/0092-8674(93)90585-E.

Maher, F.O., Nolan, Y., Lynch, M.A., 2005. Downregulation of IL-4-induced signalling in hippocampus contributes to deficits in LTP in the aged rat. Neurobiol. Ageing 26 (5), 717–728. https://doi.org/10.1016/j.neurobiolageing.2004.07.002.

Meltzer, M.D.,C., 1998. Serotonin in ageing, late-life depression, and Alzheimer's disease: the emerging role of functional imageing. Neuropsychopharmacology 18 (6), 407–430. https://doi.org/10.1016/S0893-133X(97)00194-2.

Mildner, A., Schmidt, H., Nitsche, M., Merkler, D., Hanisch, U.-K., Mack, M., Heikenwalder, M., Brück, W., Priller, J., Prinz, M., 2007. Microglia in the adult brain arise from Ly-6ChiCCR2+ monocytes only under defined host conditions. Nat. Neurosci. 10 (12), 1544–1553. https://doi.org/10.1038/nn2015.

Mittelbronn, M., Dietz, K., Schluesener, H.J., Meyermann, R., 2001. Local distribution of microglia in the normal adult human central nervous system differs by up to one order of magnitude. Acta Neuropathol. 101 (3), 249–255. https://doi.org/10.1007/s004010000284.

Morabito, G., Giannelli, S.G., Ordazzo, G., Bido, S., Castoldi, V., Indrigo, M., Cabassi, T., Cattaneo, S., Luoni, M., Cancellieri, C., Sessa, A., Bacigaluppi, M., Taverna, S., Leocani, L., Lanciego, J.L., Broccoli, V., 2017. AAV-PHP.B-Mediated global-scale expression in the mouse nervous system enables GBA1 gene therapy for wide protection from synucleinopathy. Mol. Ther. 25 (12), 2727–2742. https://doi.org/10.1016/j.ymthe.2017.08.004.

Morris, G., Walker, A.J., Berk, M., Maes, M., Puri, B.K., 2018. Cell death pathways: a novel therapeutic approach for neuroscientists. Mol. Neurobiol. 55 (7), 5767–5786. https://doi.org/10.1007/s12035-017-0793-y.

Mosher, K.I., Wyss-Coray, T., 2014. Microglial dysfunction in brain ageing and Alzheimer's disease. Biochem. Pharmacol. 88 (4), 594–604. https://doi.org/10.1016/j.bcp.2014.01.008.

Obermeier, B., Verma, A., Ransohoff, R.M., 2016. The Blood–Brain Barrier, pp. 39–59. https://doi.org/10.1016/B978-0-444-63432-0.00003-7.

Orihuela, R., McPherson, C.A., Harry, G.J., 2016. Microglial M1/M2 polarization and metabolic states. Br. J. Pharmacol. 173 (4), 649–665. https://doi.org/10.1111/bph.13139.

Patterson, S.L., 2015. Immune dysregulation and cognitive vulnerability in the ageing brain: interactions of microglia, IL-1β, BDNF and synaptic plasticity. Neuropharmacology 96, 11–18. https://doi.org/10.1016/j.neuropharm.2014.12.020.

Petralia, R.S., Mattson, M.P., Yao, P.J., 2014. Communication breakdown: the impact of ageing on synapse structure. Ageing Res. Rev. 14, 31–42. https://doi.org/10.1016/j.arr.2014.01.003.

Porges, E.C., Woods, A.J., Edden, R.A.E., Puts, N.A.J., Harris, A.D., Chen, H., Garcia, A.M., Seider, T.R., Lamb, D.G., Williamson, J.B., Cohen, R.A., 2017. Frontal gamma-aminobutyric acid concentrations are associated with cognitive performance in older adults. Biol. Psychiatr. 2 (1), 38–44. https://doi.org/10.1016/j.bpsc.2016.06.004.

Pösel, C., Möller, K., Boltze, J., Wagner, D.-C., Weise, G., 2016. Isolation and flow cytometric analysis of immune cells from the ischemic mouse brain. J. Vis. Exp. 108 https://doi.org/10.3791/53658.

Puhlmann, L.M.C., Valk, S.L., Engert, V., Bernhardt, B.C., Lin, J., Epel, E.S., Vrticka, P., Singer, T., 2019. Association of short-term change in leukocyte telomere length with cortical thickness and outcomes of mental training among healthy adults. JAMA Netw. Open 2 (9), e199687. https://doi.org/10.1001/jamanetworkopen.2019.9687.

Raefsky, S.M., Mattson, M.P., 2017. Adaptive responses of neuronal mitochondria to bioenergetic challenges: roles in neuroplasticity and disease resistance. Free Radical Biol. Med. 102, 203–216. https://doi.org/10.1016/j.freeradbiomed.2016.11.045.

Rankin, L.C., Artis, D., 2018. Beyond host defense: emerging functions of the immune system in regulating complex tissue physiology. Cell 173 (3), 554–567. https://doi.org/10.1016/j.cell.2018.03.013.

Ransohoff, R.M., 2016. A polarizing question: do M1 and M2 microglia exist? Nat. Neurosci. 19 (8), 987–991. https://doi.org/10.1038/nn.4338.

Regulski, M.J., 2017. Cellular senescence: what, why, and how. Wounds 29 (6), 168–174. http://www.ncbi.nlm.nih.gov/pubmed/28682291.

Rogaeva, E., Meng, Y., Lee, J.H., Gu, Y., Kawarai, T., Zou, F., Katayama, T., Baldwin, C.T., Cheng, R., Hasegawa, H., Chen, F., Shibata, N., Lunetta, K.L., Pardossi-Piquard, R., Bohm, C., Wakutani, Y., Cupples, L.A., Cuenco, K.T., Green, R.C., St George-Hyslop, P., 2007. The neuronal sortilin-related receptor SORL1 is genetically associated with Alzheimer disease. Nat. Genet. 39 (2), 168–177. https://doi.org/10.1038/ng1943.

Ross, C.A., Tabrizi, S.J., 2011. Huntington's disease: from molecular pathogenesis to clinical treatment. Lancet Neurol. 10 (1), 83–98. https://doi.org/10.1016/S1474-4422(10)70245-3.

Salim, S., 2017. Oxidative stress and the central nervous system. J. Pharmacol. Exp. Therapeut. 360 (1), 201–205. https://doi.org/10.1124/jpet.116.237503.

Shay, J.W., Wright, W.E., 2000. Hayflick, his limit, and cellular ageing. Nat. Rev. Mol. Cell Biol. 1 (1), 72–76. https://doi.org/10.1038/35036093.

Sierra, A., Abiega, O., Shahraz, A., Neumann, H., 2013. Janus-faced microglia: beneficial and detrimental consequences of microglial phagocytosis. Front. Cell. Neurosci. 7 https://doi.org/10.3389/fncel.2013.00006.

Spierings, E., Fleischhauer, K., 2019. Histocompatibility. In: The EBMT Handbook: Hematopoietic Stem Cell Transplantation and Cellular Therapies. http://www.ncbi.nlm.nih.gov/pubmed/32091748.

Streit, W.J., Sammons, N.W., Kuhns, A.J., Sparks, D.L., 2004. Dystrophic microglia in the ageing human brain. Glia 45 (2), 208–212. https://doi.org/10.1002/glia.10319.

Sudhakar, V., Richardson, R.M., 2019. Gene therapy for neurodegenerative diseases. Neurotherapeutics 16 (1), 166–175. https://doi.org/10.1007/s13311-018-00694-0.

Tang, Y., Le, W., 2016. Differential roles of M1 and M2 microglia in neurodegenerative diseases. Mol. Neurobiol. 53 (2), 1181–1194. https://doi.org/10.1007/s12035-014-9070-5.

Thion, M.S., Low, D., Silvin, A., Chen, J., Grisel, P., Schulte-Schrepping, J., Blecher, R., Ulas, T., Squarzoni, P., Hoeffel, G., Coulpier, F., Siopi, E., David, F.S., Scholz, C., Shihui, F., Lum, J., Amoyo, A.A., Larbi, A., Poidinger, M., Garel, S., 2018. Microbiome influences prenatal and adult microglia in a sex-specific manner. Cell 172 (3), 500–516.e16. https://doi.org/10.1016/j.cell.2017.11.042.

Trist, B.G., Hare, D.J., Double, K.L., 2019. Oxidative stress in the ageing substantia nigra and the etiology of Parkinson's disease. Ageing Cell 18 (6). https://doi.org/10.1111/acel.13031.

Ulland, T.K., Song, W.M., Huang, S.C.-C., Ulrich, J.D., Sergushichev, A., Beatty, W.L., Loboda, A.A., Zhou, Y., Cairns, N.J., Kambal, A., Loginicheva, E., Gilfillan, S., Cella, M., Virgin, H.W., Unanue, E.R., Wang, Y., Artyomov, M.N., Holtzman, D.M., Colonna, M., 2017. TREM2 maintains microglial metabolic fitness in Alzheimer's disease. Cell 170 (4), 649–663.e13. https://doi.org/10.1016/j.cell.2017.07.023.

Vandamme, C., Adjali, O., Mingozzi, F., 2017. Unraveling the complex story of immune responses to AAV vectors trial after trial. Hum. Gene Ther. 28 (11), 1061–1074. https://doi.org/10.1089/hum.2017.150.

Venegas, C., Heneka, M.T., 2017. Danger-associated molecular patterns in Alzheimer's disease. J. Leukoc. Biol. 101 (1), 87–98. https://doi.org/10.1189/jlb.3MR0416-204R.

von Bartheld, C.S., Bahney, J., Herculano-Houzel, S., 2016. The search for true numbers of neurons and glial cells in the human brain: a review of 150 years of cell counting. J. Comp. Neurol. 524 (18), 3865–3895. https://doi.org/10.1002/cne.24040.

Wang, D., Tai, P.W.L., Gao, G., 2019. Adeno-associated virus vector as a platform for gene therapy delivery. Nat. Rev. Drug Discov. 18 (5), 358–378. https://doi.org/10.1038/s41573-019-0012-9.

Wohleb, E.S., Powell, N.D., Godbout, J.P., Sheridan, J.F., 2013. Stress-induced recruitment of bone marrow-derived monocytes to the brain promotes anxiety-like behavior. J. Neurosci. 33 (34), 13820–13833. https://doi.org/10.1523/JNEUROSCI.1671-13.2013.

Wolf, S.A., Boddeke, H.W.G.M., Kettenmann, H., 2017. Microglia in physiology and disease. Annu. Rev. Physiol. 79 (1), 619–643. https://doi.org/10.1146/annurev-physiol-022516-034406.

Xiang, C., Zhang, Y., Guo, W., Liang, X.-J., 2020. Biomimetic carbon nanotubes for neurological disease therapeutics as inherent medication. Acta Pharm. Sin. B 10 (2), 239–248. https://doi.org/10.1016/j.apsb.2019.11.003.

Yin, F., Sancheti, H., Patil, I., Cadenas, E., 2016. Energy metabolism and inflammation in brain ageing and Alzheimer's disease. Free Radic. Biol. Med. 100, 108–122. https://doi.org/10.1016/j.freeradbiomed.2016.04.200.

Zhang, Y., Sloan, S.A., Clarke, L.E., Caneda, C., Plaza, C.A., Blumenthal, P.D., Vogel, H., Steinberg, G.K., Edwards, M.S.B., Li, G., Duncan, J.A., Cheshier, S.H., Shuer, L.M., Chang, E.F., Grant, G.A., Gephart, M.G.H., Barres, B.A., 2016. Purification and characterization of progenitor and mature human astrocytes reveals transcriptional and functional differences with mouse. Neuron 89 (1), 37–53. https://doi.org/10.1016/j.neuron.2015.11.013.

CHAPTER 3

Genetic Aspects of Early-Onset Alzheimer's Disease

VIJAY R. BOGGULA, MSC, PHD

3.1 BACKGROUND

Neurodegenerative disorders are caused due to defects in either neuronal transmission or neuronal cell deterioration. This group is also defined in terms of neuro-progressive disorders because of their manifestations seen in late ages (age-related disorders). One among them is Alzheimer's disease (AD), which includes heterogeneous and broad symptoms with dementia being prominent form seen (seen in 50%−75% individuals). Other common symptoms are attention deficit, lack of ability to perform daily activities, and disorientation ('2020 Alzheimer's disease facts and figures', 2020). Brain autopsy shows neurofibrillary tangles and senile plaques in neurons, which are characterized by the accumulation of hyperphosphorylated tau protein. These histological findings are due to failure in signaling cascade involved in phosphorylation process (Price et al., 1991). Based on the age of development of symptoms, it has been classified into two categories: early-onset ($\leq$60 years, EOAD) and late-onset (>60 years, LOAD). The National Institute on Aging recognizes AD in three groups: preclinical AD, mild cognitive impairment due to AD, and dementia due to AD (Sperling et al., 2011). Being a complex disorder, the cause depends on lifestyle, environmental factors, and genetic factors. At genome level, variations could be sequence to structural levels. The sequence variations can be studied in two broad groups: rare variants and common variants. These include mutations, polymorphisms occurring at gene, and genomic loci either affecting normal gene function involved in biological pathways or helping in knowing the risk toward susceptibility of AD. This chapter discusses genes involved in EOAD.

3.2 INTERPRETATION AND METHODOLOGY

The American College of Medical Genetics (ACMG) identifies mutations (permanent and rare change) and polymorphisms collectively as "variants." Based on their outcomes or effect, they are classified into five groups: (1) pathogenic, (2) likely pathogenic, (3) uncertain significance, (4) likely benign, and (5) benign (Richards et al., 2015). Various online tools (in silico) are available to check the functional effect of genomic variants such as PolyPhen-2 (polymorphism phenotyping version 2), SIFT (sorting intolerant from tolerant), Mutation Taster 2, and LRT (likelywood ratio test) (Ng, 2003; Chun and Fay, 2009; Adzhubei et al., 2010; Schwarz et al., 2014). EOAD mainly exists in monogenic (single gene) form; "mutation" term is used whereever applicable throughout the text.

Genetic analysis applies a broad range of techniques to find variants at gene and genome levels. In the past, linkage analysis was instrumental in identifying chromosomal regions in EOAD. It was very useful in identifying genes following Mendelian mode of inheritance in EOAD families but revealed less information in LOAD due to its complexity. Further candidate genes genotyping between individual with AD and controls revealed many important risk genes responsible for LOAD (Driscoll et al., 2019). With the understanding of genome variations, structure, and organization using modern techniques, many genetic variants have been found. From the past decade, researchers have been using SNP (single nucleotide polymorphism)-based genome-wide association studies (GWAS) (Kunkle et al., 2019). Using this approach, apart from APOE gene, researchers have identified many other loci or genes with reference to LOAD. This methodology can identify susceptibility genes, and rare variants cannot be identified. NGS (next-generation sequencing) is fruitful in identifying new rare variant genes responsible for EOAD. Whole-exome sequencing has been performed in large AD families with EOAD. Using this technique, only the genes are investigated, and many new variants can be identified (Novarino et al., 2014).

The Molecular Immunology of Neurological Diseases. https://doi.org/10.1016/B978-0-12-821974-4.00013-3

3.3 EARLY-ONSET ALZHEIMER'S DISEASE

Earlier studies on families with EOAD phenotype have identified genes responsible for the cause. Linkage analysis method was first used in one or more individuals with EOAD noticed in first-degree relatives. In brief, methodology involves phenotype association study using polymorphic markers that cotransmit. The inherited blocks encompassing markers adjacent to genes can be further studied by means of positional cloning strategy. The familiar gene mutations associated with EOAD are *APP* (OMIM 104760), *PSEN1* (OMIM 104311), and *PSEN2* (OMIM 600759) (Levy-Lahad et al., 1995; Sherrington et al., 1995; Mann et al., 1996). Mutations in these genes majorly exhibit autosomal dominant inheritance, and very few sporadic cases were also reported.

3.3.1 Amyloid-β Precursor Protein

APP gene structure consists of 18 exons spanning over 290 kilo bair pair (kbp) (genomic coordinates: 21: 25,880,549-26,171,127, https://www.ncbi.nlm.nih.gov/351). It is localized to long arm of 22 chromosomes (22q21.3). *APP* gene encodes amyloid precursor protein involved in various cellular processes such as neuronal proliferation, migration, differentiation, plasticity, and synaptogenesis (Young-Pearse et al., 2007; Joo et al., 2010; Rama et al., 2012; Vasques et al., 2017). This gene encodes several sizes of APP protein. Among them APP695 (central nervous system), APP751, and APP770 (peripheral and central nervous system) are predominately expressed (Tanaka et al.,

1989; Golde et al., 1990). APP protein undergoes proteolytic cleavage by means of α-, β-, and γ-secretase complexes (Fig. 3.1). The γ-secretase cleavage process can happen in either nonamyloidogenic or amyloidogenic pathways giving rise to Aβ peptides with different lengths: Aβ40 (90%) and Aβ 42(10%) (Sun et al., 2017). The Aβ42 are more prone to form aggregates when present in high proportion. Pathogenic mutations cause abnormal accumulation of these peptides in neuronal cell. Usually, change in the amino acid sequence at the site of cleavage leads to increased ratio of Aβ42/Aβ40. Commonly found mutations are more close to either β or γ cleavage sites. Most of the pathogenic mutations are reported in exons 16 and 17. According to LOVD database (Leiden open variation database, http://www.lovd.nl/APP), a more number of missense mutations are reported followed by gross insertions or duplications of genomic DNA region encompassing APP gene. Over 80 pathogenic mutations were reported till date, among these, 26 are unique pathogenic variants (also refer to http://alzforum.org/mutations/app). The common mutations seen in different ethnic population are in Aβ domain of β cleavage secretase site such as Arctic (E693G), Dutch (E693Q), Felmish (A692G), Iowa (D694N), Italian (E693K), and Piedmont (L705V) are associated with EOAD (Levy et al., 1990; Hendriks et al., 1992; Kamino et al., 1992; Miravalle, 2000; Grabowski et al., 2001; Obici et al., 2005). The most frequent mutations found among different ethnic population show change of

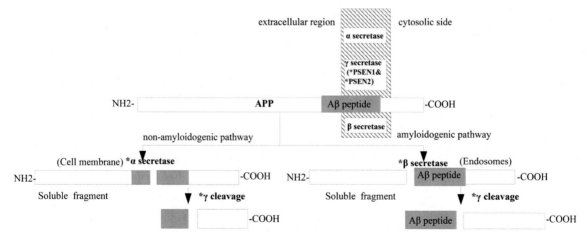

FIG. 3.1 Graphical representation of APP, PSEN1, and PSEN2 transmembrane proteins. APP protein undergoes proteolytic cleavage through amyloidogenic (full-length Aβ peptide) and nonamyloidogenic pathways (partial Aβ peptide). "*" denotes that mutations in or around cleavage sites results in abnormal Aβ peptide ratio. Also PSEN1 and PSEN2 gene mutations lead to abnormal cleavage efficiency. γ cleavage is final and common process in both pathways.

amino acid valine to isoleucine at position 717 of protein (Val717Ile, V717I) located in protein region encoded by exon 17 near to γ-secretase cleavage site (Talarico et al., 2010). The other missense mutations affecting the same amino acid are V717F, V717G, and V717L. This amino acid is located in intramembrane, not present in Aβ region, but affects relative levels of Aβ peptides. This missense mutation is seen in approximately 50% of APP-linked EOAD families and first to be reported (Chartier-Harlin et al., 1991; Murrell et al., 1991). Till date, no mutations are reported at C-terminal region (intracellular side). Near to β-secretase site, four missense mutations are reported: E682K, D678, A673, and double mutation KM670/671NL. All of them lead to variable levels of abnormal Aβ depositions (Mullan et al., 1992; Di Fede et al., 2009; Zhou et al., 2011; Chen et al., 2012; Jonsson et al., 2012). At α-secretase site, four pathogenic mutations are reported, leading to similar consequences like at β-secretase site (Van Broeckhoven et al., 1990; Tomiyama et al., 2008; Kaden et al., 2012). Mutations reported near to α- and β-secretase sites are present in Aβ regions. Also noncoding region mutations of *APP* gene are in intron-17 (IVS17 83-88delAAGTAT) and 3' UTR region (C.*331_*332del) associated with AD and CAA (cerebral amyloid angiopathy) phenotypes (Kamino et al., 1992; Nicolas et al., 2016a). Duplication of *APP* gene is also reported, which results in overexpression in Down syndrome individuals who are more prone to develop EOAD phenotypes (Rovelet-Lecrux et al., 2006; Sleegers et al., 2006; McNaughton et al., 2012). Initial studies have shown very small fraction of APP mutations in familial and sporadic cases. Later a study finding revealed that mutations in APP can be found in less than 20% with mean onset age of 51.2 years in mutation carriers (Tanzi et al., 1992; Raux, 2005). With reference to mutation rate in familial EOAD, APP stands second to *PSEN1*.

3.3.2 Presenilin 1

The PSEN1 encodes presenilin 1 protein, which is subunit of γ-secretase complex. The gene is comprised of 12 exons (87 Kbp) and encodes a 467-amino-acid (aa)-length protein (genomic coordinates:14:73,136,435-73,223,690, https://www.ncbi.nlm.nih.gov/gene/5663www.ncbi.nlm.nih.gov/gene/5663). This gene is localized to long arm of chromosome 14 (14q24.2). The heterologous proteolysis gives rise to 30 kDa (kilodalton) N-terminal (NTF) and 20 kDa C-terminal fragments (Fraering et al., 2004). It has nine transmembrane domains with large loop between seven and

eight domains having N-terminal facing cytosol and C-terminal toward the lumen. It is found to be localized in endoplasmic reticulum involved in processing of different proteins and cleaves many different proteins involved in cell signaling process from outside to the cell nucleus (e.g., Notch signaling) (Ray et al., 1999; Yang et al., 2019). As mentioned earlier, it is known to play a prominent role in processing of APP protein in neuronal cell. The γ-secretase complex is predominantly composed of presenilin (PSEN1 or PSEN2), nicastrin (NCSTN), and APN1. These four components are minimal requirement for its secretase activity (Lu et al., 2014; Bai et al., 2015). Several isoforms are encoded by *PSEN1* gene by means of alternative splicing mechanism; among them, isoform 3 is predominantly expressed. Till date, more than 200 unique pathogenic mutations have been reported; mostly mutations occur in exons 5, 6, 7, and 9 following autosomal dominant inheritance (LOVD database). Similar to APP gene, majority of are missense mutations with very less insertions, and deletions are also reported in the database. It is observed that approximately 50% of EOAD individuals have mutations in this gene (http://alzforum.org/mutations/psen-1). Mutations in the promoter region have also been reported, which leads to abnormal expression (Duijn et al., 1999). At protein domain levels, mutations in transmembranes 2 and 4 are more associated with early onset compared with those in transmembranes 6 and 8 (Lippa et al., 2000). The onset age of *PSEN1* mutations carriers is earlier when compared with APP carriers. The onset begins before the 35 years termed as very early-onset Alzheimer's disease (VEOAD) (Campion et al., 1996; Selkoe, 2001). The most common mutation in this gene found among different ethnic population is H163R, which is encoded by exon 5 and is predicted to be possibly or probably damaging (Campion et al., 1995; Gómez-Tortosa et al., 2010; Lohmann et al., 2012; Yagi et al., 2014; Shi et al., 2015). The other mutations reported in the same protein site are H163P and H163Y. All of these mutations are reported to cause increase in Aβ42/Aβ40 ratio (Li et al., 2016; Giau et al., 2019a). The only mutation reported in N-terminal domain is Q15H encoded from exon 3 is associated with frontotemporal dementia. This mutation does not have experimental evidence, but it is classified as possibly deleterious (Koriath et al., 2018). Deletions are observed from exons 8 and 9 (5 mutations), which are heterogeneous even within family members but with one similarity being deletion of exon 9. The common amino acid change in the protein seen is serine replaced by cysteine (S290C) at the splice junction, which further results in exon 9 skipping (Hutton et al., 1996; Crook et al., 1998; Smith et al., 2001; Rovelet-Lecrux et al., 2015). In another study, exon 9 to exon 10 (S290W, serine is replaced by tryptophan) are reported to be deleted in familial EOAD (Lanoiselée et al., 2017; Le Guennec et al., 2017). The deletion of exon 9 (Δ E9) results in absence of 30 amino acids from its original protein sequence and is associated with neuropathological features such as peculiar cotton wool appearance due to deposition and shows more immunopositive toward Aβ42 peptide. All the six deletions are commonly associated with spastic paraparesis and AD. Collectively, mutations in this gene provide significant proof as Aβ accumulation is the root cause for AD manifestation.

3.3.3 Presenilin 2

PSEN2 gene comprising of 13 exons spanning 25.483 kbp in length (genomic coordinates: 1:226,870,571-226,903,828, http://ncbi.nlm.nih.gov/gene/5664) is localized on long arm of chromosome 1; 1q42.13. It has 70% homology with PSEN1 protein structure and is composed of 448 amino acids (Levy-Lahad et al., 1995; Rogaev et al., 1995). Its product comprised of nine transmembrane domains having long cytosolic loop is encoded by exons 8, 9, and 10. This gene transcribes two isoforms, where the isoform 2 is predominantly present in brain, kidney, liver, skeletal muscle, and cell type such as fibroblast (Prihar et al., 1996). PSEN2 protein is one of the four components of γ-secretase complex. The PSEN2 protein subcellular localization is not yet well studied. However, its presence is observed more in endosome and lysosome (Sannerud et al., 2016). Studies on PSEN2 protein revealed that it may play a role in various cell physiological processes such as apoptosis, neurodegeneration, neuroinflammation, and inducer of immunological mediators through Aβ peptides (Araki et al., 2002; Ghidoni et al., 2007; Nguyen et al., 2007; Agrawal et al., 2016; Qin et al., 2017). Mutations in *PSEN2* are found to be very rare; over 56 unique public variants have been reported (http://databases.lovd.nl/shared/genes/PSEN2). Among them, 15 mutations are reported to be pathogenic (https://www.alzforum.org/mutations/psen-2). A more number of missense mutations are reported such as *APP* and *PSEN1* genes. Though being the rare mutated gene, amino acid change at 141-position encoded by exon 5 is reported by 3 studies and found to be pathogenic. This N141 amino acid is present in transmembrane 2 of PSEN2 protein, and substitutions reported are N141D, N141I, and N141Y (Rogaev et al., 1995; Niu et al., 2014; Wang et al., 2019). Till date, four frame shift mutations are reported in coding region: K115Efs, E126fs, K306fs, and K82fs. Among these, K115Efs is caused due to deletion of 2 nucleotides in exon 5, leading to premature stop codon in exon 6. In patient fibroblasts, Aβ40 levels are considerably decreased to that of control, and Aβ42 is undetectable (Jayadev et al., 2010). Remaining three frame shift mutations are reported in exon 9 but do not have evidence of pathogenicity (El Kadmiri et al., 2014). The K306fs mutation is identified in long cytosolic loop encoded by exon 9. Few intronic region mutations are also reported, which include fame shift and point mutations with unclear panthogenicity (Perrone et al., 2018; Wang et al., 2019). Mutations leading to T122P, N141I, M239I, and M239V cause increase in Aβ peptide levels (Walker et al., 2005). Altogether, data indicate that *PSEN2* gene mutations cause alterations in Aβ levels. The age of onset of AD is found to be variable ranging from 40 to 80 years (Ryan and Rossor, 2010; Youn et al., 2014). It is observed that individuals with the *PSEN2* mutation develop AD in much older ages than *PSEN1* mutation carriers. Apart from mutations associated with EOAD, this gene mutation was also seen in other disorders such as LOAD, frontotemporal dementia (FTD), dementia with Lewy bodies, breast cancer, and dilated cardiopathy (An et al., 2015).

3.3.4 Other Genes Related to Early-Onset Alzheimer's Disease

The EOAD is heritable; mostly, *APP*, *PSEN1*, and *PSEN2* mutations are only associated with familial cases. However, there are studies where sporadic cases have been reported. Considering the complexity of the familial EOAD, mutations in these genes can be found in only 50% of cases with reference to cause; still, other half is unexplained. With the introduction of next-generation sequencing techniques such as whole-genome sequencing and whole-exome sequencing, more genes related to EOAD are expected. Consequently, since the past decade, many studies used whole-exome sequencing in familial and sporadic cases of EOAD in an attempt to explore other gene mutations or new mutations in known genes. The results of these finding varied among different studies.

A recent study using Sanger's sequencing has shown that 13% of mutations in sporadic individuals (n = 129) whose parents did not have mutations for the same are found to be de novo. The limitation of this study is lack of functional assessment of possibly and probable pathogenic variants, which further help in classification (Lanoiselée et al., 2017). A deep sequencing study on 100 brains and 355 blood samples from sporadic EOAD individuals found that mutation rate to be 10.8%. In addition to *APP* gene mutations, other mutations found are *SORL1, NCSTN, and MARK4*. Out of five mutations identified in *SORL1* gene, two were found to be pathogenic. However, when compared with routine *APP, PSEN1*, and *PSEN2* mutations in familial EOAD, this figure is low (Nicolas et al., 2018). Contrarily, other study did not find any mutations in sporadic cases (Nicolas et al., 2016b). Some important whole-exome sequencing studies identified mutations in other genes, such as *MAPT, CSF1R, NOTCH3, SEZ6, TYROBP*, and *TREM2* in EOAD individuals. These are likely to be pathogenic mutations associated with neurological disorders such as FTD, Parkinson's disease, amyotrophic lateral sclerosis, and febrile seizures (Guerreiro et al., 2012; Jin et al., 2014; Pottier et al., 2016; Sims et al., 2017; Paracchini et al., 2018; Cochran et al., 2019; Giau et al., 2019b). Although aforementioned disorders are not related to AD, they have some overlapping features.

The *SORL1* (sorting protein–related receptor) gene product is involved in recycling of APP protein from the cell surface, regulates its intracellular trafficking, and plays an important role in formation of Aβ peptides (Fig. 3.2) (Andersen et al., 2005; Caglayan et al., 2014). Furthermore, *MAPT* (microtubule-associated protein tau) mutations are responsible for neurodegenerative disorders implicating complexity of AD. Mutations are referred to be risk factors. Its protein is implicated in microtubule stability and neuronal polarity (Yoshida and Goedert, 2012). Two studies identified mutations in *CSF1R* (colony-stimulating factor 1 receptor), which is associated with hereditary diffuse leucoencephalopathy with axonal spheroids. This disorder can have variable phenotype including AD, FTD, atypical parkinsonism, and multiple sclerosis (Sundal et al., 2012). Other dementia-related gene is *NOTCH3* (neurogenic locus notch homolog protein 3); its mutations are associated with cerebral arteriopathy with subcortical infarcts and leucoencephalopathy 1 (CADASIL). Notch 3 protein being one of the members of Notch signaling pathway is cleaved by γ-secretase complex at cell membrane and released into cell for transcription activation

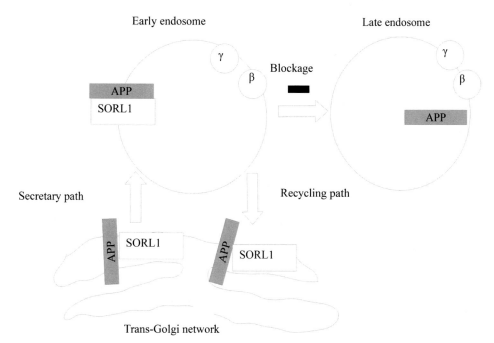

FIG. 3.2 Representative figure of SORL1 role together with APP protein from trans-Golgi network to early endosome and involvement in blockage to late endosome, which will further lead to either nonamylodogenic and amyloidogenic pathways. Collectively, SORL1 protein is involved in retaining of APP protein. Rare variant or pathogenic mutation in SORL1 gene leads to dominant amyloidogenic pathway.

and thus plays an important role in cell-to-cell communication (Peters et al., 2004). Mutations reported in coding part of *TREM2* (triggering receptor expressed on myeloid receptor 2) and *TRYOBP* (tyro protein tyrosine kinase—binding protein) could modify the risk of developing AD (Guerreiro et al., 2013; Pottier et al., 2016). TREM2 protein receptor complex with TYROBP protein forms a signaling cascade (Paloneva et al., 2002). TREM2 is an immunoreceptor expressed on wide variety of activated immunological cells including microglia (Fig. 3.3) (Colonna, 2003). Recent in vivo study has shown that both proteins play an important role in Aβ peptide turn over (Haure-Mirande et al., 2017).Therefore, further work on aforementioned genes and studies on their mutations among large sample size with different ethnic background are needed to have clear etiological contributors of EOAD. The genes reported till date with reference to EOAD and their biological functions are given Table 3.1.

3.4 CONCLUSION

AD is a complex disorder with reference to phenotype and genotype. EOAD is associated with *APP*, *PSEN1*, and *PSEN2* mutations that mostly follow Mendelian inheritance identified in families. Therefore, familial cases can be identified earlier, which helps in disease management. Previously studied mutations in classical genes revealed that their proteins interact at cellular levels and importance in disease manifestation. Since remaining half of etiology is unknown, it is expected that still underlying genetic cause is related to EOAD. Introduction of next-generation sequencing (specifically WES) technology using panel of additional genes related to neurological disorders revealed not only new mutations but also new genes related to neuronal degeneration. The new mutations, which are classified as "likely pathogenic," needed to be functionally validated for their importance. In sporadic EOAD, identified genes such as *SORL1*, *CSF1R*, *NOTCH3*, *TYROBP*, and *TREM2* could be potential candidate genes, as they interact with proteins of already known genes (classical) in neuronal cells. Furthermore, these gene mutations are required to be studied in various populations to have deeper knowledge of EOAD. In general, studying rare variants or mutation can help us in developing potential drug targets, which will ameliorate AD manifestations and further understanding of much more complex LOAD (seen in 90%).

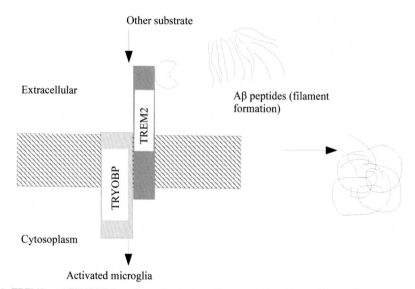

FIG. 3.3 TREM2 and TRYOBP in cell membrane together recognize Aβ peptides and also other substrate. These proteins are expressed in immunoactive cell surface of cells such as microglia. In normal condition, Aβ peptide activation of microglia cells leads to compaction reducing the abnormal accumulation. Mutation in these genes might affect the above molecular process.

TABLE 3.1
Summary of Genes Related to Early-Onset Alzheimer's Disease and Their Functions.

Gene	Chromosomal Location	Nucleotide Position (GRCh38.p13)	Biological Function of Protein
[a]APP	21q21.3	25880550 to 26171128	Neuronal migration during early development, might help in cell growth promotion before and after birth
[a]PSEN1	14q24.2	73136436 to 73223692	Processing of APP protein, being subunit of γ-secretase complex involved in processing several proteins that are essential for transmitting extracellular signal to nucleus
[a]PSEN2	1q42.13	226870572 to 226903829	Processing of APP protein, helps in neuronal growth before and after birth
TREM2	6p21.2	41158508 to 41163116	Expressed on many immunological cells, especially dendritic cells. Activates several cells in central nervous system. Thought to play an important role in microglial function
SORL1	11q24.1	121452314 to 121633763	Endocytosis and protein sorting process
MAPT	17q21.31	45894382 to 46028334	Cell shape maintenance, cell division, transport of several cellular components within the cell
NOTCH3	19p13.12	15159038 to 15200995	Receptor protein on cell surface involved in transmitting signal from outside to nucleus

[a] these genes are primarily or exclusively mutated (rare variants) in EOAD. Remaining gene mutations are rarely reported.

REFERENCES

Alzheimer's disease facts and figures', 2020. Alzheimer's dement. United States 16 (3), 391–460.

Adzhubei, I.A., Schmidt, S., Peshkin, L., Ramensky, V.E., Gerasimova, A., Bork, P., et al., 2010. A method and server for predicting damaging missense mutations. Nat. Methods 7 (4), 248–249.

Agrawal, V., Sawhney, N., Hickey, E., et al., 2016. Loss of presenilin 2 function is associated with defective LPS-mediated innate immune responsiveness. Mol. Neurobiol. 53 (5), 3428–3438.

An, S.S., Cai, Y., Kim, S., 2015. Mutations in presenilin 2 and its implications in Alzheimer's disease and other dementia-associated disorders. Clin. Interv. Aging 10, 1163–1172.

Andersen, O.M., Reiche, J., Schmidt, V., Gotthardt, M., Spoelgen, R., Behlke, J., et al., 2005. Neuronal sorting protein-related receptor sorLA/LR11 regulates processing of the amyloid precursor protein. Proc. Natl. Acad. Sci. U.S.A. 102 (38), 13461–13466.

Araki, W., Yuasa, K., Takeda, S., Takeda, K., Shirotani, K., Takahashi, K., et al., 2002. Pro-apoptotic effect of presenilin 2 (PS2) overexpression is associated with down-regulation of Bcl-2 in cultured neurons. J. Neurochem. 79 (6), 1161–1168.

Bai, X., Yan, C., Yang, G., Lu, P., Ma, D., Sun, L., et al., 2015. An atomic structure of human γ-secretase. Nature 525 (7568), 212–217.

Van Broeckhoven, C., Haan, J., Bakker, E., Hardy, J.A., Van Hul, W., Wehnert, A., et al., 1990. Amyloid beta protein precursor gene and hereditary cerebral hemorrhage with amyloidosis (Dutch). Science 248 (4959), 1120–1122.

Caglayan, S., Takagi-Niidome, S., Liao, F., Carlo, A.-S., Schmidt, V., Burgert, T., et al., 2014. Lysosomal sorting of amyloid- by the SORLA receptor is impaired by a familial alzheimer's disease mutation. Sci. Transl. Med. 6 (223), 223ra20.

Campion, D., Flaman, J.-M., Brice, A., Hannequin, D., Dubois, B., Martin, C., et al., 1995. Mutations of the presenilin I gene in families with early-onset Alzheimer's disease. Hum. Mol. Genet. 4 (12), 2373–2377.

Campion, D., Brice, A., Dumanchin, C., Puel, M., Baulac, M., De La Sayette, V., et al., 1996. A novel presenilin 1 mutation resulting in familial Alzheimer's disease with an onset age of 29 years. Neuroreport 7 (10), 1582–1584.

Chartier-Harlin, M.-C., Crawford, F., Houlden, H., Warren, A., Hughes, D., Fidani, L., et al., 1991. Early-onset Alzheimer's disease caused by mutations at codon 717 of the β-amyloid precursor protein gene. Nature 353 (6347), 844–846.

Chen, W.T., Hong, C.J., Lin, Y.T., Chang, W.H., Huang, H.T., Liao, J.Y., et al., 2012. Amyloid-Beta (Aβ) D7H mutation increases oligomeric Aβ42 and alters properties of Aβ-Zinc/copper assemblies. PLoS One 7 (4), e35807.

Chun, S., Fay, J.C., 2009. Identification of deleterious mutations within three human genomes. Genome Res. 19 (9), 1553–1561.

Cochran, J.N., McKinley, E.C., Cochran, M., Amaral, M.D., Moyers, B.A., Lasseigne, B.N., et al., 2019. Genome sequencing for early-onset or atypical dementia: high diagnostic yield and frequent observation of multiple contributory alleles. Cold Spring Harb. Mol. Case Stud. 5 (6), a003491.

Colonna, M., 2003. TREMs in the immune system and beyond. Nat. Rev. Immunol. 3 (6), 445–453.

Crook, R., Verkkoniemi, A., Perez-Tur, J., Mehta, N., Baker, M., Houlden, H., et al., 1998. A variant of Alzheimer's disease with spastic paraparesis and unusual plaques due to deletion of exon 9 of presenilin 1. Nat. Med. 4 (4), 452–455.

Driscoll, I., Snively, B.M., Espeland, M.A., Shumaker, S.A., Rapp, S.R., Goveas, J.S., et al., 2019. A candidate gene study of risk for dementia in older, postmenopausal women: results from the women's health initiative memory study. Int. J. Geriatr. Psychiatr. 34 (5), 692–699.

Duijn, C. M. van, Cruts, M., Theuns, J., Van Gassen, G., Backhovens, H., van den Broeck, M., et al., 1999. Genetic association of the presenilin-1 regulatory region with early-onset Alzheimer's disease in a population-based sample. Eur. J. Hum. Genet. 7 (7), 801–806.

Di Fede, G., Catania, M., Morbin, M., Rossi, G., Suardi, S., Mazzoleni, G., et al., 2009. A recessive mutation in the APP gene with dominant-negative effect on amyloidogenesis. Science 323 (5920), 1473–1477.

Fraering, P.C., Ye, W., Strub, J.-M., Dolios, G., LaVoie, M.J., Ostaszewski, B.L., et al., 2004. Purification and characterization of the human γ-secretase complex. Biochemistry 43 (30), 9774–9789.

Ghidoni, R., Paterlini, A., Benussi, L., et al., 2007. Presenilin 2 is secreted in mouse primary neurons: a release enhanced by apoptosis. Mech. Ageing Dev. 128 (4), 350–353.

Giau, V. Van, Bagyinszky, E., Youn, Y.C., et al., 2019a. APP, PSEN1, and PSEN2 mutations in Asian patients with early-onset Alzheimer disease. Int. J. Mol. Sci. 20 (19), 4757.

Giau, V. Van, Bagyinszky, E., Yang, Y.S., et al., 2019b. Genetic analyses of early-onset Alzheimer's disease using next generation sequencing. Sci. Rep. 9 (1), 8368.

Golde, T.E., Estus, S., Usiak, M., et al., 1990. Expression of β amyloid protein precursor mRNAs: recognition of a novel alternatively spliced form and quantitation in Alzheimer's disease using PCR. Neuron 4 (2), 253–267.

Gómez-Tortosa, E., Barquero, S., Barón, M., Gil-Neciga, E., Castellanos, F., Zurdo, M., et al., 2010. Clinical-genetic correlations in familial Alzheimer's disease caused by presenilin 1 mutations. J. Alzheim. Dis. 19 (3), 873–884.

Grabowski, T.J., Cho, H.S., Vonsattel, J.P.G., et al., 2001. Novel amyloid precursor protein mutation in an Iowa family with dementia and severe cerebral amyloid angiopathy. Ann. Neurol. 49 (6), 697–705.

Le Guennec, Veugelen, S., Quenez, O., Szaruga, M., Rousseau, S., Nicolas, G.K., et al., 2017. Deletion of exons 9 and 10 of the presenilin 1 gene in a patient with early-onset Alzheimer disease generates longer amyloid seeds. Neurobiol. Dis. 104, 97–103.

Guerreiro, R.J., Lohmann, E., Kinsella, E., Brás, J.M., Luu, N., Gurunlian, N., et al., 2012. Exome sequencing reveals an unexpected genetic cause of disease: NOTCH3 mutation in a Turkish family with Alzheimer's disease. Neurobiol. Aging 33 (5), 1008.e17–1008.e23.

Guerreiro, R.J., Lohmann, E., Brás, J.M., Gibbs, J.R., Rohrer, J.D., Gurunlian, N., et al., 2013. Using exome sequencing to reveal mutations in TREM2 presenting as a frontotemporal dementia–like syndrome without bone involvement'. JAMA Neurol. 70 (1), 78.

Haure-Mirande, J.-V., Audrain, M., Fanutza, T., Kim, S.H., Klein, W.L., Glabe, C., et al., 2017. Deficiency of TYROBP, an adapter protein for TREM2 and CR3 receptors, is neuroprotective in a mouse model of early Alzheimer's pathology. Acta Neuropathol. 134 (5), 769–788.

Hendriks, L., van Duijn, C.M., Cras, P., Cruts, M., Van Hul, W., van Harskamp, F., et al., 1992. Presenile dementia and cerebral haemorrhage linked to a mutation at codon 692 of the β–amyloid precursor protein gene. Nat. Genet. 1 (3), 218–221.

Hutton, M., Busfield, F., Wragg, M., Crook, R., Perez-Tur, J., Clark, R.F., et al., 1996. Complete analysis of the presenilin 1 gene in early onset Alzheimer's disease. Neuroreport 7 (3), 801–805.

Jayadev, S., Leverenz, J.B., Steinbart, E., Stahl, J., Klunk, W., Yu, C.-E., et al., 2010. Alzheimer's disease phenotypes and genotypes associated with mutations in presenilin 2. Brain 133 (4), 1143–1154.

Jin, S.C., Benitez, B.A., Karch, C.M., Cooper, B., Skorupa, T., Carrell, D., et al., 2014. Coding variants in TREM2 increase risk for Alzheimer's disease. Hum. Mol. Genet. 23 (21), 5838–5846.

Jonsson, T., Atwal, J.K., Steinberg, S., Snaedal, J., Jonsson, P.V., Bjornsson, S., et al., 2012. A mutation in APP protects against Alzheimer's disease and age-related cognitive decline. Nature 488 (7409), 96–99.

Joo, Y., Ha, S., Hong, B.H., Kim, J.A., Chang, K.A., Liew, H., et al., 2010. Amyloid precursor protein binding protein-1 modulates cell cycle progression in fetal neural stem cells. PLoS One 5 (12), e14203.

Kaden, D., Harmeier, A., Weise, C., Munter, L.M., Althoff, V., Rost, B.R., et al., 2012. Novel APP/Aβ mutation K16N produces highly toxic heteromeric Aβ oligomers. EMBO Mol. Med. 4 (7), 647–659.

El Kadmiri, N., Zaid, N., Zaid, Y., Tadevosyan, A., Hachem, A., Dubé, M.-P., et al., 2014. Novel presenilin mutations within Moroccan patients with early-onset Alzheimer's disease. Neuroscience 269, 215–222.

Kamino, K., Orr, H.T., Payami, H., Wijsman, E.M., Alonso, M.E., Pulst, S.M., et al., 1992. Linkage and mutational analysis of familial Alzheimer disease kindreds for the APP gene region. Am. J. Hum. Genet. 51 (5), 998–1014.

Koriath, C., Kenny, J., Adamson, G., Druyeh, R., Taylor, W., Beck, J., et al., 2018. Predictors for a dementia gene mutation based on gene-panel next-generation sequencing of a large dementia referral series'. Mol. Psychiatr. 2018 https://doi.org/10.1038/s41380-018-0224-0.

Kunkle, B.W., Grenier-Boley, B., Sims, R., Bis, J.C., Damotte, V., Naj, A.C., et al., 2019. Genetic meta-analysis of diagnosed Alzheimer's disease identifies new risk loci and implicates Aβ, tau, immunity and lipid processing. Nat. Genet. 51 (3), 414–430.

Lanoiselée, H.M., Nicolas, G., Wallon, D., Rovelet-Lecrux, A., Lacour, M., Rousseau, S., et al., 2017. APP, PSEN1, and PSEN2 mutations in early-onset Alzheimer disease: a genetic screening study of familial and sporadic cases. PLoS Med. 14 (3), e1002270.

Levy-Lahad, E., Wasco, W., Poorkaj, P., Romano, D., Oshima, J., Pettingell, W., et al., 1995. Candidate gene for the chromosome 1 familial Alzheimer's disease locus. Science 269 (5226), 973–977.

Levy, E., Carman, M., Fernandez-Madrid, I., Power, M., Lieberburg, I., van Duinen, S., et al., 1990. Mutation of the Alzheimer's disease amyloid gene in hereditary cerebral hemorrhage. Dutch type. Science 248 (4959), 1124–1126.

Li, N., Liu, K., Qiu, Y., Ren, Z., Dai, R., Deng, Y., et al., 2016. Effect of presenilin mutations on APP cleavage; insights into the pathogenesis of FAD. Front. Aging Neurosci. 8, 51.

Lippa, C.F., Swearer, J.M., Kane, K.J., Nochlin, D., Bird, T.D., Ghetti, B., et al., 2000. Familial Alzheimer's disease: site of mutation influences clinical phenotype. Ann. Neurol. 48 (3), 376–379.

Lohmann, E., Guerreiro, R.J., Erginel-Unaltuna, N., Gurunlian, N., Bilgic, B., Gurvit, H., et al., 2012. Identification of PSEN1 and PSEN2 gene mutations and variants in Turkish dementia patients. Neurobiol. Aging 33 (8), 1850.e17–1850.e27.

Lu, P., Bai, X., Ma, D., Xie, T., Yan, C., Sun, L., et al., 2014. Three-dimensional structure of human γ-secretase. Nature 512 (7513), 166–170.

Mann, D.M.A., Iwatsubo, T., Ihara, Y., Cairns, N.J., Lantos, P.L., Bogdanovic, N., et al., 1996. Predominant deposition of amyloid-beta 42(43) in plaques in cases of Alzheimer's disease and hereditary cerebral hemorrhage associated with mutations in the amyloid precursor protein gene. Am. J. Pathol. 148 (4), 1257–1266.

McNaughton, D., Knight, W., Guerreiro, R., Ryan, N., Lowe, J., Poulter, M., et al., 2012. Duplication of amyloid precursor protein (APP), but not prion protein (PRNP) gene is a significant cause of early onset dementia in a large UK series. Neurobiol. Aging 33 (2), 426.e13–426.e21.

Miravalle, L., 2000. Substitutions at codon 22 of Alzheimer's A beta peptide induce conformational changes and diverse apoptotic effects in human cerebral endothelial cells'. J. Biol. Chem. 275 (35), 27110–27116.

Mullan, M., Crawford, F., Axelman, K., Houlden, H., Lilius, L., Winblad, B., et al., 1992. A pathogenic mutation for

probable Alzheimer's disease in the APP gene at the N-terminus of β–amyloid. Nat. Genet. 1 (5), 345–347.

Murrell, J., Farlow, M., Ghetti, B., et al., 1991. A mutation in the amyloid precursor protein associated with hereditary Alzheimer's disease. Science 254 (5028), 97–99.

Ng, P.C., 2003. SIFT: predicting amino acid changes that affect protein function. Nucleic Acids Res. 31 (13), 3812–3814.

Nguyen, H.N., Lee, M.S., Hwang, D.Y., Kim, Y.K., Yoon, D.Y., Lee, J.W., et al., 2007. Mutant presenilin 2 increased oxidative stress and p53 expression in neuronal cells. Biochem. Biophys. Res. Commun. 357 (1), 174–180.

Nicolas, G., Wallon, D., Goupil, C., et al., 2016a. Mutation in the 3'untranslated region of APP as a genetic determinant of cerebral amyloid angiopathy. Eur. J. Hum. Genet. 24 (1), 92–98.

Nicolas, G., Wallon, D., Charbonnier, C., et al., 2016b. Screening of dementia genes by whole-exome sequencing in early-onset Alzheimer disease: input and lessons. Eur. J. Hum. Genet. 24 (5), 710–716.

Nicolas, G., Acuña-Hidalgo, R., Keogh, M.J., Quenez, O., Steehouwer, M., Lelieveld, S., et al., 2018. Somatic variants in autosomal dominant genes are a rare cause of sporadic Alzheimer's disease. Alzheimer's Dementia 14 (12), 1632–1639.

Niu, F., Yu, S., Zhang, Z., Yi, X., Ye, L., Tang, W., et al., 2014. A novel mutation in the PSEN2 gene (N141Y) associated with early-onset autosomal dominant Alzheimer's disease in a Chinese Han family. Neurobiol. Aging 35 (10), 2420.e1–2420.e5.

Novarino, G., Fenstermaker, A.G., Zaki, M.S., Hofree, M., Silhavy, J.L., Heiberg, A.D., et al., 2014. Exome sequencing links corticospinal motor neuron disease to common neurodegenerative disorders. Science 343 (6170), 506–511.

Obici, L., Demarchi, A., de Rosa, G., Bellotti, V., Marciano, S., Donadei, S., et al., 2005. A novel AβPP mutation exclusively associated with cerebral amyloid angiopathy. Ann. Neurol. 58 (4), 639–644.

Paloneva, J., Manninen, T., Christman, G., Hovanes, K., Mandelin, J., Adolfsson, R., et al., 2002. Mutations in two genes encoding different subunits of a receptor signaling complex result in an identical disease phenotype. Am. J. Hum. Genet. 71 (3), 656–662.

Paracchini, L., Beltrame, L., Boeri, L., Fusco, F., Caffarra, P., Marchini, S., et al., 2018. Exome sequencing in an Italian family with Alzheimer's disease points to a role for seizure-related gene 6 (SEZ6) rare variant R615H. Alzheimer's Res. Ther. 10 (1), 106.

Perrone, F., Cacace, R., Van Mossevelde, S., Van den Bossche, T., De Deyn, P.P., Cras, P., et al., 2018. Genetic screening in early-onset dementia patients with unclear phenotype: relevance for clinical diagnosis. Neurobiol. Aging 69, 292.e7–292.e14.

Peters, N., Opherk, C., Zacherle, S., et al., 2004. CADASIL-associated Notch3 mutations have differential effects both on ligand binding and ligand-induced Notch3 receptor signaling through RBP-Jk. Exp. Cell Res. 299 (2), 454–464.

Pottier, C., Ravenscroft, T.A., Brown, P.H., Finch, N.A., Baker, M., Parsons, M., et al., 2016. TYROBP genetic variants in early-onset Alzheimer's disease. Neurobiol. Aging 48, 222.e9—222.e15.

Price, J.L., Davis, P.B., Morris, J.C., et al., 1991. The distribution of tangles, plaques and related immunohistochemical markers in healthy aging and Alzheimer's disease. Neurobiol. Aging 12 (4), 295—312.

Prihar, G., Fuldner, R.A., Perez-Tur, J., Lincoln, S., Duff, K., Crook, R., et al., 1996. Structure and alternative splicing of the presenilin-2 gene. Neuroreport 7 (10), 1680—1684.

Qin, J., Zhang, X., Wang, Z., Li, J., Zhang, Z., Gao, L., et al., 2017. Presenilin 2 deficiency facilitates Aβ-induced neuroinflammation and injury by upregulating P2X7 expression. Sci. China Life Sci. 60 (2), 189—201.

Rama, N., Goldschneider, D., Corset, V., et al., 2012. Amyloid precursor protein regulates netrin-1-mediated commissural axon outgrowth. J. Biol. Chem. 287 (35), 30014—30023.

Raux, G., 2005. Molecular diagnosis of autosomal dominant early onset Alzheimer's disease: an update. J. Med. Genet. 42 (10), 793—795.

Ray, W.J., Yao, M., Mumm, J., Schroeter, E.H., Saftig, P., Wolfe, M., et al., 1999. Cell surface presenilin-1 participates in the γ-secretase-like proteolysis of notch. J. Biol. Chem. 274 (51), 36801—36807.

Richards, S., Aziz, N., Bale, S., Bick, D., Das, S., Gastier-Foster, J., et al., 2015. Standards and guidelines for the interpretation of sequence variants: a joint consensus recommendation of the American college of medical genetics and genomics and the association for molecular pathology. Genet. Med. 17 (5), 405—423.

Rogaev, E.I., Sherrington, R., Rogaeva, E.A., Levesque, G., Ikeda, M., Liang, Y., et al., 1995. Familial Alzheimer's disease in kindreds with missense mutations in a gene on chromosome 1 related to the Alzheimer's disease type 3 gene. Nature 376 (6543), 775—778.

Rovelet-Lecrux, A., Hannequin, D., Raux, G., Meur, N Le, Laquerrière, A., Vital, A., et al., 2006. APP locus duplication causes autosomal dominant early-onset Alzheimer disease with cerebral amyloid angiopathy. Nat. Genet. 38 (1), 24—26.

Rovelet-Lecrux, A., Charbonnier, C., Wallon, D., Nicolas, G., Seaman, M.N.J., Pottier, C., et al., 2015. De novo deleterious genetic variations target a biological network centered on Aβ peptide in early-onset Alzheimer disease. Mol. Psychiatr. 20 (9), 1046—1056.

Ryan, N.S., Rossor, M.N., 2010. Correlating familial Alzheimer's disease gene mutations with clinical phenotype. Biomarkers Med. 4 (1), 99—112.

Sannerud, R., Esselens, C., Ejsmont, P., Mattera, R., Rochin, L., Tharkeshwar, A.K., et al., 2016. Restricted location of PSEN2/γ-secretase determines substrate specificity and generates an intracellular Aβ pool. Cell 166 (1), 193—208.

Schwarz, J.M., Cooper, D.N., Schuelke, M., et al., 2014. MutationTaster2: mutation prediction for the deep-sequencing age. Nat. Methods 11 (4), 361—362.

Selkoe, D.J., 2001. Alzheimer's disease: genes, proteins, and therapy. Physiol. Rev. 81 (2), 741—766.

Sherrington, R., Rogaev, E.I., Liang, Y., Rogaeva, E.A., Levesque, G., Ikeda, M., et al., 1995. Cloning of a gene bearing missense mutations in early-onset familial Alzheimer's disease. Nature 375 (6534), 754—760.

Shi, Z., Wang, Y., Liu, S., Liu, M., Liu, S., Zhou, Y., et al., 2015. Clinical and neuroimaging characterization of Chinese dementia patients with PSEN1 and PSEN2 mutations. Dement. Geriatr. Cognit. Disord. 39 (1—2), 32—40.

Sims, R., van der Lee, S.J., Naj, A.C., Bellenguez, C., Badarinarayan, N., Jakobsdottir, J., et al., 2017. Rare coding variants in PLCG2, ABI3, and TREM2 implicate microglial-mediated innate immunity in Alzheimer's disease. Nat. Genet. 49 (9), 1373—1384.

Sleegers, K., Brouwers, N., Gijselinck, I., Theuns, J., Goossens, D., Wauters, J., et al., 2006. APP duplication is sufficient to cause early onset Alzheimer's dementia with cerebral amyloid angiopathy. Brain 129 (11), 2977—2983.

Smith, M.J., Kwok, J.B.J., McLean, C.A., Kril, J.J., Broe, G.A., Nicholson, G.A., et al., 2001. Variable phenotype of Alzheimer's disease with spastic paraparesis. Ann. Neurol. 49 (1), 125—129.

Sperling, R.A., Aisen, P.S., Beckett, L.A., Bennett, D.A., Craft, S., Fagan, A.M., et al., 2011. Toward defining the preclinical stages of Alzheimer's disease: recommendations from the National Institute on Aging-Alzheimer's Association workgroups on diagnostic guidelines for Alzheimer's disease. Alzheimer's Dementia 7 (3), 280—292.

Sun, L., Zhou, R., Yang, G., et al., 2017. Analysis of 138 pathogenic mutations in presenilin-1 on the in vitro production of Aβ42 and Aβ40 peptides by γ-secretase. Proc. Natl. Acad. Sci. U.S.A. 114 (4), E476—E485.

Sundal, C., Lash, J., Aasly, J., Øygarden, S., Roeber, S., Kretzschman, H., et al., 2012. Hereditary diffuse leukoencephalopathy with axonal spheroids (HDLS): a misdiagnosed disease entity. J. Neurol. Sci. 314 (1—2), 130—137.

Talarico, G., Piscopo, P., Gasparini, M., Salati, E., Pignatelli, M., Pietracupa, S., et al., 2010. The london APP mutation (Val717Ile) associated with early shifting abilities and behavioral changes in two Italian families with early-onset Alzheimer's disease. Dement. Geriatr. Cognit. Disord. 29 (6), 484—490.

Tanaka, S., Shiojiri, S., Takahashi, Y., Kitaguchi, N., Ito, H., Kameyama, M., et al., 1989. Tissue-specific expression of three types of β-protein precursor mRNA: enhancement of protease inhibitor-harboring types in Alzheimer's disease brain. Biochem. Biophys. Res. Commun. 165 (3), 1406—1414.

Tanzi, R.E., Vaula, G., Romano, D.M., Mortilla, M., Huang, T.L., Tupler, R.G., et al., 1992. Assessment of amyloid beta-protein precursor gene mutations in a large set of familial and sporadic Alzheimer disease cases. Am. J. Hum. Genet. 51 (2), 273—282.

Tomiyama, T., Nagata, T., Shimada, H., Teraoka, R., Fukushima, A., Kanemitsu, H., et al., 2008. A new amyloid β variant favoring oligomerization in Alzheimer's-type dementia. Ann. Neurol. 63 (3), 377—387.

Vasques, J.F., Heringer, P.V.B., Jesus Gonçalves, R.G., et al., 2017. Monocular denervation of visual nuclei modulates APP processing and sAPPα production: a possible role on neural plasticity. Int. J. Dev. Neurosci. 60 (1), 16—25.

Walker, E.S., Martinez, M., Brunkan, A.L., et al., 2005. Presenilin 2 familial Alzheimer's disease mutations result in partial loss of function and dramatic changes in Abeta 42/40 ratios. J. Neurochem. 92 (2), 294−301.

Wang, G., Zhang, D.-F., Jiang, H.-Y., Fan, Y., Ma, L., Shen, Z., et al., 2019. Mutation and association analyses of dementia-causal genes in Han Chinese patients with early-onset and familial Alzheimer's disease. J. Psychiatr. Res. 113, 141−147.

Yagi, R., Miyamoto, R., Morino, H., Izumi, Y., Kuramochi, M., Kurashige, T., et al., 2014. Detecting gene mutations in Japanese Alzheimer's patients by semiconductor sequencing. Neurobiol. Aging 35 (7), 1780.e1−1780.e5.

Yang, G., Zhou, R., Zhou, Q., Guo, X., Yan, C., Ke, M., et al., 2019. Structural basis of Notch recognition by human γ-secretase. Nature 565 (7738), 192−197.

Yoshida, H., Goedert, M., 2012. Phosphorylation of microtubule-associated protein tau by AMPK-related kinases. J. Neurochem. 120 (1), 165−176.

Youn, Y.C., Bagyinszky, E., Kim, H., et al., 2014. Probable novel PSEN2 Val214Leu mutation in Alzheimer's disease supported by structural prediction. BMC Neurol. 14 (1), 105.

Young-Pearse, T.L., Bai, J., Chang, R., et al., 2007. A critical function for -amyloid precursor protein in neuronal migration revealed by in utero RNA interference. J. Neurosci. 27 (52), 14459−14469.

Zhou, L., Brouwers, N., Benilova, I., Vandersteen, A., Mercken, M., Van Laere, K., et al., 2011. Amyloid precursor protein mutation E682K at the alternative β-secretase cleavage β′-site increases Aβ generation. EMBO Mol. Med. 3 (5), 291−302.

CHAPTER 4

Role of Neuroinflammation in Neurodegenerative Disorders

VISHWA MOHAN, BSC, MSC, PHD •
CHANDRAKANTH REDDY EDAMAKANTI, MSC, PHD • VYOM SHARMA, MSC, PHD

4.1 INTRODUCTION

The prevalence of neurodegenerative diseases (alzheimer's disease [AD], parkinson's disease [PD], amyotrophic lateral sclerosis [ALS], frontotemporal dementia [FTD], and huntington's disease [HD]) is increasing due to the substantial increase in average life expectancy compared with the past century. Neurodegenerative diseases affect millions of people worldwide; Alzheimer's disease and Parkinson's disease are the most common neurodegenerative diseases. Neurodegenerative disorders are characterized by progressive loss of neuronal population; most neurodegenerative disorders share the common fact that they are derived from aberrant proteins that undergo a misfolding event and a tendency to form aggregate in the cells. These aggregation products are toxic to these cells, and toxicity spreads all over different brain areas. Diseases can be classified according to primary molecular abnormality (amyloidoses, tauopathies, α-synucleinopathies, and TDP-43 proteinopathies), clinical features (e.g., dementia, parkinsonism, or motor neuron disease), and regional distribution of neurodegeneration (e.g., frontotemporal degenerations, extrapyramidal disorders, or spinocerebellar degenerations) (Kovacs, 2017).

Other than these most common neurodegenerative disorders, there is a different group of diseases called Polyglutamine (PolyQ) repeat disorders also known as trinucleotide repeat expansion disorders (TREDs), dominantly inherited neurological diseases caused by the expansion of unstable repeats in specific regions of the associated genes. CAG repeat expansion in translated regions of the respective genes results in PolyQ-rich proteins that form intracellular aggregates that affect numerous cellular activities. Currently, there are nine known polyQ-expansion diseases that comprise HD, spinocerebellar ataxias type 1, 2, 3, 6, 7, and 17 (SCAs),

dentatorubropallidoluysian atrophy, and spinal and bulbar muscular atrophy (SBMA).

For decades, it was accepted that the central nervous system (CNS) is an immune-privileged site; however, the identification of meningeal lymphatic system in brain has challenged the idea of the brain being an immune-privileged site (Brioschi and Colonna, 2019; Louveau et al., 2018). These studies suggest that innate and adaptive immune responses could be involved in the CNS in neurodegenerative disease.

There are three different anatomical regions in brain (i.e., cerebrospinal fluid [CSF], meninges, and parenchyma of the brain), and immune cells have access to all of three regions under physiological circumstances and disease states. Microglia are the resident macrophages of the CNS that has roles in maintaining homeostasis and immune responses; major functions of microglia include phagocytosis, extracellular signaling, and antigen presentation. The microglia are the primary phagocytic cells within the brain that act by engulfing misfolded proteins, cell debris, aggregated proteins, and toxic lipid products (Brioschi and Colonna, 2019). The blood—brain barrier (BBB), formed by multiple layers of tightly packed endothelial cells, prevents entry of immune cells in the CNS, and leakiness in this barrier is associated with pathologies. Under normal physiological conditions, a small number of leukocytes are present in the cerebrospinal fluid (CSF), and various immune cell types are constitutively present in the blood—CSF barrier and in the space between meninges. The blood—CSF barrier controls the trafficking of leukocytes from the blood to the CSF (Brioschi and Colonna, 2019; Kunis et al., 2013). Brain and the spinal cord encapsulated by meningeal spaces and various immune cell types are present in the CSF. Meningeal leukocytes, as well as brain antigens, potentially drain via the dural

The Molecular Immunology of Neurological Diseases. https://doi.org/10.1016/B978-0-12-821974-4.00001-7

lymphatics, where they communicate with the peripheral immune system (Brioschi and Colonna, 2019; Louveau et al., 2018).

In this chapter, we present the current state of knowledge regarding the role of the immune system in the pathogenesis of AD, PD, and Poly-Q disease. We focused on how neuroinflammatory processes are directly related to the neurodegenerative processes and involvement of both astrocytes and microglia in the neuroinflammatory processes in neurodegenerative diseases.

4.2 NEUROINFLAMMATION IN ALZHEIMER'S DISEASE

Neuroinflammation is a process implicated in several neurodegenerative disorders, and it is an important contributor to AD pathogenesis and progression. Several damaging signals appear to induce neuroinflammation, such as trauma, infection, oxidative agents, redox iron, oligomers of tau, and β-amyloid (Aβ). AD is characterized by widespread neuronal degeneration and synaptic loss affecting mainly the hippocampus and cortex, resulting in brain atrophy. Intracellular neurofibrillary tangles (NFTs) that are formed due to the aggregation of hyperphosphorylated tau protein along with the extracellular plaques comprise oligomers of Aβ peptide that became the major hallmarks of AD; more recently, neuroinflammation has emerged as a third hallmark of the disease (Heneka et al., 2014). Due to increase in average life expectancy, AD has become a major health problem, with an estimated 50 million people all over the world having it (Bettens et al., 2010). According to the World Health Organization (WHO), AD progressively affects learning and memory, and behavior. The prevalence of AD constantly increasing (Guzmán-Martinez et al., 2013; Maccioni, 2012), microglia are the key cell type which regulate the innate immune response in CNS and self/non–self-recognizing receptors expressed by microglia but also present on astrocytes and oligodendrocytes (Hernangomez et al., 2014).

Microglial cells regulate the immune functions of astrocytes; the inflammatory factors released by activated microglia can induce intracellular signaling in astrocytes and make them reactive. However, the reactive astrocytes release factors that cause permeability changes in BBB, resulting in the recruitment of immune cells in the brain parenchyma. This amplifies the initial innate immune response, and these reactive astrocytes secrete a wide range of factors, such as neurotrophic factors, growth factors, and cytokines, promoting neuronal survival, neurite growth, and neurogenesis. Both the microglia and the astrocytes release various signaling molecules, establishing an autocrine feedback.

There are several neuroinflammatory factors secreted by microglia and astrocyte that are known to involve in onset and the progression of AD (Maccioni et al., 2009). Microglial cells are one of the major regulators of innate immunity and the main source of proinflammatory factors in the brain (Guzman-Martinez et al., 2019). The involvement of microglia in AD pathogenesis was studied in relation to the Aβ (Fang et al., 2018; Guillot-Sestier et al., 2015; Heneka et al., 2015) as well as in the context of tau oligomerization (Maccioni, 2012; Maccioni et al., 2009; Morales et al., 2010, 2014).

Astrocytes are also known to be involved in the neuroinflammation process; they have roles in metabolic regulation, neuronal scaffold, and synaptogenesis. Astrocytes are key modulators of synaptic structure and function (Allen et al., 2012; Christopherson et al., 2005; Chung et al., 2013; Kucukdereli et al., 2011; Smith et al., 2020), and their activation to a "reactive" state is a common feature of neurodegenerative disorders. Degenerating neurons release aggregates and toxic tau proteins that create a positive feedback loop to generate more reactive cells, fueling this vicious cycle of neuroinflammation.

Activated microglia release proinflammatory cytokines to their cellular niche, among which IL-1β, IL-6, IL-12, IFN-γ, and TNF-α are few to name. On the other hand, the astrocytes are the most abundant of the glial cells; they are fundamental to maintain homeostasis and support the function of neurons. Reactive astrocyte can cause a condition known as astrogliosis, i.e., marked by the aberrant increase in astrocytes that can also be considered as an important factor in chronic neuroinflammation affecting considerably the neuron and its integrity (Sofroniew and Vinters, 2010). Along with microglia, astrocytes also synthesize and secrete proinflammatory cytokines. In addition, experimental evidence has indicated that interleukins (ILs) such as IL-1β, TNF-a, IL-6, and C1q coactivate the astrocytes, resulting in neuronal malfunction and degeneration (Jacobs and Tavitian, 2012).

Genome-wide association studies of late-onset AD have identified genetic risk factors, which can be divided into distinct functional classes of genes involved in innate immune response genes, cholesterol metabolism, endocytosis, and processing of Aβ precursor protein (Webers et al., 2020).

Microglial cells become the focal point in the development of the neuroinflammatory process. Activated microglia can increase the permeability of the BBB and promote an increase in infiltration of peripheral immune cells and contribute to neuronal dysfunction, thus accelerating the neurodegenerative process (Zarrouk et al., 2018). The importance of microglia in AD pathogenesis is also underscored by the presence of activated microglia around amyloid plaques from AD patients and in AD animal models (Frautschy et al., 1998; McGeer et al., 1987). Microglia are involved in Aβ clearance, which can be beneficial, to inhibit Aβ buildup and damaging, when levels of Aβ are elevated that result in chronic inflammation (Webers et al., 2020). Microglia express a range of different receptors that can bind Aβ and trigger inflammation, such as different Toll-like receptors and NACHT, LRR, and pyrin domains—containing protein 3 (NLRP3); these receptors induce release of TNFα and IL-1β, which mediate neuroinflammation and neurotoxicity and cause sustained low-grade inflammation. Deletion of receptors such as TLR4 reduces Aβ-induced cytokine production (Heneka et al., 2013, 2015; Fiebich et al., 2018). Unfolded, misfolded, and aggregated proteins are recognized by the damage-associated molecular pattern (DAMP) receptors found on the surface of innate immune cells. It has been shown that aggregated Aβ acts as a DAMP, resulting in the activation of the innate immune system in the brain with subsequent proinflammatory cytokine production (Heneka, 2017).

There are now multiple studies that link neuroinflammation to the progression of AD, and it can now be considered as a chronic inflammatory disease. However, there remains considerable gap in our knowledge on the interactions between the different cell types involved in AD and the molecular details of the pathways, which link Aβ accumulation and ongoing inflammation. Deciphering the underlying mechanisms of how neuroinflammation pathways help in progression of AD pathology could help to identify new therapeutic targets and to provide an understanding as to why current therapeutic strategies are often failing.

There is still no cure for AD; however, six therapeutic approaches are underway, reducing MAPT (microtubule associated protein tau) gene expression, modulating post-translational modification, preventing tau aggregation, immune neutralization, or clearance of different tau, and stabilizing microtubules (VandeVrede et al., 2020).

4.3 NEUROIMMUNOLOGY OF PARKINSON'S DISEASE

PD was first described by James Parkinson in 1817; it is one of the most common neurodegenerative disorders. Among people over 50 years, about 2.0% is affected with PD (Maarouf et al., 2012). Currently, there are more than six million people affected by this worldwide, and numbers are constantly increasing due to increasing life expectancy. Hallmarks of pathology include intra-neuronal and intra-axonal α-synuclein-positive inclusions (Lewy bodies) and loss of dopaminergic neurons in the substantia nigra; reduction in dopamine

concentration in the striatum gives rise to the hallmark motor symptoms of the PD (Tan et al., 2020).

The clinical signature consists of motor, cognitive, and neuropsychiatric symptoms. The motor symptoms include rest tremor, bradykinesia (impairment in voluntary movements), postural instability, and rigidity (Thomas and Beal, 2007). There are fewer dopaminergic neurons in the substantia nigra, which causes low levels of dopamine in the striatum, which in turn result in malfunctioning in the basal ganglia.

The relationship of PD progression with neuroinflammation is still unclear, and exact mechanism must be elucidated for the better understanding of PD.

Autoimmune disorders have been associated with PD as risk factor, and studies have shown that autoantibodies to α-synuclein are present in the serum and cerebrospinal fluid (CSF) of patients with PD (Akhtar et al., 2018; Bach and Falkenburger, 2018; Wu et al., 2017). In one large Swedish epidemiological study which includes 33 different autoimmune disorders, it has been concluded that people with autoimmune disorder have 33% more risk of developing PD (Li et al., 2012).

The innate and adaptive immune systems both generate cellular immune responses, which involve activity of T cells, neutrophils, monocytes, dendritic cells, natural killer cells, and humoral immune responses, which are mediated by molecular components such as antibodies and complement system. Dysregulation of both cellular and humoral immune responses in the periphery has been observed in patients with PD.

Microglia are the primary innate immune cells in the brain and help maintaining CNS homeostasis. Chemokines direct the movement of leukocytes to areas of insult and can increase the permeability of the BBB, thereby enabling infiltration of immune cells in the CNS (Cardona et al., 2008). In PD, the increased permeability of BBB results in infiltration of inflammatory molecules to the brain and microglia activation. In PD, patients elevated serum cytokine (IL-2, IL-6, TNF-α) levels, and peripheral lymphocyte activation promotes systemic immune response (Collins et al., 2012). In PD patients, it has been shown that levels of proinflammatory cytokines, including transforming growth factor-β1, IL-6, and IL-1β, are increased in the brain and CSF (Brodacki et al., 2008; Lin et al., 2019).

After inflammatory processes, induction-elevated levels of MHC II in astrocytes and microglia in a mice PD model indicate the involvement of adaptive immune response (Martin et al., 2016). One of the hallmarks of PD pathology is accumulation of α-synuclein-positive cytoplasmic inclusions in neurons. α-Synuclein is encoded by the SNCA gene. Expression of SNCA in astrocytes is low, whereas it is well expressed in neurons. α-Synuclein-positive inclusions have been found in PD postmortem brain in astrocytes as well as in neurons, which suggests that α-synuclein secreted by neurons is also entering astrocytes. Multiple studies have now revealed that astrocytes can take up α-synuclein, and this has been shown to occur via a TLR4-independent endocytosis pathway (Fellner et al., 2013; Rannikko et al., 2015). This suggests that astrocytes have a role in its removal and degradation, potentially maintaining a healthy environment in which neurons can grow well.

Activation of microglia by classical inflammatory mediators can convert astrocytes to a neurotoxic A1 phenotype in a variety of neurological diseases (Liddelow et al., 2017). Glucagon-like peptide-1 receptor (GLP-1R) agonists are known as potential neuroprotective agents for neurologic disorders such as AD and PD. Recently, it has been shown that GLP-1R agonist prevents the microglial-mediated conversion of astrocytes to an A1 neurotoxic phenotype and provides neuroprotection in PD mouse model (Yun et al., 2018). Since the past few years, a new angle has been added in PD pathology by linking the gut–brain axis regulation to the PD (Chapelet et al., 2019; Klingelhoefer and Reichmann, 2015). Common early manifestations in PD that do not involve movement or motor skills are olfactory impairment and constipation. This is shared with the Braak hypothesis that α-synuclein pathology spreads from the gastrointestinal tract to the brain stem through the vagus nerve and reaches to the substantia nigra (Braak et al., 2003; Kim et al., 2019). The neuropathological process leading to the disease seems to start in the enteric nervous system or the olfactory bulb, spreading to the substantia nigra and finally to the CNS, suggesting that the environment could be part of the disease onset and development. Several studies indicate that immune system malfunction has a role in PD; this includes associations between autoimmune disease and PD, and impaired cellular and humoral immune responses. The mechanisms that link the immune system with PD remain unclear, but in recent years, there are several immune-based therapeutic that showed some potential in PD. These approaches include treatment with antiinflammatory drugs, clearing α-synuclein aggregates with antibodies, restoring lysosome function to accelerate α-synuclein clearance, inhibiting proinflammatory cytokine release with immunosuppressants, blocking antigen presentation to T cells, modulating CD4+ and CD8+ T cell activity, and inducing immune tolerance by modulating activity of regulatory T cells (Liu et al., 2017; Savitt and Jankovic, 2019).

4.4 NEUROINFLAMMATION IN POLYQ DISORDERS

Polyglutamine (polyQ) repeat–containing proteins are widespread in the human proteome, but only nine of them are associated with highly debilitating neurodegenerative disorders (see Table 4.1). The genetic expansion of the polyQ tract in disease-related proteins triggers a series of events, resulting in neurodegeneration. The polyQ repeat length is variable in the normal population, but expansion beyond a certain threshold, distinct for each disease, enhances the tendency to associate into β-rich amyloid-like fibrils both in vitro and in vivo (Silva et al., 2018). The common feature of polyQ diseases is the loss of neurons in specific brain regions accompanied by reactive gliosis and astrocytosis, which may suggest involvement of inflammation in pathogenesis. The inflammatory response on the CNS level is reflected by the interplay between neurons, microglial cells, and astrocytes.

In triplet repeat expansion diseases, chronic stimulation of the immune system by a mutant protein/transcript may play an important role in disease progression. Immune response markers such as elevated cytokine, reactive oxygen species (ROS), and nitric oxide (NO) levels, activation of caspases, and changes in gene expression have been observed in the blood and CNS of patients and mouse models (Björkqvist et al., 2008; Tai et al., 2007). Aggregation of polyQ proteins that affect various cellular functions and cause selective neurodegeneration in specific brain regions is a common feature of polyQ diseases (Klement et al., 1998; Paulson et al., 1997; Scherzinger et al., 1997). Most studies have focused on protein toxicity, but studies also showed that mutant transcripts also contribute to the disease by an RNA gain-of-function mechanism (Nalavade et al., 2013; Tsoi and Chan, 2014). This hypothesis assumes that nuclear foci formed by expanded CAG repeat transcripts sequester specific RNA-binding proteins and might lead to loss of their normal function.

HD is the most prevalent and most studied polyQ disorder, and alleles with repeat lengths from 36 to 39 show less penetrance, whereas 40 or more repeats are fully penetrant and associated with the development of HD (Duyao et al., 1993; Myers et al., 1993). Dementia and movement phenotype resulted by selective loss of neurons in the striatum and cortex, and death occurs around the fifth decade of life. Microglia are the resident macrophages in brain, surrounded by astrocytes and neurons and form the first line of defense of the innate immune system. Under normal conditions, microglia play roles in eliminating neuronal cells during development and in maintaining their survival by removing

TABLE 4.1

The Polyglutamine Family of Neurodegenerative Diseases, their Causative Genes and Protein Function.

S. No.	Disease	Gene	Pathogenic Repeat Length	Protein Name and Function	Affected Site
1	Huntington's disease	HTT	≥37 CAG repeats	Huntingtin/early development, neurogenesis	Striatum, globus pallidus, substantia nigra
2	Dentatorubropallidoluysian atrophy	ATN1	≥38 CAG repeats	Atrophin-1/ transcription regulation	Globus pallidus, subthalamic and dentate nucleus, white matter
3	Spinal and bulbar muscular atrophy	AR	≥38 CAG repeats	Androgen receptor/ transcriptional regulation	Skeletal muscle, anterior horn
4	Spinocerebellar ataxia 1 (SCA1)	ATXN1	≥39 CAG repeats	Ataxin-1/ transcriptional regulation	Cerebellum, dentate nucleus, brain stem
5	SCA2	ATXN2	≥32 CAG repeats	Ataxin-2/mRNA maturation, translation	Cerebellum, pons, thalamus, substantia nigra
6	SCA3/Machado–Joseph disease	ATXN3/ MJD	≥44 CAG repeats	Ataxin-3/ubiquitin-mediated proteolysis, chromatin remodeling	Substantia nigra, striatum
7	SCA6	CACNA1A	≥19 CAG repeats	Alpha-1A calcium channel Protein/calcium-dependent processes	Cerebellum, dentate nucleus, inferior olive
8	SCA7	ATXN7	≥37 CAG repeats	Ataxin-7/transcription regulation	Cerebellum, pons, inferior olive, retina
9	SCA17	TBP	≥45 CAG repeats	TATA-binding protein/ transcriptional regulation	Cerebellum, striatum, cerebral cortex

toxic cellular debris. In response to a stimulus, microglia proliferate, migrate toward site, and start producing cascade of proinflammatory cytokines (e.g., IL-6, IL-12, TNF-α, and IL-1β). These could lead to caspase activation, changes in intracellular calcium levels, and free radical production. Overstimulation of these pathways may lead to neurodegeneration (Hanisch, 2002; Lobsiger and Cleveland, 2007).

Polyglutamine diseases are typically mid to late onset and involve accumulation of the disease-specific protein. Out of nine PolyQ disorders, six are ataxias; i.e., spinocerebellar ataxia (SCA) types 1, 2, 3, 6, 7, and 17 are among the most understood ataxias. Even though the mutant protein expresses all over brain, the predominant affected region is cerebellum. The mechanism behind the regional vulnerability of cerebellum is not known. Recent evidence suggests that

cerebellar microglia and astrocytes (including Bergmann glia) have a unique vigilant immune reactivity phenotype compared with microglia and astrocytes from other brain regions. These findings may indicate that the selective vulnerability comes from glial cells of cerebellum. In fact, glia are shown to contribute to the neurodegeneration through cell-autonomous and non–cell-autonomous mechanism (Crotti et al., 2014; Efthymiou and Goate, 2017; Albanito et al., 2011). Moreover, the glial cells act as primary mediators of neuroinflammation. Microglia and astrocytes act as resident macrophages of CNS. These glial cells are known for the policing brain insults to which they react with process of activation, which included changes in proliferation, function, morphology, and gene expression. Postmortem analysis of patient and staining of mouse cerebellum suggest that the microglia activation is a very prominent feature of several ataxias including ataxia—telangiectasia, SCA6, SCA21, multiple system atrophy, and Friedreich's ataxia (Aikawa et al., 2015; Seki et al., 2018; Shen et al., 2016; Quek et al., 2017; Nykjaer et al., 2017). Moreover, the PolyQ ataxia mouse models demonstrated that the activation of microglia occurs presymptomatically, well before behavioral symptoms and cerebellar degeneration appear (Aikawa et al., 2015; Seki et al., 2018; Cvetanovic et al., 2015; Qu et al., 2017; Edamakanti et al., 2018). With the current research knowledge, understanding of neuroinflammation in PolyQ disorders exposes a crucial role of astrocytes and microglia to the disease progression. In many PolyQ disorders, the early activation of microglia and astrocytes with active participation in immune defense system like release of proinflammatory factors makes glial cells as a promising therapeutic target. In recent times, the ASO (antisense oligonucleotide) therapy is most effective in PolyQ neurodegenerative disorders (Scoles et al., 2017; Friedrich et al., 2018). Since the polyQ aggregates express within the glial cells and lead to non—cell-autonomous toxicity of neurons, it is important to target the antisense oligonucleotides specific to the glial population to reduce the neuroinflammation, thereby ameliorating the neurodegenerative phenotype. These experiments will be able to tease out the individual role of glial cells in pathogenesis of neurovegetative disorders.

4.5 CONCLUSION

Nervous and immune systems are interlinked, which is highlighted by multiple studies revealing meningeal vessels that directly link the brain with the lymphatic system. In many neurodegenerative diseases, the innate immune response in the CNS plays an important role in the onset and progression of disease. During development microglia perform functions such as synaptic pruning and serve as principal macrophage in CNS, which is important for neuronal development. Adaptive and innate immune cells that enter the CNS trigger damage but are crucial for immune regulation. There are multiple studies that suggest that microglia, astrocytes, neurons, and oligodendrocytes also contribute to immune surveillance in the CNS. There is now substantial evidence that neuroinflammation contributes to the progression of AD, PD, and polyQ diseases. There is common pattern of pathological events seen in various neurodegenerative diseases in which aggregating protein results in the activation of microglia and astrocytes that in turn release proinflammatory signals that drive the degeneration of neurons (Fig. 4.1).

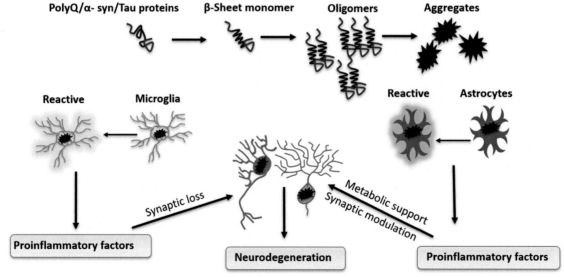

FIG. 4.1 Role of glial cells in neurodegenerative disorders: Mutant monomeric proteins can miss fold and form beta sheets and thereby adopt an oligomer confirmation. These oligomers inturn make nuclear aggregates in most of the brain cell types, including neurons and glial cells of the brain. These mutant protein aggregates in glial population contribute to the neurodegeneration through cell-autonomous and non—cell-autonomous mechanism. The early activation of microglia and astrocytes (reactive) leads to the release proinflammatory factors that in turn contribute to the regional specific toxicity the brain. Moreover, the non—cell-autonomous toxicity of neurons comes from their role in altered synaptic and metabolic support.

AD can now be considered as a chronic inflammatory disease. However, in spite of the recent advances, there remain substantial gaps in our knowledge on the interactions between the different cell types involved in these degenerative disorders and the molecular details of the pathways that link aggregating protein and ongoing inflammation. Deciphering the underlying mechanisms of this system could help to identify new therapeutic targets and to provide a better understanding as to why current therapeutic strategies are often failing.

REFERENCES

Aikawa, T., Mogushi, K., Iijima-Tsutsui, K., Ishikawa, K., Sakurai, M., Tanaka, H., et al., 2015. Loss of MyD88 alters neuroinflammatory response and attenuates early Purkinje cell loss in a spinocerebellar ataxia type 6 mouse model. Hum. Mol. Genet. 24 (17), 4780–4791.

Akhtar, R.S., Licata, J.P., Luk, K.C., Shaw, L.M., Trojanowski, J.Q., Lee, V.M., 2018. Measurements of auto-antibodies to alpha-synuclein in the serum and cerebral spinal fluids of patients with Parkinson's disease. J. Neurochem. 145 (6), 489–503.

Albanito, L., Reddy, C.E., Musti, A.M., 2011. c-Jun is essential for the induction of Il-1β gene expression in in vitro activated Bergmann glial cells. Glia 59 (12), 1879–1890.

Allen, N.J., Bennett, M.L., Foo, L.C., Wang, G.X., Chakraborty, C., Smith, S.J., et al., 2012. Astrocyte glypicans 4 and 6 promote formation of excitatory synapses via GluA1 AMPA receptors. Nature 486 (7403), 410–414.

Bach, J.P., Falkenburger, B.H., 2018. What autoantibodies tell us about the pathogenesis of Parkinson's disease: an Editorial for 'Measurements of auto-antibodies to α-synuclein in the serum and cerebral spinal fluids of patients with Parkinson's disease' on page 489. J. Neurochem. 145 (6), 433–435.

Bettens, K., Sleegers, K., Van Broeckhoven, C., 2010. Current status on Alzheimer disease molecular genetics: from past, to present, to future. Hum. Mol. Genet. 19 (R1), R4–r11.

Björkqvist, M., Wild, E.J., Thiele, J., Silvestroni, A., Andre, R., Lahiri, N., et al., 2008. A novel pathogenic pathway of immune activation detectable before clinical onset in Huntington's disease. J. Exp. Med. 205 (8), 1869–1877.

Braak, H., Rüb, U., Gai, W.P., Del Tredici, K., 2003. Idiopathic Parkinson's disease: possible routes by which vulnerable neuronal types may be subject to neuroinvasion by an unknown pathogen. J. Neural. Transm. 110 (5), 517–536.

Brioschi, S., Colonna, M., 2019. The CNS immune-privilege goes down the drain(age). Trends Pharmacol. Sci. 40 (1), 1–3.

Brodacki, B., Staszewski, J., Toczyłowska, B., Kozłowska, E., Drela, N., Chalimoniuk, M., et al., 2008. Serum interleukin (IL-2, IL-10, IL-6, IL-4), TNFalpha, and INFgamma concentrations are elevated in patients with atypical and idiopathic parkinsonism. Neurosci. Lett. 441 (2), 158–162.

Cardona, A.E., Li, M., Liu, L., Savarin, C., Ransohoff, R.M., 2008. Chemokines in and out of the central nervous system: much more than chemotaxis and inflammation. J. Leukoc. Biol. 84 (3), 587–594.

Chapelet, G., Leclair-Visonneau, L., Clairembault, T., Neunlist, M., Derkinderen, P., 2019. Can the gut be the missing piece in uncovering PD pathogenesis? Park. Relat. Disord. 59, 26–31.

Christopherson, K.S., Ullian, E.M., Stokes, C.C., Mullowney, C.E., Hell, J.W., Agah, A., et al., 2005. Thrombospondins are astrocyte-secreted proteins that promote CNS synaptogenesis. Cell 120 (3), 421–433.

Chung, W.S., Clarke, L.E., Wang, G.X., Stafford, B.K., Sher, A., Chakraborty, C., et al., 2013. Astrocytes mediate synapse elimination through MEGF10 and MERTK pathways. Nature 504 (7480), 394–400.

Collins, L.M., Toulouse, A., Connor, T.J., Nolan, Y.M., 2012. Contributions of central and systemic inflammation to the pathophysiology of Parkinson's disease. Neuropharmacology 62 (7), 2154–2168.

Crotti, A., Benner, C., Kerman, B.E., Gosselin, D., Lagier-Tourenne, C., Zuccato, C., et al., 2014. Mutant Huntingtin promotes autonomous microglia activation via myeloid lineage-determining factors. Nat. Neurosci. 17 (4), 513–521.

Cvetanovic, M., Ingram, M., Orr, H., Opal, P., 2015. Early activation of microglia and astrocytes in mouse models of spinocerebellar ataxia type 1. Neuroscience 289, 289–299.

Duyao, M., Ambrose, C., Myers, R., Novelletto, A., Persichetti, F., Frontali, M., et al., 1993. Trinucleotide repeat length instability and age of onset in Huntington's disease. Nat. Genet. 4 (4), 387–392.

Edamakanti, C.R., Do, J., Didonna, A., Martina, M., Opal, P., 2018. Mutant ataxin1 disrupts cerebellar development in spinocerebellar ataxia type 1. J. Clin. Invest. 128 (6), 2252–2265.

Efthymiou, A.G., Goate, A.M., 2017. Late onset Alzheimer's disease genetics implicates microglial pathways in disease risk. Mol. Neurodegener. 12 (1), 43.

Fang, Y., Wang, J., Yao, L., Li, C., Wang, J., Liu, Y., et al., 2018. The adhesion and migration of microglia to beta-amyloid (Abeta) is decreased with aging and inhibited by Nogo/NgR pathway. J. Neuroinflammation 15 (1), 210.

Fellner, L., Irschick, R., Schanda, K., Reindl, M., Klimaschewski, L., Poewe, W., et al., 2013. Toll-like receptor 4 is required for α-synuclein dependent activation of microglia and astroglia. Glia 61 (3), 349–360.

Fiebich, B.L., Batista, C.R.A., Saliba, S.W., Yousif, N.M., de Oliveira, A.C.P., 2018. Role of microglia TLRs in neurodegeneration. Front. Cell. Neurosci. 12, 329.

Frautschy, S.A., Yang, F., Irrizarry, M., Hyman, B., Saido, T.C., Hsiao, K., et al., 1998. Microglial response to amyloid plaques in APPsw transgenic mice. Am. J. Pathol. 152 (1), 307–317.

Friedrich, J., Kordasiewicz, H.B., O'Callaghan, B., Handler, H.P., Wagener, C., Duvick, L., et al., 2018. Antisense oligonucleotide-mediated ataxin-1 reduction prolongs survival in SCA1 mice and reveals disease-associated transcriptome profiles. JCI Insight 3 (21).

Guillot-Sestier, M.V., Doty, K.R., Town, T., 2015. Innate immunity fights Alzheimer's disease. Trends Neurosci. 38 (11), 674–681.

Guzmán-Martinez, L., Farías, G.A., Maccioni, R.B., 2013. Tau oligomers as potential targets for Alzheimer's diagnosis and novel drugs. Front. Neurol. 4, 167.

Guzman-Martinez, L., Maccioni, R.B., Andrade, V., Navarrete, L.P., Pastor, M.G., Ramos-Escobar, N., 2019. Neuroinflammation as a common feature of neurodegenerative disorders. Front. Pharmacol. 10, 1008.

Hanisch, U.K., 2002. Microglia as a source and target of cytokines. Glia 40 (2), 140–155.

Heneka, M.T., Kummer, M.P., Stutz, A., Delekate, A., Schwartz, S., Vieira-Saecker, A., et al., 2013. NLRP3 is activated in Alzheimer's disease and contributes to pathology in APP/PS1 mice. Nature 493 (7434), 674–678.

Heneka, M.T., Kummer, M.P., Latz, E., 2014. Innate immune activation in neurodegenerative disease. Nat. Rev. Immunol. 14 (7), 463–477.

Heneka, M.T., Golenbock, D.T., Latz, E., 2015. Innate immunity in Alzheimer's disease. Nat. Immunol. 16 (3), 229–236.

Heneka, M.T., 2017. Inflammasome activation and innate immunity in Alzheimer's disease. Brain Pathol. 27 (2), 220–222.

Hernangomez, M., Carrillo-Salinas, F.J., Mecha, M., Correa, F., Mestre, L., Loria, F., et al., 2014. Brain innate immunity in the regulation of neuroinflammation: therapeutic strategies by modulating CD200-CD200R interaction involve the cannabinoid system. Curr. Pharmaceut. Des. 20 (29), 4707–4722.

Jacobs, A.H., Tavitian, B., 2012. Noninvasive molecular imaging of neuroinflammation. J. Cerebr. Blood Flow Metabol. 32 (7), 1393–1415.

Kim, S., Kwon, S.H., Kam, T.I., Panicker, N., Karuppagounder, S.S., Lee, S., et al., 2019. Transneuronal propagation of pathologic α-synuclein from the gut to the brain models Parkinson's disease. Neuron 103 (4), 627-641.e7.

Klement, I.A., Skinner, P.J., Kaytor, M.D., Yi, H., Hersch, S.M., Clark, H.B., et al., 1998. Ataxin-1 nuclear localization and aggregation: role in polyglutamine-induced disease in SCA1 transgenic mice. Cell 95 (1), 41–53.

Klingelhoefer, L., Reichmann, H., 2015. Pathogenesis of Parkinson disease—the gut-brain axis and environmental factors. Nat. Rev. Neurol. 11 (11), 625–636.

Kovacs, G.G., 2017. Concepts and classification of neurodegenerative diseases. Handb. Clin. Neurol. 145, 301–307.

Kucukdereli, H., Allen, N.J., Lee, A.T., Feng, A., Ozlu, M.I., Conatser, L.M., et al., 2011. Control of excitatory CNS synaptogenesis by astrocyte-secreted proteins Hevin and SPARC. Proc. Natl. Acad. Sci. U.S.A. 108 (32), E440–E449.

Kunis, G., Baruch, K., Rosenzweig, N., Kertser, A., Miller, O., Berkutzki, T., et al., 2013. IFN-gamma-dependent activation of the brain's choroid plexus for CNS immune surveillance and repair. Brain 136 (Pt 11), 3427–3440.

Li, X., Sundquist, J., Sundquist, K., 2012. Subsequent risks of Parkinson disease in patients with autoimmune and related disorders: a nationwide epidemiological study from Sweden. Neurodegener. Dis. 10 (1–4), 277–284.

Liddelow, S.A., Guttenplan, K.A., Clarke, L.E., Bennett, F.C., Bohlen, C.J., Schirmer, L., et al., 2017. Neurotoxic reactive astrocytes are induced by activated microglia. Nature 541 (7638), 481–487.

Lin, C.H., Chen, C.C., Chiang, H.L., Liou, J.M., Chang, C.M., Lu, T.P., et al., 2019. Altered gut microbiota and inflammatory cytokine responses in patients with Parkinson's disease. J. Neuroinflammation 16 (1), 129.

Liu, Y., Xie, X., Xia, L.P., Lv, H., Lou, F., Ren, Y., et al., 2017. Peripheral immune tolerance alleviates the intracranial lipopolysaccharide injection-induced neuroinflammation and protects the dopaminergic neurons from neuroinflammation-related neurotoxicity. J. Neuroinflammation 14 (1), 223.

Lobsiger, C.S., Cleveland, D.W., 2007. Glial cells as intrinsic components of non-cell-autonomous neurodegenerative disease. Nat. Neurosci. 10 (11), 1355–1360.

Louveau, A., Herz, J., Alme, M.N., Salvador, A.F., Dong, M.Q., Viar, K.E., et al., 2018. CNS lymphatic drainage and neuroinflammation are regulated by meningeal lymphatic vasculature. Nat. Neurosci. 21 (10), 1380–1391.

Maarouf, C.L., Beach, T.G., Adler, C.H., Shill, H.A., Sabbagh, M.N., Wu, T., et al., 2012. Cerebrospinal fluid biomarkers of neuropathologically diagnosed Parkinson's disease subjects. Neurol. Res. 34 (7), 669–676.

Maccioni, R.B., Rojo, L.E., Fernandez, J.A., Kuljis, R.O., 2009. The role of neuroimmunomodulation in Alzheimer's disease. Ann. N.Y. Acad. Sci. 1153, 240–246.

Maccioni, R.B., 2012. Introductory remarks. Molecular, biological and clinical aspects of Alzheimer's disease. Arch. Med. Res. 43 (8), 593–594.

Martin, H.L., Santoro, M., Mustafa, S., Riedel, G., Forrester, J.V., Teismann, P., 2016. Evidence for a role of adaptive immune response in the disease pathogenesis of the MPTP mouse model of Parkinson's disease. Glia 64 (3), 386–395.

McGeer, P.L., Itagaki, S., Tago, H., McGeer, E.G., 1987. Reactive microglia in patients with senile dementia of the Alzheimer type are positive for the histocompatibility glycoprotein HLA-DR. Neurosci. Lett. 79 (1–2), 195–200.

Morales, I., Farías, G., Maccioni, R.B., 2010. Neuroimmunomodulation in the pathogenesis of Alzheimer's disease. Neuroimmunomodulation 17 (3), 202–204.

Morales, I., Guzmán-Martínez, L., Cerda-Troncoso, C., Farías, G.A., Maccioni, R.B., 2014. Neuroinflammation in the pathogenesis of Alzheimer's disease. A rational framework for the search of novel therapeutic approaches. Front. Cell. Neurosci. 8, 112.

Myers, R.H., MacDonald, M.E., Koroshetz, W.J., Duyao, M.P., Ambrose, C.M., Taylor, S.A., et al., 1993. De novo expansion of a (CAG)n repeat in sporadic Huntington's disease. Nat. Genet. 5 (2), 168–173.

Nalavade, R., Griesche, N., Ryan, D.P., Hildebrand, S., Krauss, S., 2013. Mechanisms of RNA-induced toxicity in CAG repeat disorders. Cell Death Dis. 4 (8), e752.

Nykjaer, C.H., Brudek, T., Salvesen, L., Pakkenberg, B., 2017. Changes in the cell population in brain white matter in multiple system atrophy. Mov. Disord. 32 (7), 1074–1082.

Paulson, H.L., Perez, M.K., Trottier, Y., Trojanowski, J.Q., Subramony, S.H., Das, S.S., et al., 1997. Intranuclear inclusions of expanded polyglutamine protein in spinocerebellar ataxia type 3. Neuron 19 (2), 333–344.

Qu, W., Johnson, A., Kim, J.H., Lukowicz, A., Svedberg, D., Cvetanovic, M., 2017. Inhibition of colony-stimulating factor 1 receptor early in disease ameliorates motor deficits in SCA1 mice. J. Neuroinflammation 14 (1), 107.

Quek, H., Luff, J., Cheung, K., Kozlov, S., Gatei, M., Lee, C.S., et al., 2017. A rat model of ataxia-telangiectasia: evidence for a neurodegenerative phenotype. Hum. Mol. Genet. 26 (1), 109–123.

Rannikko, E.H., Weber, S.S., Kahle, P.J., 2015. Exogenous α-synuclein induces toll-like receptor 4 dependent inflammatory responses in astrocytes. BMC Neurosci. 16, 57.

Savitt, D., Jankovic, J., 2019. Targeting α-synuclein in Parkinson's disease: progress towards the development of disease-modifying therapeutics. Drugs 79 (8), 797–810.

Scherzinger, E., Lurz, R., Turmaine, M., Mangiarini, L., Hollenbach, B., Hasenbank, R., et al., 1997. Huntingtin-encoded polyglutamine expansions form amyloid-like protein aggregates in vitro and in vivo. Cell 90 (3), 549–558.

Scoles, D.R., Meera, P., Schneider, M.D., Paul, S., Dansithong, W., Figueroa, K.P., et al., 2017. Antisense oligonucleotide therapy for spinocerebellar ataxia type 2. Nature 544 (7650), 362–366.

Seki, T., Sato, M., Kibe, Y., Ohta, T., Oshima, M., Konno, A., et al., 2018. Lysosomal dysfunction and early glial activation are involved in the pathogenesis of spinocerebellar ataxia type 21 caused by mutant transmembrane protein 240. Neurobiol. Dis. 120, 34–50.

Shen, Y., McMackin, M.Z., Shan, Y., Raetz, A., David, S., Cortopassi, G., 2016. Frataxin deficiency promotes excess microglial DNA damage and inflammation that is rescued by PJ34. PLoS One 11 (3), e0151026.

Silva, A., de Almeida, A.V., Macedo-Ribeiro, S., 2018. Polyglutamine expansion diseases: more than simple repeats. J. Struct. Biol. 201 (2), 139–154.

Smith, H.L., Freeman, O.J., Butcher, A.J., Holmqvist, S., Humoud, I., Schätzl, T., et al., 2020. Astrocyte unfolded protein response induces a specific reactivity state that causes non-cell-autonomous neuronal degeneration. Neuron 105 (5), 855-866.e5.

Sofroniew, M.V., Vinters, H.V., 2010. Astrocytes: biology and pathology. Acta Neuropathol. 119 (1), 7–35.

Tai, Y.F., Pavese, N., Gerhard, A., Tabrizi, S.J., Barker, R.A., Brooks, D.J., et al., 2007. Microglial activation in presymptomatic Huntington's disease gene carriers. Brain 130 (Pt 7), 1759–1766.

Tan, E.K., Chao, Y.X., West, A., Chan, L.L., Poewe, W., Jankovic, J., 2020. Parkinson disease and the immune system - associations, mechanisms and therapeutics. Nat. Rev. Neurol. 16 (6), 303–318.

Thomas, B., Beal, M.F., 2007. Parkinson's disease. Hum. Mol. Genet. 16 (Spec No. 2), R183–R194.

Tsoi, H., Chan, H.Y., 2014. Roles of the nucleolus in the CAG RNA-mediated toxicity. Biochim. Biophys. Acta 1842 (6), 779–784.

VandeVrede, L., Boxer, A.L., Polydoro, M., 2020. Targeting tau: clinical trials and novel therapeutic approaches. Neurosci. Lett. 134919.

Webers, A., Heneka, M.T., Gleeson, P.A., 2020. The role of innate immune responses and neuroinflammation in amyloid accumulation and progression of Alzheimer's disease. Immunol. Cell Biol. 98 (1), 28–41.

Wu, M.C., Xu, X., Chen, S.M., Tyan, Y.S., Chiou, J.Y., Wang, Y.H., et al., 2017. Impact of Sjogren's syndrome on Parkinson's disease: a nationwide case-control study. PLoS One 12 (7), e0175836.

Yun, S.P., Kam, T.I., Panicker, N., Kim, S., Oh, Y., Park, J.S., et al., 2018. Block of A1 astrocyte conversion by microglia is neuroprotective in models of Parkinson's disease. Nat. Med. 24 (7), 931–938.

Zarrouk, A., Debbabi, M., Bezine, M., Karym, E.M., Badreddine, A., Rouaud, O., et al., 2018. Lipid biomarkers in Alzheimer's disease. Curr. Alzheimer Res. 15 (4), 303–312.

CHAPTER 5

Interconnectivity of Gene, Immune System, and Metabolism in the Muscle Pathology of Duchenne Muscular Dystrophy (DMD)

NIRAJ KUMAR SRIVASTAVA, MSC, PHD • RAMAKANT YADAV, MD, DM[a] • SOMNATH MUKHERJEE, MSC, PHD[a]

5.1 MUSCULAR DYSTROPHY: NEUROMUSCULAR DISORDER

Neuromuscular disorders (muscular dystrophies/myopathies), especially muscular dystrophies, are significant health problems worldwide. Muscular dystrophy is a genetic disease and characterized by progressive muscle wasting and weakness with variable distribution and severity. The essential features of muscular dystrophy are selective involvement, significant wasting, and weakness of muscles. This is in distinction with other types of myopathies, in which muscle weakness is diffused, relatively more than wasting, and enlargement in muscle is exceptional. Several types of muscular dystrophy have been illustrated in the literature on the basis of the age, progress, site of involvement, and the inheritance pattern. The most important types are Duchenne muscular dystrophy (DMD), Becker muscular dystrophy (BMD), Emery-Dreifuss muscular dystrophy (EDMD), facioscapulohumeral dystrophy (FSHD), limb girdle muscular dystrophy (LGMD), myotonic dystrophy (DM), and congenital muscular dystrophy (CMD).The genes and their respective protein, which are responsible for these diseases, have now been identified (Emery, 1998, 2000).

Study of the progressive degenerative muscle diseases was started in the mid-nineteenth century, particularly in France and Germany. Aran (1850) and Wachsmuth (1855) reconsidered the subject with modest effort at classification of disease states or the differentiation of neurogenic from myopathic disease.

In 1852, Meryon described the comprehensible explanation of progressive muscular paralysis in young boys and demonstrated that it was observed due to "granular degeneration" of the muscles exclusive of alterations in the anterior horns of the spinal cords or in the motor roots. Dr. Gullaume Benjamin Amand Duchenne (1868) was a French neurologist and performed the milestone work in the field of muscular dystrophy. He recognized and illustrated the pseudo-hypertrophy muscular dystrophy as a primary muscle disease. Afterward, this disease is known as "**Duchenne muscular dystrophy**" (Arikawa-Hirasawa et al., 1995; Amato and Russell, 2008; Schapira and Griggs, 1999).

Muscular dystrophy (muscle diseases) has been described in ancient Indian medical treatise "Charak Samhita" (100 BC−200 CE). Charak Samhita had been written by Maharishi Charak in Sanskrit language, who was a physician of ancient India (Meulenbeld, 1999; Samhita, 1959). Such description has been analyzed by modern Indian physicians and neurologists in the light of present clinical observations to understand the concept of ancient physicians (Gouri-Devi and Venkatram, 1983). The ancient physicians had the knowledge of proximal pseudohypertrophic muscle weakness, and they recognized or expressed it to excessive accumulation of morbid humor (provoked by its interaction with fat) in the hip, thigh, and calf regions. The Sanskrit description is given in the following:

Sanehachamam chitam kosthe vatadeenmedasa saha I

Rudhwashu gouravdooru yatuadhogaih siradibhih II

---

[a]Both authors are considered as second author.

The Molecular Immunology of Neurological Diseases. https://doi.org/10.1016/B978-0-12-821974-4.00007-8

Poorayan sakthijanghoru dosho medo balotkatah I

Avidheya parispandam janayatyalpa vikramam II

This description is matched to the DMD. The inheritance of the disease was also described, and sense of this description was simply that limb or organ becomes defective, whose unique representative part in the germoplasm has been caused to be defective. This is also the concept of X-linked inheritance of DMD (Samhita, 1959; Gouri-Devi and Venkatram, 1983). This is apparent from their knowledge and intelligence that "among others, 'flesh' (muscle) is one of the 'mother engendered' parts, which pass from mother to embryo during its formation." The management of such patients had been performed by physiotherapy and possibly acupressure. This sense was made by the description that exercise whenever possible should be prescribed or the patient may be made to walk on uneven ground covered with gravel and sand. The Sanskrit version is describe in the following:

Kaphkshayartham shakyeshu vyayameshwanuyojayet I

Sthalanyakramayet kalyam sharkarah sikatastatha II

The disease was well recognized in several centuries ago but still not much has happened in the treatment of DMD (Samhita, 1959; Gouri-Devi and Venkatram, 1983).

5.2 DUCHENNE MUSCULAR DYSTROPHY

DMD is a lethal and the most rapidly progressive form of dystrophy. This dystrophy occurs due to mutation in dystrophin gene. The observed incidence of this dystrophic form is 1 in 3500 live male births, and a reported prevalence is $50-70 \times 10^{-6}$ of total male population. In India, the hospital-based report showed the relative prevalence of DMD is 30%. The most consistent estimates of the DMD cases have been situated in the range of 18–30 per 1, 00,000 live male birth and its prevalence in the population as a whole from 1.9 to 4.8 per 100 000. In all these cases of DMD, one-third of cases are new mutants, one-third have a prior family history, and one-third are born to unsuspecting and frequently mutant carriers. The mutation rate is about $7-10 \ 1 \times 10^{-5}$ per gene per generation (Emery, 1998, 2000; Arikawa-Hirasawa et al., 1995; Amato and Russell, 2008; Schapira and Griggs, 1999)

Clinical symptoms in DMD patients are started in between the age of 3–5 years. These are difficulty in standing, walking, running, and climbing the stairs. Frequent fall is also observed in these patients. Muscle wasting and weakness are symmetrical and selective.

The selective involvement of muscles is the quadriceps, iliopsoas, and glutei in the lower limb and the latissimus dorsi, serrate, sternocostal head of pectoralis major, biceps, triceps, and brachioradialis in the upper limb. Decrease in muscle tone (hypotonia), prominence of the calves, and abnormalities of gait are also observed in these patients. Progressive muscle wasting and weakness lead to inability to walk by the age of 8–12 years of the onset (patient becomes chair bound) and later to contracture and thoracic deformity. Cardiac involvement is invariable, and intellectual impairment is frequent, but not invariable. Death of these patients is occurred in second or third decade caused by respiratory or cardiac failure, often associated with respiratory infection (Emery, 1998, 2000; Arikawa-Hirasawa et al., 1995; Amato and Russell, 2008; Schapira and Griggs, 1999; Lane, 1996; Stålberg, 2003).

Two important clinical signs are helpful for the diagnostic purpose. These are Gower's sign (Fig. 5.1) "Pradhan sign" (Fig. 5.2), or "valley sign." Gower's sign is observed due to weakness of the knees and hip extensors. This sign is visualized, when the child attempts to rise from the floor. The another important clinical sign "Pradhan sign" or "valley sign" is visualized in patients with DMD as a linear groove or sometimes an oval depression due to the wasting of the muscles participating in the formation of the posterior axillary fold, the teris major, teris minor, posterior-most part of the deltoid, and the lateral one-third of the infraspinatus. On either side of the depression, two prominent mounts are visible: the inferomedial formed by the hypertrophied infraspinatus muscle and the superolateral by the hypertrophied deltoid muscle. The whole

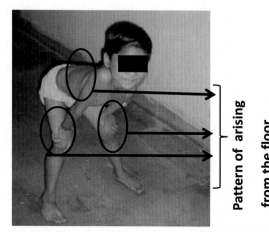

FIG. 5.1 Gower's maneuver or sign in DMD patient. *DMD,* Duchenne muscular dystrophy.

appearance is like a "valley between two mounts" (Amato and Russell, 2008; Emery, 2002; Pradhan, 1994).

Electromyography (EMG) is a laboratory examination, which refers to methods of studying the electrical activity of muscle. EMG examination is performed for confirming the diagnosis of myopathy in DMD patients by using concentric bipolar needle. Myopathy is established after the expression of the myopathic EMG pattern. The myopathic EMG pattern appeared in the form of predominately small amplitude (200 µv per division) and short duration (10 ms/cm) motor units

potential (MUPs) with increase in the proportion of polyphasia (Fig. 5.3) (DeLucchi, 1976; Mishra and Kalita, 2006).

The value of serum CK (creatine kinase) has been paid to its diagnostic assist in clinical medicine. The enzyme is present in high concentrations in skeletal and cardiac muscle and a lesser extent in the brain tissue but is only found in the relatively small concentration in other tissues (Colombo et al., 1962). The serum CK is 10−20 times higher in DMD patients as compared with normal individuals (Emery, 2002). Value of serum CK in DMD patients also depends on the age. Below 5 year age, it is higher and then subsequently decreases according to age. Serum CK level is strikingly elevated, even in preclinical stages of the disease (Bushby and Anderson, 2001). Serum CK measurement is a very simple diagnostic tool for DMD patients, and average CK value is found to be 9387.51 IU/L. The minimum and maximum values of CK has been estimated 1183 IU/L and 29 000 IU/L, respectively (Srivastava et al., 2010).

Multiplex PCR (mPCR) is a common method for the DMD diagnosis in India (Fig. 5.4). The mPCR method is principally qualitative or semiquantitative, and this permits the revealing of approximately 98% of all the deletions, which covers the identification of all the 65% mutations. MLPA (multiplex ligation dependent probe amplification) has facilitated more reliable and quicker quantitative deletion of the complete dystrophin gene containing 79 exons to study the deletions and duplications. Following multiplex PCR by MLPA

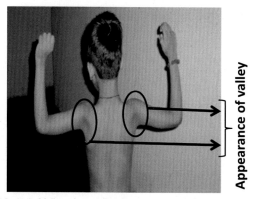

FIG. 5.2 Valley sign or Pradhan sign in DMD patient. *DMD*, Duchenne muscular dystrophy.

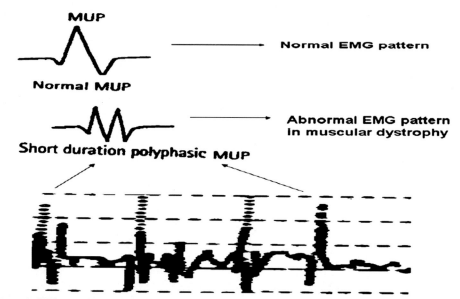

FIG. 5.3 EMG examination of a DMD patient showed the myopathic pattern. *DND*, Duchenne muscular dystrophy; *EMG*, electromyography.

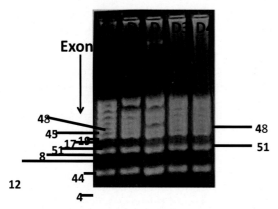

FIG. 5.4 Analysis of the exons deletion in dystrophin gene (performed by mPCR) showed deletion of exon 48 in D1 (DMD patient) and exons 48 and 51 in D2 (DMD patient). C represents the normal subject. *DMD*, Duchenne muscular dystrophy; *mPCR*, multiplex polymerase chain reaction.

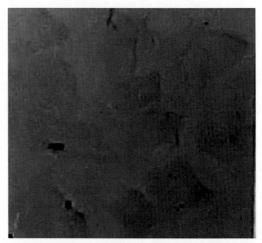

FIG. 5.5 Dystrophin staining showed the complete absent of dystrophin in DMD muscle specimen. *DMD*, Duchenne muscular dystrophy.

increases the percentage of patients with a precise diagnosis to 75%. MLPA-based diagnostic method is used to diagnose DMD in the United States, China, and several European countries (Chaturvedi et al., 2001; Murugan et al., 2010; Potnis-Lele, 2012; Srivastava et al., 2016a,b).

Immunohistochemical-based muscle biopsy examination has diagnostic significance in the case where gene mutation or mPCR-based examination failed to provide the diagnosis. In such cases, immunohistochemical analysis, i.e., dystrophin analysis provided the confirm result. Dystrophin is completely absent in the muscle biopsy specimen of DMD patients (Fig. 5.5) (Srivastava et al., 2010; Hilton-Jones et al., 2014; Woolf et al., 2002; Swash and Schwartz, 2013).

5.3 MUSCLE PATHOLOGY IN DUCHENNE MUSCULAR DYSTROPHY

The primary contribution of muscle pathology in DMD is the genetic defect or gene mutation. This genetic defect or gene mutation is further connected to immune-mediated mechanism and metabolic abnormalities. The interconnectivity of all these events is responsible for loss of muscle regenerative capacity, enhancing the rate of muscle degeneration (Nigro and Piluso, 2015).

5.3.1 Molecular Genetics Defects: Initiation of Muscle Pathology in Duchenne Muscular Dystrophy

5.3.1.1 Dystrophin gene

DMD is caused by mutations of the dystrophin gene. This gene is located on the short arm of the

X-chromosome (Xp21.2). According to the first molecular report, the gene responsible for DMD was referred to as the "Duchenne gene." However, this gene produces the dystrophin protein, which is a normal cytoskeletal protein and required for the structural and functional integrity of the muscle membrane. In this regard, the accurate and appropriate terminology "dystrophin gene" is established in the place of the "Duchenne gene." In another molecular discovery, a mouse model (mdx mouse) was found to lack dystrophin. Such study was also performed on several other animal models of muscular dystrophy such as a golden retriever dog, cat, and Rottweiler dog and observed that the completely lacks dystrophin. So it has been absolutely well established that DMD arises due to mutation in the dystrophin gene. Such genetic defect or defect in dystrophin gene is the base for the initiation of muscle pathology because dystrophin gene produces the defective dystrophin protein of muscle membrane (Emery, 1998, 2000; Kakulas, 1996; Pozzoli et al., 2003).

The dystrophin gene is a very largest human gene (0.1% of the entire human genome), and it contains 2.4–3.0 megabases (Mb) of DNA. The coding region of this gene contains 1–79 exons, which is separated by introns of 200 kb size. The exons of this gene are represented by 14 kDa of mRNA (Pozzoli et al., 2003; Lee et al., 2012).

5.3.1.1.1 Mutation in dystrophin gene. Due to a very largest size of the dystrophin gene, the rate of mutation in this gene is very high (Bennett et al., 2001). Several studies have shown mutation at the dystrophin locus of the X-chromosome in patients with DMD (severe and lethal form) and milder allelic form, Becker muscular dystrophy (BMD) (Le Rumeur, 2015).With extensive analysis and study, Monaco et al. (1986) postulated that the absence of dystrophin in DMD was due to the "frame shift" deletions. In BMD, the mainstream deletions are in frame, and in this regard, a truncated protein is produced. This postulation puts forward the explanation for difference in severity between DMD and BMD and is also known as the "reading frame" hypothesis of Monaco (Monaco et al., 1986). At the same time, other investigators publicized important information concerning the gene and the gene product. By using the technique of immunoblotting, Hoffman et al. (1988) demonstrated absence of dystrophin in DMD muscles (Kunkel et al., 1985; Ray et al., 1985–1986; Worton and Thompson, 1988; Hoffman et al., 1987). Afterward, this technique had been developed to recognize and distinguish DMD/BMD from other related diseases. Later, this had been observed that the expression of dystrophin gene was regulated in the course of different promoters in different tissues (Amato and Russell, 2008; Schapira and Griggs, 1999). According to the suggestions of Hoffman et al. (1987, 1988), DMD and BMD can be divided into four clinical categories on the basis of the presence of dystrophin quantity. DMD (with the patient becoming wheelchair bound at the age of 11 years) occurs when the quantity of dystrophin is less than 3% of normal; in severe BMD (when the patient becomes wheelchair bound from 13 to 20 year of age) dystrophin, quantities are between 3% and 10%; and moderate or mild BMD (with ambulation beyond the age of 20 years) is related to a dystrophin quantity of 20% or more. These strategies have established and applied in the clinical investigation of the intermittent cases (Hoffman et al., 1987). However, there is no strong or strict correlation between dystrophin quantity and the severity of the disease due to occurrence of several exceptional cases of DMD. Some cases of DMD with abundant dystrophin may still show the destructive DMD phenotype, whereas in some milder cases, patients have no confirmable dystrophin (Gao and McNally, 2015).

5.3.1.2 Pattern of mutation in dystrophin gene
Several mutation patterns have been observed in the dystrophin gene. These patterns of mutations have been described in the following:

Large intragenic exons deletions: Southern blot analysis has publicized that 50%–70% of the dystrophin mutations arise from large intragenic deletions involving several exons. Major and minor hot spots are located in the gene. The major hot spot is located near the middle of the gene and encircling exons 44 and 45. The minor hot spot covers a broad region near the 5′ ends of the gene, and several deletions occupy the first 20 exons. This is very rare to get large deletions of more than 60 exons. Southern blot analysis has now been largely replaced by mPCR method (Fig. 5.6), and it identifies 98% of the deletion detected by Southern blotting and is easier and cheaper as compared with Southern blotting (Worton and Thompson, 1988; Hoffman et al., 1987; Danieli et al., 1993).

Duplications: Dystrophin gene duplications account for 5%–10% of all the mutations. Duplications in DMD patients have been identified with both genomic probes and cDNA probes. According to the study of Hu et al. (1988), duplications are tandem repeats and can result in a genetic disorder through the disruption of exons organization. Hu et al. (1990) studied the gene duplication in 72 nondeleted DMD/BMD patients. Ten patients had a duplication of part of the gene of which six had a novel restriction fragment that spanned the duplication junction. Kodaira et al. (1993) studied the 21 nondeleted Japanese DMD patients by PFGE and found that seven had partial duplications spanning 50–400 kb, four had duplications corresponding to the major hot spot (7.5–8.5 kb from the 5′ end of the cDNA), and two had duplications in the region about 10 kb from the 5′ end of the cDNA. (At this point, the causative mutations are rare.) One patient, however, had duplication in the duplication-prone region, i.e., 1.0 kb from the 5′ end of the cDNA (Hu et al., 1988, 1990; Kodaira et al., 1993).

Point mutation and microdeletions: Point mutations and small deletions or microdeletions were discovered at a point near the beginning stage. The identification of point mutations in the dystrophin gene represents a challenge due to large size and complexity of the dystrophin gene. The first nonsense mutation was reported by Bulman et al. (1991) in exon 26 of a DMD patient, where immunological analysis of the truncated dystrophin from muscle biopsy specimen allowed prior localization of the mutation. Detection of point mutation became easier after single-strand conformational polymorphism analysis (SSCP), the protein truncation test, heteroduplex analysis, and other available analytical methods. In most of the cases of DMD/BMD, deletion of one to a few nucleotides generates a stop codon except the deletion is in frame. Missense mutation involving a single nucleotide

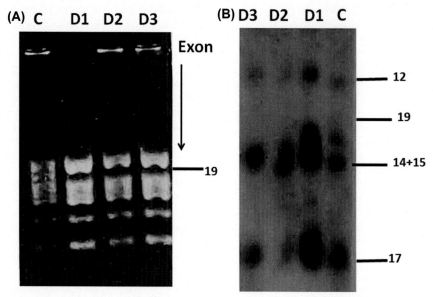

FIG. 5.6 Analysis of the exons deletion in dystrophin gene has been performed by **(A)** mPCR, and **(B)** Southern hybridization showed deletion of exon 19 in D1,D2 and D3 (DMD patient). C represents the normal subject. Both techniques showed the same results and presently mPCR replaced the southern hybridization based analysis. *DMD*, Duchenne muscular dystrophy; *mPCR*, multiplex polymerase chain reaction.

infrequently produces pathologic phenotype. Occasional nonsense mutations in the cysteine-rich and C-terminal regions result in a disease phenotype, most likely owing to nonsense-mediated mRNA crumble (Bulman et al., 1991; Roberts et al., 1992; Mittal et al., 1997; Singh et al., 1997). Kneppers et al. (1995) used mPCR products in SSCP for screening of point mutations in a set of 70 nondeleted DMD/BMD patients and identified six patients with band shift. In all these six band shift patients, five were the result of a frame shift or termination mutation, while the other band shift was found to be a rare polymorphism (Kneppers et al., 1995).

Frame shift rule: When the number of nucleotides removed by a deletion or duplication is an integral multiple of 3N (where N is an integer), the mutation is "in frame," and the reading frame remains open; but if the number is 3N+1 or 3N+2, then the mutation is out of frame. This usually generates a stop codon after a variable number of missense codons. In general, frame shift mutations cause a DMD phenotype, whereas in-frame mutations result in a milder BMD phenotype. This rule is valid for deletions and duplications regardless of the size of the deletion (Mittal et al., 1997; Koenig et al., 1989; Gillard et al., 1989). However, in a few cases, an in-frame deletion generates a DMD phenotype, and

out-of-frame deletion causes a BMD phenotype (Malhotra et al., 1988).

5.3.2 Immune System—Mediated Muscle Pathology of Duchenne Muscular Dystrophy

5.3.2.1 Dystrophin protein

Dystrophin gene produces the dystrophin protein with a molecular weight of 427 kDa and possesses 3685 amino acids. This protein is found in skeletal muscle (0.002% of the entire muscle protein) and also found in heart, brain, and smooth muscle with a molecular weight of 405 kDa. Four domains of dystrophin protein are identified. The first is the actin-binding N-terminal domain, the second is the rod domain, the third is the cysteine-rich domain, and the fourth is the C-terminal domain (Amato and Russell, 2008; Kakulas, 1996; Gao and McNally, 2015; Brown et al., 2012).

These domains and related exons are described in the following:

Actin-binding domain (exons 1—8): The close resemblance of 240 N-terminal amino acids of dystrophin to the actin-binding domains of spectrin and α-actinin predicted that dystrophin would bind to actin filaments.

Rod or triple helical repeat domain (exons 9—62): The next and largest dystrophin domain has a mass of

300 kDa and 125 nm in length. This is encoded by exons 9–62 and contains 2400 amino acid residues.

Cysteine-rich domain (exons 63–69): This domains has 15 cysteine residues and a stretch of 142 residues that show 24% homology to the C-terminus of α-actinin.

C-terminal domain (exons 70–79): The fourth domain encompasses the last 420 amino acid residues of dystrophin and is highly conserved across species (Amato and Russell, 2008; Kakulas, 1996; Gao and McNally, 2015; Koenig et al., 1988; Brown et al., 1997).

5.3.2.1.1 Dystrophin: localization and functions.
Dystrophin protein is located at the cytoplasmic membrane of skeletal, cardiac, and smooth muscles (at the skeletal myotendinous junction and synapses in the central nervous system). The C-terminal region (containing the cysteine-rich and C-terminal domains) of dystrophin is linked with a complex of at least six proteins, called the dystrophin-associated glycoproteins (DAGs) or dystrophin-associated proteins (DAPs) or dystrophin–glycoprotein complexes (DGCs) [Fig. 5.7]. The cysteine-rich domain may be most critical because it interacts with the intracellular tail of β-dystroglycan, which links dystrophin to the sarcolemmal membrane and preserves

the entire DGC. The C-terminal domain binds additional proteins of DGC (Amato and Russell, 2008; Kakulas, 1996; Gao and McNally, 2015; Tiidus, 2008).

Dystrophin is considered an essential element in the muscle fiber in that it links cytoskeletal actin to the DGC, which in turns forms a connection through the sarcolemmal membrane to the basal lamina. The DGC along with additional proteins forms riblike lattice on the cytoplasmic face of the sarcolemma, known as costameres that facilitate force transmission between active and nonactive fibers. This complex of protein also likely acts as a signaling conduct between outside and inside of the fiber. Dystrophin is also found in abundance in the myotendinous junction and is thought to facilitate the transmission of forces from the muscle fibers to the tendon (Amato and Russell, 2008; Kakulas, 1996; Gao and McNally, 2015; Tiidus, 2008).

The DGC is made up of several subcomplexes, including the dystroglycan complex and the sarcoglycan (sarcoglycan complex and the peripheral proteins of the cytoplasmic dystrophin-containing domain). The amino or N-terminal of dystrophin binds cytoskeletal actin, and its carboxy- or C-terminal region binds the dystroglycan complex. The dystroglycan complex is composed of α- and β-dystroglycans. β-dystroglycan,

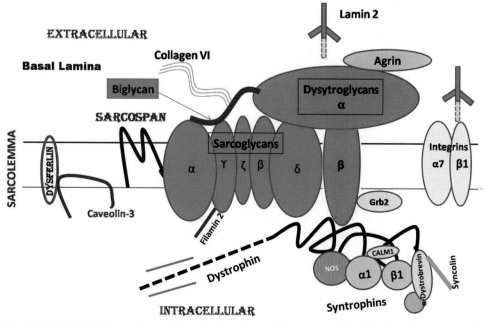

FIG. 5.7 Localization of muscle membrane protein: the cysteine-rich and C-terminal domains of dystrophin are linked with a complex of at least six proteins, called the dystrophin-associated glycoproteins (DAGs) or dystrophin-associated proteins (DAPs) or dystrophin–glycoprotein complexes (DGC).

an integral membrane protein, interacts with dystrophin in the cytosol and with α-dystroglycan, which in turn binds to laminin-2, in the basal lamina. The sarcoglycan complex, composed of α-, β-, γ-, and δ-sarcoglycan, is important for stabilizing the dystroglycan complex in the sarcolemma. The cytoplasmic dystrophin-containing domain contains α-dystrobervin that binds directly to dystrophin. β-Dystrobervin is found in the DGC of nonmuscle tissues. Another component of this domain is the syntrophin, adapter protein that links membrane-associated proteins, dystrophin, and dystrobervin to the DGC. Syntrophin is associated with the neuronal nitric oxide synthase (nNOS). Several other proteins with undefined functions are associated with the complex (Amato and Russell, 2008; Kakulas, 1996; Gao and McNally, 2015; Tiidus, 2008).

5.3.2.1.2 Dystrophin: Duchenne muscular dystrophy pathology.

The complete absence of dystrophin protein in DMD leads to the loss of DGC and thus a disrupted costameric lattice. Loss of the essential structural DGC elements results in the characteristic progressive muscle degeneration seen in DMD. The mechanism of which is still not clearly defined (Tiidus, 2008).

5.3.2.2 Immunocytochemical-based analysis of dystrophin protein in Duchenne muscular dystrophy

Immunocytochemical-based analysis of dystrophin protein in DMD is performed on muscle biopsy specimen. The dystrophin is located at the plasma membrane of skeletal muscle fibers, where it can be demonstrated by a range of immunocytochemical techniques. The monoclonal antibodies of dystrophin have been produced that recognize epitopes within various domains (N-terminal, rod domain, and C-terminal). Antibodies against dystrophin-associated proteins (such as spectrin, α-sarcoglycan, and β-dystroglycan) are also used in this technique to perform the detection of these proteins (Vogel and Zamecnik, 2005; Dubowitz et al., 2007).

In muscle biopsy specimen of DMD patients, three different labeling patterns are observed through immunocytochemical-based analysis for dystrophin protein: first, absent from most fibers with all antibodies (Fig. 5.8)]; second, present only on small clusters or a few isolated fibers (revertant fibers) (Fig. 5.9); and third, very weak labeling on most fibers (occasional fibers) (Fig. 5.10). Normal or control muscle showed the uniform labeling pattern on sarcolemma of all fibers with all antibodies (Vogel and Zamecnik, 2005; Dubowitz et al., 2007).

Immunocytochemical-based analysis of β-spectrin in muscle biopsy specimen of DMD patients showed the membrane loss due to necrosis and presence of abnormally immuno-reacted fibers (Fig. 5.11). Normal or control muscle showed the uniform labeling pattern on sarcolemma of all fibers (Vogel and Zamecnik, 2005; Dubowitz et al., 2007).

Immunocytochemical-based analysis of α-sarcoglycan and β-dystroglycan (Fig. 5.12) in muscle biopsy specimen of DMD patients showed the large level reduction on most of the fibers, whereas normal or control muscle showed the uniform labeling pattern on sarcolemma of all fibers (Vogel and Zamecnik, 2005; Dubowitz et al., 2007).

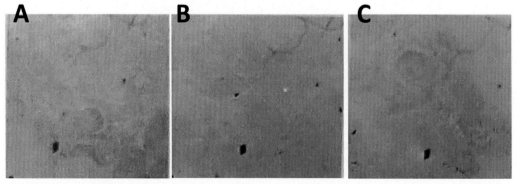

FIG. 5.8 Immunocytochemical analysis of dystrophin protein in the skeletal muscle of DMD patient has been performed with monoclonal antibody against **(A)** N-terminal of dystrophin, **(B)** rod domain of dystrophin, and **(C)** C-terminal of dystrophin showed the complete absence of dystrophin in sarcolemma of all muscle fibers. *DMD*, Duchenne muscular dystrophy.

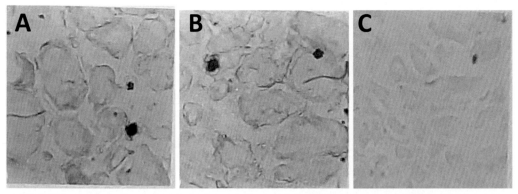

FIG. 5.9 Immunocytochemical analysis of dystrophin protein in the skeletal muscle of DMD patient has been performed with monoclonal antibody against **(A)** N-terminal of dystrophin, **(B)** rod domain of dystrophin, and **(C)** C-terminal of dystrophin showed the presence only on small clusters or a few isolated fibers (revertant fibers) of dystrophin in sarcolemma of all muscle fibers. *DMD*, Duchenne muscular dystrophy.

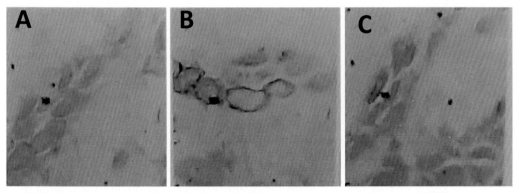

FIG. 5.10 Immunocytochemical analysis of dystrophin protein in the skeletal muscle of DMD patient has been performed with monoclonal antibody against **(A)** N-terminal of dystrophin, **(B)** rod domain of dystrophin, and **(C)** C-terminal of dystrophin showed the very weak labeling on most fibers (occasional fibers) of dystrophin in sarcolemma of all muscle fibers. *DMD*, Duchenne muscular dystrophy.

5.3.2.3 Immune-mediated muscle membrane damage in the course of defective dystrophin protein in Duchenne muscular dystrophy

5.3.2.3.1 Oxidative stress–induced muscle damage.
Oxidative stress has a central role in the pathogenesis of patients with DMD and mdx mouse model. The resultant of oxidative stress event comes in the form of the abnormal production of superoxide anion, hydrogen peroxide, or nitric oxide. These primary oxidative species are converted to secondary reactive oxygen species (such as hydroxyl radical [OH·]) and reactive nitrogen species (such as peroxynitrite [ONOO-]). Excessive levels of these secondary reactive oxygen species and reactive nitrogen species are responsible for damaging of membrane lipids, proteins (structural and regulatory), and DNA. Amplified levels of lipid and protein oxidation had been observed in dystrophic muscle of patients with DMD and animal models. These observations evidently proved that the oxidative stress is one of the important factors of muscle wasting and weakness in patients with DMD (Srivastava et al., 2015; Villalta et al., 2015; Nitahara-Kasahara et al., 2016; Shin et al., 2013; Rosenberg et al., 2015).

Oxidative stress–induced damage is responsible for the deterioration of skeletal muscle and may also be involved in the pathogenesis of the heart failure in DMD patients and mdx mice. The functional impairment of muscle force generating capacity in DMD patients happens due to the oxidative stress–dependable events (Srivastava et al., 2015; Villalta et al., 2015;

Nitahara-Kasahara et al., 2016; Shin et al., 2013; Rosenberg et al., 2015).

Excessive levels of oxidative stress are also responsible for generating reactive oxygen species—based NADPH oxidase complex [Nox2] in sarcolemma of dystrophic muscle. This fact has been proved by the upregulation of NADPH complex, and its inhibition prevents muscle damage in tibial anterior muscle of

mdx mice. Other causes for oxidative stress in dystrophic muscle are infiltrating neutrophils (which include elevated activity of myeloperoxidase), mitochondria, and upregulation of inducible nitric oxide synthase. Dystrophin-deficient myofibers come into sight to be more susceptible to excessive levels of oxidative stress. These raised levels of oxidative stress are responsible for muscle wasting with further enhancing the inflammatory response and interfering with the process of muscle regeneration (Srivastava et al., 2015; Villalta et al., 2015; Nitahara-Kasahara et al., 2016; Shin et al., 2013; Rosenberg et al., 2015).

5.3.2.3.2 Instigation of inflammatory process. Several studies utilize microarray techniques and have investigated the genome-wide gene expression in muscle specimens of DMD patients and in myofibers of mdx mice. These studies demonstrated a higher expression of genes concerned in inflammatory responses in the affected muscles (diaphragm and limb muscles), but not in the secure extraocular muscle. Other sovereign studies have confirmed the enhancement in the production of numerous inflammatory molecules including cytokines (e.g., TNF-α, IL-1β, and IL-6) and cell adhesion molecules (e.g., intracellular adhesion molecule-1) in the myofibers of dystrophic muscle (Amato and Russell, 2008; Schapira and Griggs, 1999; Villalta et al., 2015; Nitahara-Kasahara et al., 2016; Shin et al., 2013; Rosenberg et al., 2015).

Dystrophin gene mutations or gene defect in DMD patient is the major cause for chronic inflammation, fibrosis, fat infiltration, and impaired vasoregulation,

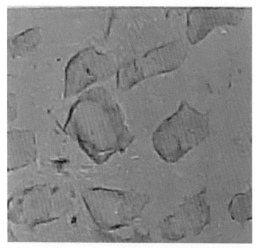

FIG. 5.11 Immunocytochemical analysis of β-spectrin protein in the skeletal muscle of DMD patient performed with monoclonal antibody against β-spectrin showed the membrane loss due to necrosis and presence of abnormally immuno-reacted fibers. *DMD*, Duchenne muscular dystrophy.

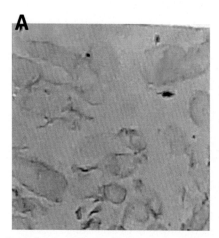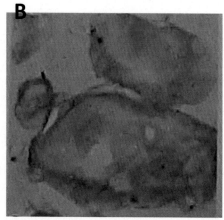

FIG. 5.12 Immunocytochemical analysis of **(A)** α-sarcoglycan and **(B)** β-dystroglycan protein in the skeletal muscle of DMD patient performed with monoclonal antibody against α-sarcoglycan and β-dystroglycan showed the large level reduction of these proteins on most of the fibers. *DMD*, Duchenne muscular dystrophy.

which is evident as muscle weakness ultimately directs to skeletal and cardiac muscle atrophy. Although involuntary injury and membrane defects are essential features that encourage dystrophic disease pathology, inflammation performs a major function in the muscle pathology of DMD. The long-standing inflammation and large numbers of inflammatory cells are promoting the muscle pathogenesis in DMD (Amato and Russell, 2008; Schapira and Griggs, 1999; Villalta et al., 2015; Nitahara-Kasahara et al., 2016; Shin et al., 2013; Rosenberg et al., 2015).

5.3.2.3.3 Immune-mediated muscle pathology.
Inflammatory cytokines are the major source for emergence of immune-mediated muscle pathology in DMD. Definitely, immune cells infiltrations lead to the onset of preliminary disease, and continual through later stages directly contributes to the enhancement and progression of the inflammation and muscle lesion in mdx mice and patients with DMD. Macrophages, T cells, and neutrophils are the primary immune cells and responsible for invading the muscle cells between the ages of 2–4 weeks. These cells play a major role in development of muscle lesions. Activated macrophages perform the lysis of muscle cells with the production of nitric oxide (NO). Macrophages are also a major source of proinflammatory cytokines and responsible for muscle wasting. This muscle wasting process is carried out by the activation of proteolytic systems and hindering the muscle regeneration. Moreover, macrophages are promoting the phagocytosis of injured myofibers and then present antigens to T cells to induce their activation. Indeed, T cells, mast cells, eosinophils, and neutrophils are also present in plenty and contribute to fiber wasting in dystrophic muscle (Amato and Russell, 2008; Schapira and Griggs, 1999; Villalta et al., 2015; Nitahara-Kasahara et al., 2016; Shin et al., 2013; Rosenberg et al., 2015).

Earlier studies showed the expression of proinflammatory factors (e.g., TNF-α, IFN-γ, IL-1, TGF-β, and MCP-1) proceeding to the onset of muscle degeneration in both DMD patients and mdx mice. These factors are also providing the damage signals that have an insightful impact on satellite cell performance throughout the repair process. There is a continually altered and reorganizing extracellular matrix (ECM) that encourages damage and dysfunction in an inflamed muscle of DMD. Intensifying deposition of fibrin within the ECM encourages inflammation-mediated muscle degeneration and regeneration with the association of $\alpha M\beta 2$ integrin engagement on macrophages, which is responsible for fibrosis development and loss of normal muscle architecture. M1 macrophages encourage the expression of proinflammatory cytokines (IL-1β, TNF-α, and IL-6), which in turn may negatively regulate satellite cell functions. In this way, these proinflammatory cytokines are regarded as a key factor in mediating the muscle damage carried out by M1 macrophages (Amato and Russell, 2008; Schapira and Griggs, 1999; Villalta et al., 2015; Nitahara-Kasahara et al., 2016; Shin et al., 2013; Rosenberg et al., 2015).

5.3.2.3.4 Innate immunity–based response in muscle pathology.
Lack of dystrophin in myofibers is responsible to contraction-induced membrane damage with release of cytoplasmic contents and stimulation of innate immunity, cycles of myofiber degeneration/regeneration, age-related replacement of muscle by fibro fatty connective tissue, muscle weakness, and ultimately death in patients with DMD. The degree of muscle pathology is usually correlated with reduction of muscle function. DMD fetal muscle represents the modest support of pathology in spite of the noticeable dystrophin deficiency at the myofiber plasma membrane. On the other hand, rapidly subsequent to birth, there is an initiation of the innate immune system with strongly activated numerous components ahead of the onset of clinical symptoms, together with altered signaling through Toll-like receptors (TLR4, TLR7) and nuclear factor κB (NF-κB) as well as expression of major histocompatibility complex (MHC) class I molecules on muscle cells (which do not normally express MHC class I). There is a strong substantiation that membrane instability and associated discharge of cytoplasmic contents into the extracellular gap arbitrate the chronic creation of the innate immune system and linked inflammatory response. The next pathological progression, which is overlapped on the chronic proinflammatory status, is that of segmental degeneration and regeneration of myofibers. Under this progression, fibers (singly or in groups) are infiltrated by neutrophils and phagocytosed by macrophages. In the meantime, occupant myogenic stem cells are triggered and distinguish into myoblasts, and regeneration of the myofiber takes place within the preexisting basal lamina. This is a well-established and well-known fact that the regenerated myofibers are dystrophin deficient. This directs to consecutive focal stretch of degeneration and regeneration, with a specific chronological staged prototype of inflammatory infiltrates. While such stretch of degeneration and regeneration is successful

in the healing of wild-type muscle, it is unable to heal DMD muscle. Eventually, with growing age, the relationship between chronic activation of innate immunity and asynchronous adjacent stretch of degeneration and regeneration yield an inadequately coordinated repair response that may itself drive disease progression (Amato and Russell, 2008; Schapira and Griggs, 1999; Villalta et al., 2015; Nitahara-Kasahara et al., 2016; Shin et al., 2013; Rosenberg et al., 2015).

Crucial features of the typical biology of muscle require its immune-restricted status, a phenomenon that is highlighted by its collapse in DMD. In the condition of normal exhaustive muscle activity, huge syncytial myofibers demonstrate outflow of cytoplasmic contents into the extracellular environment. The muscle cytoplasmic enzymes (creatine kinase) are appearing in the blood. The microscopic and cellular evidence showed the unhindered flow of cytoplasmic content across membranes. It is well established that leakage of cell cytoplasm into the extracellular environment is an effective elicit of innate immune responses, together with the binding of damage-associated molecular pattern (DAMP) molecules (for example, heat shock proteins and nucleic acids) to TLRs with subsequent inflammasome configuration. While leakage of cytoplasmic contents in vigorously exercising normal muscle (particularly eccentric or lengthening contractions) triggers TLRs, the comparative immune system advantage united with quick membrane repair and intact dystrophin emerges to boundary and hurriedly resolves inflammation. Additionally, it must be noted that there is no constitutive MHC class I expression on skeletal muscle cells, and also there is no succession from innate to adaptive immune responses with exercise-induced injury (Schapira and Griggs, 1999; Villalta et al., 2015; Nitahara-Kasahara et al., 2016; Shin et al., 2013; Rosenberg et al., 2015).

Specific muscles demonstrate noticeable variability in pathology. Such pathological variability is observed within an individual muscle, between muscles in a human patient and among species. One probable explanation for this variability is the extent to which inflammation is aggravated. The relationship between the possessions of chronic stimulation of innate immunity by membrane leakage (primary deficiency) and stretches of asynchronous degeneration/regeneration may be reserved at a low level in a few muscles, but in others may synergize to become destructive. Several immunopathological characters of dystrophin-deficient muscle diverge depending on muscle group and age. Remarkably, there are supplementary features of pathophysiology, larger than inflammation, which probably add to variability including induction of compensatory pathways (utrophin, myostatin, and others), type of activity of specific muscle groups (for example, stress on myofibers), and others (Amato and Russell, 2008; Villalta et al., 2015; Nitahara-Kasahara et al., 2016; Shin et al., 2013; Rosenberg et al., 2015).

In DMD, on the other hand, dystrophin deficiency directs to chronic membrane instability with an uninterrupted discharge of TLR ligands into the extracellular environment. Continual membrane instability directs to self-sufficient activation of the innate immune response. Proinflammatory cytokines provoke constitutive MHC class I and II expression on muscle cells, recruitment of T and B cells, and generation of an adaptive immune response in the muscle surroundings. This proinflammatory microenvironment is frequently overlaid on the neutrophil and macrophage infiltrations provoked by consecutive routes of myofiber degeneration and regeneration. In contrast to the complete resolution of a solitary stretch of inflammation and repair of normal muscle, dystrophin-deficient muscle loses the stretch consequence. Adjacent fibers or groups of fibers penetrate the necrotic phase at diverse times in the 2-week degeneration/regeneration cycle (asynchronous regeneration), consequently sustaining a chronic inflammatory condition, which, in turn, generates an additional proinflammatory environment with establishment of innate immune pathways and confirmation of amplified antigen presentation (Schapira and Griggs, 1999; Villalta et al., 2015; Nitahara-Kasahara et al., 2016; Shin et al., 2013; Rosenberg et al., 2015).

5.3.2.3.5 Molecular biology of the immune system.

In dystrophin-deficient muscle, myofiber-generated RNA molecules may be the most effective DAMPs such as TLR7 (specific for single-stranded RNA) which appearance is upregulated at exceptionally beginning phases of the disease in muscle, in infiltrating mononuclear cells, and in blood vessels of presymptomatic DMD patients (<1 year of age). TLR7 upregulation is associated with creation of inflammatory signaling pathways, leading to prominent appearance of MHC class I and II molecules and complement factors D and H, as identified by mRNA profiling and immunohistochemistry base analysis. Appearance of this early inflammatory gene cluster (HLA-DQβ, HLA-DRβ, and complement factor D) is amplified with age in presymptomatic DMD patients but largely remains steadily upregulated during the disease course. TLR1−TLR4 and TLR7−TLR9 have been found to be upregulated in the mdx murine model of DMD. TLR7-induced inflammation in DMD muscle is the extremely upregulated NF-κB signaling pathway that takes action downstream of TLR7. NF-κB establishment is recognized to encourage expression of downstream intermediaries of inflammation including cytokines, chemokines, and adhesion molecules, which recruit innate immune cells to infiltrate the tissue and trigger local cells at the damaged tissue site (Schapira and Griggs, 1999; Villalta et al., 2015; Nitahara-Kasahara et al., 2016; Shin et al., 2013; Rosenberg et al., 2015).

Allele frequencies of certain HLA types are known to differ between patients with dystrophinopathies and healthy controls. Significantly higher frequencies of HLA-A and HLA-B have been observed in patients with de novo mutations compared with controls. No difference of HLA allele's frequency is observed between patients with inherited mutation and control. This outcome indicates that HLA alleles are associated with pathogenesis of DMD especially DMD with de novo mutation (Villalta et al., 2015; Nitahara-Kasahara et al., 2016; Shin et al., 2013; Rosenberg et al., 2015).

5.3.2.3.6 Role of complement system in Duchenne muscular dystrophy.
The complement system is a component of the innate immune response. This associates the capacity of antibodies and phagocytic cells to clear opsonized cells and proteins, promotes chemotaxis, and targets cells for lysis by the membrane attack complex. Three biochemical pathways direct to activation of complement system. These are the classical, alternative, and lectin pathways. The classical pathway is triggered by the C1 complex, whereas alternative pathways are triggered by the hydrolysis of C3. Opsonins, mannose-binding lectins, and ficolins trigger the lectin pathway. Complement-mediated damage may be an underappreciated constituent of muscle and myofiber pathology in DMD. Definitely, in DMD patients, complement membrane attack complexes with complement components C5b−C9 have been identified on necrotic muscle fibers, but these are observed on nonnecrotic fibers. Immunoglobulin is not identified on such fibers, supporting the complement activation by the alternative pathway in DMD. Such finding is further supported by the upregulation of complement factors B and D and properdin in DMD. These complement factors are interacted with C3. Additionally, the C5a cleavage fragment produced during activation of the alternative pathway is chemotactic for macrophages

and neutrophils in DMD muscle. The role of C3 is not well defined and well established. While C3 is localized to all necrotic fibers in DMD, the receptor for C3 has been found to negatively regulate TLR7-mediated signaling. Consequently, C3 may be providing an antiinflammatory function in this situation. Therefore, it is confusing that ablation of C3 in mdx mice failed to have an effect on muscle pathology. In this regard, such outcome may reveal the overall slighter level of muscle pathology in mdx mice as compared with that in DMD patients. Such outcome also designates species-specific differences in activity of this complement element (Schapira and Griggs, 1999; Villalta et al., 2015; Nitahara-Kasahara et al., 2016; Shin et al., 2013; Rosenberg et al., 2015).

5.3.2.3.7 Cytokines and chemokines.
Cytokines and chemokines frequently provide a modulatory function at the boundary of innate immune creation and adaptive immunity. Several chemokines have been connected with DMD pathology and proved by studies of gene expression and immunohistochemistry. Chemokines—CCL14, CCL2, CXCL12, and CXCL14—are upregulated in DMD muscle. Immunohistochemical-based analysis showed increased expressions of CXCL12, CXCL11, and CCL2 in the blood vessel endothelium of DMD patients. Evaluation of CXCL14 is not carried out in DMD by this method. CXCL14 has been revealed to be a chemoattractant for the activation of dendritic cells with unidentified receptor. CD86, DC-LAMP, and HLA-DR are upregulated and prominent in DMD muscle and may well reflect upstream activation of CXCL14 (Villalta et al., 2015; Nitahara-Kasahara et al., 2016; Shin et al., 2013; Rosenberg et al., 2015).

The muscle of mdx mouse demonstrates the appearance of an arrangement of chemokines that are relatively different from those in patients with DMD, which may assist to elucidate the differentiation in disease pathogenesis between the two species. This fact is further supported by the finding that resident regulatory T cells (Tregs) in muscle express CCR1 (a receptor for CCL5 and CCL7) in mdx muscle, but not in DMD muscle (Villalta et al., 2015; Nitahara-Kasahara et al., 2016; Shin et al., 2013; Rosenberg et al., 2015).

Normal muscle fibers contain the CXCL12 and are highly expressed in regenerating muscle fibers in DMD. CXCL12 is activated by CXCR4 and attracts the majority of leukocytes and lymphocytes including macrophages, CD4+ T cells, and CD8+ T cells. All these leukocytes and lymphocytes are representation of the cellular infiltrate in DMD muscle (Villalta et al., 2015;

Nitahara-Kasahara et al., 2016; Shin et al., 2013; Rosenberg et al., 2015).

Tumor necrosis factor-α (TNF-α) is linked with inflammation in several autoimmune and inflammatory situations. Its contribution to inflammation in DMD is not apparent. TNF-α has not been referred as a peak inflammatory mediator in more recent gene expression profiling studies. Additionally, TNF-α is shown to have a dichotomous function in muscle physiology as demonstrated by studies in mdx mice. Mice lacking both dystrophin and TNF-α demonstrated the reduction in muscle mass and signal of accelerated pathological development in diaphragm muscles. This outcome proved that the TNF-α has a well-known function in promoting muscle maintenance and repair (Villalta et al., 2015; Nitahara-Kasahara et al., 2016; Shin et al., 2013; Rosenberg et al., 2015).

Other inflammatory cytokines including interleukin-1β (IL-1β), interferon-γ (IFN-γ), IL-6, and type I IFNs do not appear prominent in gene expression arrays, but several proteins induced by IFN-γ are certainly upregulated. Moreover, there is no straight evidence that IFN-γ contributes to inflammation in DMD. Genetic deletion of IFN-γ in mdx mice attenuated myofiber injury. In addition, specific deletions of Tregs direct to enhance expression of an IFN-γ genetic signature, signifying that IFN-γ may certainly have a function in the immunopathology of DMD. Identification of IL-1β and IL-6 in muscle tissues of DMD has been performed by immunohistochemical and reverse transcription polymerase chain reaction (RT-PCR)—based methods. The results obtained by these two methods showed the production of considerable amounts of IL-1β in dystrophin-deficient primary muscle cells (Villalta et al., 2015; Nitahara-Kasahara et al., 2016; Shin et al., 2013; Rosenberg et al., 2015).

5.3.2.3.8 The role of macrophages.
M1 macrophages (classically activated, proinflammatory) perform a strong function in muscle injury in mdx mice in initial stage of the disease due to the nitric oxide—mediated cytolytic capacity of such cells. Eradication of macrophages near the beginning in mdx disease progression created a decline in muscle lesions (Villalta et al., 2015; Nitahara-Kasahara et al., 2016; Shin et al., 2013; Rosenberg et al., 2015).

Macrophages have two subpopulations. These are M1 and M2 macrophages. M1 macrophages are proinflammatory and encourage muscle lysis with the production of nitric oxide and inflammatory cytokines. M2 macrophages are usually antiinflammatory and perform the skeletal muscle regeneration through stimulating the proliferation of satellite cells. Initial creation of proinflammatory macrophages is arbitrated by inflammatory cytokines and is not affected by muscle cells. Nevertheless, at later stages, cytokines generated by muscle cells may also contribute toward long-standing activation of macrophages. Two inflammatory cytokines (TNF-α and IFN-γ) are produced by both M1 macrophages and skeletal muscle cells. Enhanced levels of these cytokines facilitate the activation of M1 macrophages and reduce the transition from M1 to M2 macrophages. In contrast, antiinflammatory cytokines such as IL-4 and IL-10 promote activation of M2 macrophages. Deliberately, both M1 and M2 macrophages are present in dystrophic muscle of mdx mice (Villalta et al., 2015; Nitahara-Kasahara et al., 2016; Shin et al., 2013; Rosenberg et al., 2015).

5.3.2.3.9 Transforming growth factor-β in Duchenne muscular dystrophy muscle.
DMD muscle demonstrates age-linked amplification in transforming growth factor-β (TGF-β) and associated fibrotic replacement of the tissue with several chronic inflammatory situations. TGF-β-based regulation of the immune system is highly complex and circumstance dependent. Nevertheless, TGF-β is consistently upregulated in DMD patients, with expression of TGF-β and TGF-β receptors (TGFBR) connected with symptomatic disease. Patients with DMD illustrate elevated expression of TGF-β1. Expression of TGF-β1 reached to peak point in DMD patients with 2—6 years of age. The event of fibrosis is also higher in DMD patients as compared with BMD or other muscular dystrophies (Villalta et al., 2015; Nitahara-Kasahara et al., 2016; Shin et al., 2013; Rosenberg et al., 2015).

TGF-β is secreted in the form of inactive protein. Activation of this protein necessitates extra processing. Matrix metalloprotease 2 (MMP2) is upregulated in DMD muscle and has a possible role to activate TGF-β. Activation of TGF-β may also be carried out in response to pH changes, reactive oxygen species, and matrix rigidity. The mechanistic part of these factors in converting TGF-β to its active form in DMD is not well established and well defined. TGF-β1 pathways in DMD disease progression have been identified by two genetic modifier loci (SPP1 and LTBP4). SPP1 and LTBP4 are known to modify TGF-β1 arbitrated pathways and disease severity in DMD. Certainly, TGF-β1 pathways may be more significant as compared with any other downstream pathway in driving the progressive muscle wasting and weakness in DMD (Villalta et al., 2015; Nitahara-Kasahara et al., 2016; Shin et al., 2013; Rosenberg et al., 2015).

Nevertheless, with asynchronous stretches of regeneration in neighboring microenvironments, TGF-β1 becomes constitutively activated, leading to continuous connective tissue remodeling and ultimately resulting in fibrosis in DMD mouse model. Outcome of mouse model data also sustains the significance of TGF-β1 pathway modulation in the chronically inflamed muscle of DMD patients (Villalta et al., 2015; Nitahara-Kasahara et al., 2016; Shin et al., 2013; Rosenberg et al., 2015).

Myostatin or GDF8 (a member of the TGF-β superfamily) negatively regulates muscle differentiation and growth. Myostatin mRNA expression emerges to be downregulated in DMD skeletal muscle and mdx mouse. This may signify an adaptive response to sustain muscle mass and to save dystrophic muscle (Villalta et al., 2015; Nitahara-Kasahara et al., 2016; Shin et al., 2013; Rosenberg et al., 2015).

The members of TGF-β superfamily cytokines perform a key function in pathogenesis of DMD. In all of them, TGF-β1 is the most important cytokine that encourages fibrosis by enhancing the synthesis of collagen, growth, and differentiation of fibroblasts and minimizing the expression of matrix-degrading proteases. TGF-β induces cellular signaling through binding with three cell surface receptors. Type I (TβR1) and type II (TβRII) are signaling receptors and shaped the heterodimers, whereas type III (TβRIII) is a proteoglycan that regulates entry of TGF-β to signaling receptors. Binding of TGF-β to TβRII subunit directs to the phosphorylation and activation of TβRI. Activated TβRI consequently performs the phosphorylation downstream smad proteins, leading to translocation of smad complex into nucleus. In the nucleus, they bind to DNA in a sequence-specific approach and perform the transcription regulation of several target genes. In dystrophic muscles, TGF-β can be formed by several types of cells. These cells are myofibers, fibroblasts, and infiltrating immune cells. Several investigations have proved that the TGF-β1 levels are amplified in muscle tissues of DMD patients and DMD animal models (Villalta et al., 2015; Nitahara-Kasahara et al., 2016; Shin et al., 2013; Rosenberg et al., 2015).

5.3.2.3.10 Muscle fibrosis in Duchenne muscular dystrophy. Fibrosis is a pathogenic factor and characterized by chronic inflammation with continual production of profibrotic cytokines and enhanced deposition of ECM proteins (including collagens and fibronectin). This pathological factor is responsible for impairment in tissue function. Fibrosis is also a key feature in muscular dystrophy and gradually declines in the locomotor capacity, posture maintenance, and the essential function of cardiac and respiratory muscles. Progressive fibrosis is also obtained in the diaphragm of the mdx mice, which reproduces in the clinical signs of DMD patients. One of the studies in DMD patients has revealed that endomysial fibrosis is the solitary pathological feature among fiber atrophy, hypercontracted fibers, necrotic fibers, endomysial and perimysial fibrosis, and fatty degeneration, which is extensively associated with deprived motor outcome considered by muscle strength at the age of 10 years and at the age of ambulatory loss (Villalta et al., 2015; Nitahara-Kasahara et al., 2016; Shin et al., 2013; Rosenberg et al., 2015).

Enhancement in the expression of collagen species, connective tissue growth factor, osteopontin, and the tissue inhibitor of metalloproteinases-1 (TIMP1) is observed in DMD muscle. This has been connected to the enlargement of pulmonary and liver fibrosis. This may similarly contribute to fibrosis in DMD muscle. In the muscle of the mdx mouse, osteopontin is extremely expressed with infiltrating T cell, and its mRNA is upregulated in DMD patients. Knocking out osteopontin in the mdx mouse minimizes the fibrosis and increases the muscle strength through a collective consequence that resulted in the reduction of TGF-β and enhanced Treg infiltration. In the presence of other proinflammatory mediators, osteopontin may have a positive effect on muscle regeneration in vitro (Villalta et al., 2015; Nitahara-Kasahara et al., 2016; Shin et al., 2013; Rosenberg et al., 2015).

Fibrinogen is a soluble acute phase protein, and this is released into the blood in response to oxidative stress. There is a good correlation obtained in between the fibrinogen accumulation and deposition of collagen in dystrophic muscle of mdx mice and DMD patients. Fibrinogen also encourages fibrosis with enhancement of TGF-β expression in mdx diaphragm. TGF-β encourages fibrosis through increasing the expression of microribonucleic acid (miRNA)-21 in dystrophic muscle of mdx mice. The expression of miRNA-21 is extremely encouraged and associated with fibrosis in skeletal muscle of mdx mice and in muscle of DMD patients. Consistently, overexpression of miRNA-21 is adequate to further induce fibrosis in dystrophic muscle of 3-month-old mdx mice. Moreover, this has been observed that urokinase-type plasminogen activator (uPA) or plasmin-dependent mechanisms direct to proteolytic activation of TGF-β in dystrophic muscle

(Villalta et al., 2015; Nitahara-Kasahara et al., 2016; Shin et al., 2013; Rosenberg et al., 2015).

Factor XIIIa is overexpressed concomitantly with HLA-DRα by activated dendritic cells in the muscle specimens of DMD. The function of factor XIIIa (a tissue transglutaminase) in activated dendritic cells is not apparent, but it may contribute to collagen formation and fibrosis in DMD muscle (Villalta et al., 2015; Nitahara-Kasahara et al., 2016; Shin et al., 2013; Rosenberg et al., 2015).

5.3.2.3.11 Mast cells in Duchenne muscular dystrophy muscle.

Mast cells have also been concerned in DMD. Dystrophin-deficient muscle demonstrates marked proliferation of mast cells in all over the endomysial connective tissue that surrounds myofibers. These mast cells are characteristically bulky or degranulated and are chronically activated. The chronic inflammatory state in mdx mouse muscle and DMD muscle directs to recruitment of mast cells from the circulation or encourages proliferation of tissue-resident populations, which then degranulate in the proinflammatory environment (Amato and Russell, 2008; Schapira and Griggs, 1999; Villalta et al., 2015; Nitahara-Kasahara et al., 2016; Shin et al., 2013; Rosenberg et al., 2015).

The proteases and cytokines are released by degranulation of mast cells localized in endomysial connective tissue. This event might be expected to aggravate the already fragile dystrophin-deficient plasma membrane. The function of mast cells is observed to intensify both the proinflammatory state and membrane damage in mdx mice, with progression of disease pathology and increased fibrosis (Villalta et al., 2015; Nitahara-Kasahara et al., 2016; Shin et al., 2013; Rosenberg et al., 2015).

Mast cells degranulation is frequently observed in areas neighboring injured myofibers. Degranulation of mast cells discharges certain proteases (such as chymase, tryptase, and carboxypeptidase). These proteases are responsible for promoting the membrane lysis. Moreover, liberation of TNF-α and histamines from mast cell granules can encourage fiber necrosis with creating muscle microenvironment extra proinflammatory (Villalta et al., 2015; Nitahara-Kasahara et al., 2016; Shin et al., 2013; Rosenberg et al., 2015).

5.3.2.3.12 Adaptive immunity: function of T cell in Duchenne muscular dystrophy.

Numerous important studies have been performed for the establishment of the functions of CD4+ T cells and CD8+ T cells in DMD muscle pathology in humans and in animal models. CD8+ T cells present in dystrophic muscle can directly trigger muscle cell death, whereas CD4+ T cells can be an important source for inflammatory cytokines to CD8+ T cells and other immune cell types. T cells in the muscle of DMD patients expressed the Vβ2 T cell receptor (TCR) and also principally the Jβ1.3 segment. Additionally, in few DMD patient samples (all HLA-A A2- or A1-positive), there is an observation of general amino acid motif (RVSG) in the CDR3 region of Vβ2. Among CD8+ T cells (supposed to be such by HLA-A restriction) infiltrating DMD muscle, there may be selection for a subset of T cells that are specific to a restricted set of muscle antigens. In one study, this has been observed that oligoclonality of T cell subsets in the mdx mouse model, with Vβ8.1/8.2 being the predominant population in muscle, further suppors antigen restriction of the T cell response in dystrophin-deficient muscle. This is possible that these oligoclonally derived T cells are precise for dystrophin peptides in the context of MHC class I and II expressed on revertant myofibers. This is possible that long-lasting inflammation has unleashed an autoimmune response that has extended to other epitopes in diverse muscle proteins. In another study, dystrophin-specific CD8+ T cells are identified in the peripheral blood lymphocyte population of patients with DMD and are possible to originate through exposure of T cells to expression of a partial dystrophin molecule on revertant myofibers (Villalta et al., 2015; Nitahara-Kasahara et al., 2016; Shin et al., 2013; Rosenberg et al., 2015).

CD8+ T cells–dependent muscle damage is potentially contributed by recruitment of inflammatory cells (such as in eosinophils, an immune cell type is minimized in mdx mice muscle after CD8+ T cell depletion). This is very crucial that T cell depletion (by breeding mdx mice onto an SCID[severe combined immunodeficiency] background) considerably minimized the TGF-β expression as well as fibrosis in the muscles of these animals. The importance of CD8+ T cell–arbitrated pathology in DMD is also supported by the consequence of steroid treatment in DMD patients. Muscle power improvement in DMD patients treated with low-dose (0.75 mg/kg per day) or high-dose (1.5 mg/kg per day) steroids is linked with a reduction in total T cell number (predominantly in CD8+ T cells) as well as a decrease in the number of muscle fibers focally occupied by lymphocytes (Villalta et al., 2015; Nitahara-Kasahara et al., 2016; Shin et al., 2013; Rosenberg et al., 2015).

DMD is associated with chronic inflammation. This event is responsible for the infiltration of muscle tissues by a diversity of activated immune cells (such as various cytokines). The cellular sources of these cytokines contain CD4+ and CD8+ T cells, dendritic cells, B cells, neutrophils, and macrophages of both the proinflammatory M1 and the tissue regeneration–focused M2 phenotype. In particular, CD8+ T cells trigger muscle fiber death, and CD4+ T cells contribute to this procedure by supplying inflammatory cytokines to CD8+ T cells and other immune cells. Meanwhile, macrophages carry out a diversity of significant immunoregulatory and inflammatory roles and lyse muscle fibers through the generation of nitric oxide. Due to lysis of muscle fibers, large amounts of cytolytic and cytotoxic molecules are discharged (Villalta et al., 2015; Nitahara-Kasahara et al., 2016; Shin et al., 2013; Rosenberg et al., 2015).

5.3.2.3.13 Function of major histocompatibility complex in Duchenne muscular dystrophy.
Muscle cells do not express MHC class II antigens under normal physiological circumstances. They may express in a few inflammatory myopathies and also serve as significant markers of inflammation. Additionally, the dendritic cells and macrophages invading inflamed muscle do express MHC class II, and both of these populations can therefore serve as antigen-presenting cells for CD4+ MHC class II–restricted T cells that provide help to CD8+ cytolytic cells that are specific for muscle-derived proteins (dystrophin) in the context of MHC class I (Villalta et al., 2015; Nitahara-Kasahara et al., 2016; Shin et al., 2013; Rosenberg et al., 2015; Hoffman et al., 1988).

5.3.2.3.14 Function of neutrophils in Duchenne muscular dystrophy.
Neutrophils also infiltrate muscle tissues at near the beginning stages in mdx mice. The function of neutrophil in DMD has been established by the study of Hodgetts et al. (2006). Antibody-arbitrated depletion of neutrophils in mdx mice significantly minimized the fiber necrosis and successive regeneration. The major source of superoxide production is neutrophil. In this way, one of the probable mechanisms by which neutrophils perform the muscle necrosis is the superoxide generation (Villalta et al., 2015; Nitahara-Kasahara et al., 2016; Shin et al., 2013; Rosenberg et al., 2015).

5.3.2.3.15 Function of thrombospondin 4 in Duchenne muscular dystrophy.
Thrombospondin 4 (TSP-4) is upregulated in DMD by as much as 15-fold. The function of TSP-4 in inflammation has been proved in mouse models of atherosclerosis in which TSP-4 is ablated. Such types of studies also proved that TSP-4 activates endothelial cells and macrophages. TSP-4 also encourages macrophage adhesion and migration. TSP-4 may contribute to inflammation in DMD, but for well-established and well-defined function in DMD, there is a need of additional investigation (Villalta et al., 2015; Nitahara-Kasahara et al., 2016; Shin et al., 2013; Rosenberg et al., 2015).

5.3.2.3.16 Activation of signaling pathways (nuclear factor kappa B pathway).
Attack of inflammatory cells is carried out at the endomysial, perimysial, and perivascular sites of necrotizing myofibers. Moreover, several cytokines (containing IL-1α and IL-17) can exert direct effects on the muscle tissue through the activation of signaling pathways. One of the most important signaling pathways is the NF-κB pathway. This pathway performs the enhancement in the inflammatory response through upregulation of cytokine/chemokine production. Promising substantiation puts forward that NF-κB transcription factor is one of the chief mediators of muscle wasting in muscular dystrophy. The NF-κB family contains five members in the form of homo- and heterodimers. These are RelA (also known as p65), RelB, c-Rel, p105/p50, and p100/p52 ((Villalta et al., 2015; Nitahara-Kasahara et al., 2016; Shin et al., 2013; Rosenberg et al., 2015).

NF-κB is a major proinflammatory transcription factor, and its activation in skeletal muscle of mdx mice has been examined. DNA-binding activity of NF-κB and the expression of NF-κB-regulated inflammatory cytokines TNF-α and IL-1β have been found to be significantly upregulated in the skeletal muscle of mdx mice (Villalta et al., 2015; Nitahara-Kasahara et al., 2016; Shin et al., 2013; Rosenberg et al., 2015).

Activation of NF-κB can take place through canonical or noncanonical signaling pathway. Canonical NF-κB signaling engages the upstream inhibitors activation of κB (IκB) kinase-β (IKKβ) and following phosphorylation and degradation of IκB protein. Comparatively, the activation of noncanonical NF-κB pathway needs the activation of NF-κB-inducing kinase (NIK) and IKKα leading to phosphorylation and proteolytic processing of p100 subunit into p52. NF-κB is also well known for the regulation of several processes comprising the development of immune system, inflammation, acute stress response, cellular proliferation and differentiation, and also protection against cell death. NF-κB is necessary for cell survival. Abundance of literature supports the fact that the activation of NF-κB is responsible for the wasting of skeletal muscle. Three types of mechanistic pathways have been established in which activated NF-κB encourages muscle loss in both physiological and pathophysiological conditions: (a) NF-κB enhances the expression of several proteins involved in the ubiquitin–proteasome system including E3 ubiquitin ligase MuRF1 (which promotes the loss of skeletal muscle); (b) NF-κB is also responsible for the enhancement in the expression of several proinflammatory cytokines, chemokines, cell adhesion molecules, and matrix-degrading enzymes (which intensify the loss of skeletal muscle); (c) ultimately, activated NF-κB can obstruct the regeneration of myofibers in response to injury (Villalta et al., 2015; Nitahara-Kasahara et al., 2016; Shin et al., 2013; Rosenberg et al., 2015).

Muscle-specific activation of NF-κB leads to the onset of muscle degeneration in mdx mice. Several studies showed the elevated levels of activated NF-κB in skeletal muscle of diverse animal models of DMD. NF-κB activity is upregulated by proinflammatory factors. The activity of this transcription factor is inhibited by IL-10, but not by other factors such as IL-6 and AP-1, or even by NF-κB itself (Villalta et al., 2015; Nitahara-Kasahara et al., 2016; Shin et al., 2013; Rosenberg et al., 2015).

5.3.3 Metabolism-Mediated Muscle Pathology in Duchenne Muscular Dystrophy

5.3.3.1 Metabolomics (metabolism, metabolites, and omics technology)

Metabolism is the sum of many interconnected reaction sequences that interconvert cellular metabolites. Metabolites are the end products of cellular regulatory processes, and their levels can be regarded as the ultimate response of biological systems to genetic or environmental changes (Lehninger et al., 2005; Beckonert et al., 2007). The components usually considered as metabolites are molecules with a molecular weight less than 2000 Da. It could be primary metabolites such as sugars, amino acids, organic acids, fatty acids, and energetic metabolites (nucleotides) (Fillet and Frédérich, 2015).

Omics has become the new mantra in molecular research. "Omics" technologies include genomics, transcriptomics, proteomics, and metabolomics (Fig. 5.13). Genomics had proved the stationary sequence of genes and proteins. The center of attention has now been reallocated to their dynamic functions and interactions.

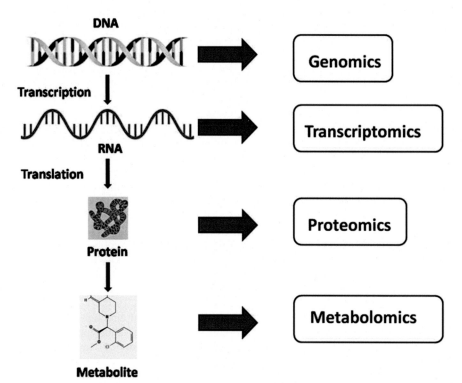

FIG. 5.13 The sequential approaches of omics-based techniques (genomics, transcriptomics, proteomics, and metabolomics) for the complete analysis of central dogma of molecular biology.

Transcriptomics, proteomics, and metabolomics disclose the biological utility of the gene product. The "omics" technologies are high-throughput technologies, and they amplify considerably several proteins or genes that can be identified concurrently to relate complex mixtures to complex effects in the form of gene or protein expression profiles. Metabolomics is a comparatively new branch of "omics" science that focuses on the system-wide characterization of metabolites. The term metabolomics was coined at the end of the 1990s to explain the development of approaches, which objective to compute all the metabolites that are present within a cell, tissue, or organism during a genetic amendment or physiological stimulus (Savorani et al., 2013; Griffin, 2006; Du et al., 2013; Young and Wallace, 2009).

5.3.3.2 Metabolomics (consequences of the activity of genes and proteins)

Although genomics engages the study of gene expression and proteomics engages the expression of proteins, metabolomics examines the consequences of the activity of these genes and proteins. Additionally, in numerous ways, transcriptomics, genomics, and proteomics are upstream of the ultimate physiology of the cell, whereas the metabolic profile is probable nearer in response to the disease process. In addition, it is gradually more the case that investigations are incorporating proteomics, genomics, transcriptomics, and metabolomics information to generate a fuller picture of the disease process. This affords an additional absolute representation of human health, including disease prediction, disease status, and therapeutic response as earlier accessible. Metabolomics has appeared as a functional genomics methodology that contributes to our understanding of the complex molecular interactions in biological systems as it affords information about both an organism's phenotype and its environment. Since metabolomics affords a distinctive window on gene—environment connections, it is performing a gradually more significant function in several quantitative phenotyping and functional genomics studies (Beckonert et al., 2007; Fillet and Frédérich, 2015; Savorani et al., 2013; Griffin, 2006; Du et al., 2013; Young and Wallace, 2009; Emwas et al., 2016, 2019).

5.3.3.3 Metabolomics (utility of analytical methods or techniques)

Two most potent and frequently utilized analytical approaches for metabolic fingerprinting are mass spectrometry (MS) and nuclear magnetic resonance (NMR) spectrometry. NMR-based metabolomics suggests numerous distinctive advantages in a clinical situation since it is a relatively quick, nondestructive, and noninvasive technique that can be carried out on standard preparations of blood cells, serum, plasma, urine, or other body fluids as well as tissue extract. To enhance the sensitivity and reduce the size of the samples required, most NMR-based metabolomic investigations generate utility of reasonably high-field NMR instruments (500—800 MHz) often equipped with a cryo-probe, which reduces the electronic noise in the detection system, thus enhancing signal from sample. In broad-spectrum, metabolomics methodologies are divided into two separate groups: untargeted metabolomics, a deliberated comprehensive investigation of the entire quantifiable characteristics in a sample comprising chemical unknowns; and targeted metabolomics, the measurement of definite groups of chemically characterized and biochemically interpreted metabolites. The outcomes of targeted metabolomics-based investigations have established both feasibility and flexibility across physiological, pathological, interventional, and epidemiological human studies. In addition to the recognition of metabolites in the disease progression, metabolomics also has the availability of the groups of well-established web-based databases. These databases have connected the recognized metabolites or components with definite diseases and physiological characteristics, and they also have well defined the biological pathways or genome, transcriptome, and proteome emphasizing the change of metabolites. As a consequence, the innovative methodologies of metabolomics have afforded very useful apparatus for the assessment of these interactions on a global scale and recommend alternative resources of investigation to that of some of the more reductionist molecular biology methods (Beckonert et al., 2007; Fillet and Frédérich, 2015; Savorani et al., 2013; Griffin, 2006; Du et al., 2013; Young and Wallace, 2009; Emwas et al., 2016, 2019).

5.3.3.4 Nuclear magnetic resonance—based metabolomics in disease diagnosis

NMR-based metabolomics can provide an efficient and rapid approach to the diagnosis of different kinds of human diseases and is also considered as a disease diagnostic tool. Several studies have repeatedly proposed on the use of NMR-based metabolic profiling to aid human disease diagnosis, such as diabetes, Alzheimer, amyotrophic lateral sclerosis (ALS), schizophrenia, inborn errors of metabolism, osteoarthritis, male infertility, meningitis, Fabry disease, hepatitis B, virus-infected cirrhosis and alcoholic cirrhosis, nonalcoholic fatty disease, Hepatitis C virus infection, HIV-1, malaria, epithelial ovarian cancer, lung cancer, breast cancer, and inflammatory bowel disease (Beckonert et al., 2007; Fillet and Frédérich, 2015; Savorani et al., 2013; Griffin, 2006; Du et al., 2013; Young and Wallace, 2009; Emwas et al., 2016, 2019).

5.3.3.5 Analysis of metabolites (abnormality in metabolism) in Duchenne muscular dystrophy

Alteration of the lipid components in muscular dystrophy was previously described in ancient Indian medical treatise "Charak Samhita" (Samhita, 1959; Gouri-Devi and Venkatram, 1983; Srivastava et al., 2010). Pearce et al. (1981) described the enhancement in the proportion of sphingomyelin in the muscle of DMD patients, but adipose tissue had higher proportions of sphingomyelin and lysophosphatidylcholine with decreased quantity of choline phosphoglyceride (Srivastava et al., 2010). Time-of-flight secondary ion MS imaging—based investigations of striated muscle specimens from DMD-affected children showed the differences in the distribution of fatty acids, phospholipids, diacylglycerols, and triglycerides in normal and dystrophic muscle tissue sections (Srivastava et al., 2010). Decline in membrane flexibility is also observed in DMD muscle due to altered proportion of unsaturated phosphatidylcholine species in muscle-degenerated areas (Srivastava et al., 2010). Lipid mapping analysis of most damaged areas of the striated muscle specimen of DMD patients revealed the enhancement in the level of intact phosphatidylcholine, cholesterol, sphingomyelin, triglyceride, and monounsaturated fatty acid species (Srivastava et al., 2010). Matrix-assisted laser desorption/ionization time-of-flight MS and tandem MS analysis of the fatty acids composition in phospholipid of skeletal muscle of mdx mice showed the enhancement in the ratio of saturated fatty acid composition/unsaturated fatty acid composition in damaged area of the skeletal muscle (Srivastava et al., 2010). Fischbeck et al. (1983), using freeze-fracture study of cholesterol—digitonin complexes, have found increased cholesterol in Duchenne sarcolemma, which may lead to increased fragility of the membrane (Srivastava et al., 2010). Biochemical, morphological, and biophysical characteristics of red blood cell (RBC) membrane established an alteration in DMD (Srivastava et al., 2010). Elevated levels of free fatty acids and ketone bodies are also observed in the serum of DMD patients (Srivastava et al., 2010). NMR spectroscopy—based analysis of lipid constituents of the serum of DMD patients showed the elevated levels of triglycerides, phospholipids, free cholesterol, cholesterol ester, and total cholesterol (Fig. 5.14) as compared with healthy subjects (Srivastava et al., 2010).

In vitro, high-resolution ^{31}P NMR spectroscopy—based analysis of whole blood of DMD patients showed (Fig. 5.15) the elevated levels of phosphatidylcholine (PC), phosphatidylethanolamine (PE), phosphatidylinositol (PI), phosphatidylserine (PS), and lysophosphatidylcholine (Lys-PC) (Srivastava et al., 2016a,b).

NMR-based quantification of the lipid components in muscle tissue of DMD also showed a marked differentiation as compared with normal subjects. There is a considerable elevation of triglycerides and total cholesterol, but a significant reduction in the level of linoleic acid is observed in the DMD muscle. So, NMR spectroscopy—based lipid analysis indicates the disturbed lipid metabolism in patients with DMD (Srivastava et al., 2017).

In, vitro, ^{1}H NMR spectroscopy was also successfully applied on the mouse model of muscular dystrophy (mdx) as compared with control subjects to observe the amendment in metabolic profiling (aqueous metabolites such as citrate and alanine) (Srivastava et al., 2018). In vitro, proton NMR spectroscopy—based quantification of metabolites in the native serum of patients with DMD showed the elevated levels of branched chain amino acids (BCAs) and acetate (Ace), whereas reduced levels of glutamine (Gln) as compared with normal subjects(Srivastava et al., 2016a,b). Analysis of perchloric acid (PCA) extract of the muscle tissue of the DMD patients by NMR spectroscopy showed the considerable difference in the quantity of aqueous metabolites as compared with control subjects (Srivastava et al., 2018). Quantity of BCAs, glutamine/glutamate (Gln/Glu), acetate (Ace), and fumarate (Fum) showed the marked decrease, whereas considerable enhancement of the level of histidine (His) is observed in muscle tissue of DMD patients as compared with control subjects. Propionate (Prop) is present in muscle tissue of DMD and absent in muscle tissue of control subjects (Srivastava et al., 2018). In vivo, ^{31}P NMR spectroscopy—based investigation showed the amendment of high-energy phosphate metabolite (phosphocreatinine [PCr]) in DMD skeletal muscle. Increased ratio of Pi (inorganic phosphate)/PCr at rest has been observed in patients with DMD indicating an impaired energy metabolism (Srivastava et al., 2018). All these

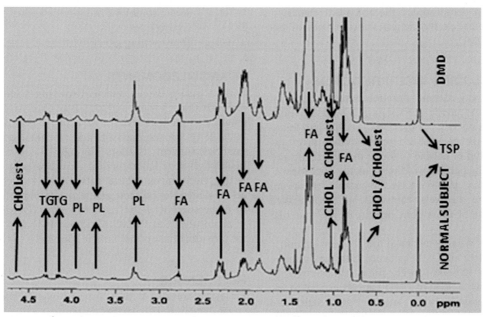

FIG. 5.14 ¹H NMR spectra of the lipid extract of the serum of DMD as compared with normal subject showed the elevated levels of TG (triglycerides), PL (phospholipids), CHOL (free cholesterol), and CHOLest (cholesterol esters). *DMD*, Duchenne muscular dystrophy; *NMR*, nuclear magnetic resonance.

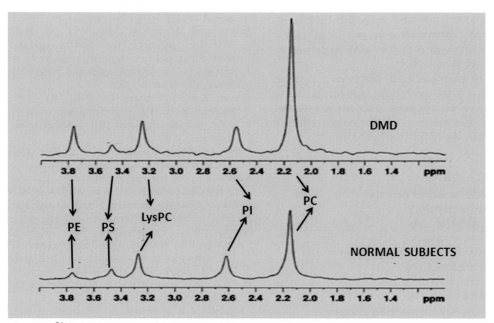

FIG. 5.15 ³¹P NMR spectra of the whole blood of DMD patients as compared with healthy subjects showed the elevated levels of phospholipid components. *DMD*, Duchenne muscular dystrophy; *Lys-PC*, Lyso-phosphatidylcholine; *NMR*, nuclear magnetic resonance; *PC*, phosphatidylcholine; *PE*, phosphatidylethanolamine; *PI*, phosphatidylinositol; *PS*, phosphatidylserine.

investigations demonstrated the abnormal metabolic pattern in muscle's bioenergetics metabolism of patients with DMD.

5.4 CONCLUDING INTERPRETATION

Interconnectivity of gene, immune system, and metabolism in the muscle pathology of DMD has been described on the basis of the several studies and investigations. There are a lot of research studies, investigations, and studies needed to know the systematic and well-established answers of the following questions:

(a) Dystrophin gene produces the protein dystrophin. What are the well-established and well-defined functions of dystrophin protein in the muscular activity?

(b) Immune system—mediated muscle degeneration takes place in DMD patients, but the immune response has not been fully characterized in dystrophic muscle of DMD patients as well as animal models. What are the systematic and well-established immune-mediated mechanistic pathways of muscle pathology in DMD patients as well as in animal models?

(c) Due to defectiveness in dystrophin gene and respective dystrophin protein, there is an alteration that takes place in muscle metabolism and metabolites. What are the systematic and well-established metabolism-mediated mechanistic pathways of muscle pathology in DMD patients as well as animal models?

Last, but not the least, DMD had been recognized centuries ago, but the therapeutic approach is still in vague and lot of research work is needed to establish a permanent treatment or to develop an effective medicine to save the life of DMD patients.

ABBREVIATIONS

BMD	Becker muscular dystrophy
CHOL	Free cholesterol
CHOLest	Cholesterol ester
CK	Creatine kinase
DMD	Duchenne muscular dystrophy
EMG	Electromyography
FA	Fatty acids
HLA	Human leukocyte antigen
MHC	Major histocompatibility complex
mPCR	Multiplex polymerase chain reaction
MUPs	Motor unit potentials
PL	Phospholipids
SCID	Severe combined immunodeficiency
TG	Triglycerides
TGF-β	Transforming growth factor-β

ACKNOWLEDGMENTS

Authors wish to thank Council of Scientific & Industrial Research (No.13 (8660-A)/2013-Pool) and University Grant Commission (No.F.4−2/2006 (BSR)/13−194/2008(BSR)), Government of India, for their generous financial support. Authors sincerely thank staff of neurophysiology laboratory for providing help in collecting electromyographical data. Senior residents are acknowledged for providing clinical data and collection of tissue specimens. Professor Rajkumar (Department of neurosurgery, SGPGIMS, Lucknow) is acknowledged for providing the normal muscle specimens.

REFERENCES

Amato, A.A., Russell, J.A., 2008. Neuromuscular Disorders, first ed. McGraw-Hill Professional.

Arikawa-Hirasawa, E., Koga, R., Tsukahara, T., Nonaka, I., Mitsudome, A., Goto, K., et al., 1995. A severe muscular dystrophy patient with an internally deleted very short (110 kDa) dystrophin: presence of the binding site for dystrophin-associated glycoprotein (DAG) may not be enough for physiological function of dystrophin. Neuromuscul. Disord. 5 (5), 429−438.

Beckonert, O., Keun, H.C., Ebbels, T.M.D., Bundy, J.G., Holmes, E., Lindon, J.C., et al., 2007. Metabolic profiling, metabolomic and metabonomic procedures for NMRspectroscopy of urine, plasma, serum and tissue extracts. Nat. Protoc. 2 (11), 2692−2703.

Bennett, R.R., den Dunnen, J., O'Brien, K.F., Darras, B.T., Kunkel, L.M., 2001. Detection of mutations in the dystrophin gene via automated DHPLC screening and direct sequencing. BMC Genet. 2, 17.

Brown, S.C., Brown, S.S., Lucy, J.A., 1997. Dystrophin: Gene, Protein and Cell Biology. Cambridge University Press.

Brown, K.J., Marathi, R., Fiorillo, A.A., Ciccimaro, E.F., Sharma, S., Rowlands, D.S., et al., 2012. Accurate quantitation of dystrophin protein in human skeletal muscle using mass spectrometry. J. Bioanal. Biomed. (Suppl. 7), 001.

Bulman, D.E., Gangopadhyay, S.B., Bebchuck, K.G., Worton, R.G., Ray, P.N., 1991. Point mutation in the human dystrophin gene: identification through western blot analysis. Genomics 10 (2), 457−460.

Bushby, K.M.D., Anderson, L.V.B., 2001. Muscular Dystrophy: Methods and Protocols. Humana Press.

Chaturvedi, L.S., Mukherjee, M., Srivastava, S., Mittal, R.D., Mittal, B., 2001. Point mutation and polymorphism in Duchenne/Becker muscular dystrophy (D/BMD) patients. Exp. Mol. Med. 33 (4), 251−256.

Colombo, J.P., Richterich, R., Rossi, E., 1962. Serum creatine phosphokinase: determination and diagnostic significance. Klin. Wochenschr. 40, 37–44.

Danieli, G.A., Mioni, F., Müller, C.R., Vitiello, L., Mostacciuolo, M.L., Grimm, T., 1993. Patterns of deletions of the dystrophin gene in different European populations. Hum. Genet. 91 (4), 342–346.

DeLucchi, M.R., 1976. Techniques and Methods of Data Acquisition of EEG and EMG. Elsevier.

Du, F., Virtue, A., Wang, H., Yang, X.F., 2013. Metabolomic analyses for atherosclerosis, diabetes, and obesity. Biomark. Res. 1 (1), 17.

Dubowitz, V., Sewry, C.A., Lane, R.J.M., 2007. Muscle Biopsy: A Practical Approach, third ed. Elsevier Health Sciences.

Emery, A.E.H., 1998. Neuromuscular Disorders: Clinical and Molecular Genetics, first ed. Wiley.

Emery, A.E.H., 2000. Muscular Dystrophy: The Facts. Oxford University Press, England.

Emery, A.E.H., 2002. The muscular dystrophies. Lancet 359 (9307), 687–695.

Emwas, A.H., Roy, R., McKay, R.T., et al., 2016. Recommendations and standardization of biomarker quantification using NMR-based metabolomics with particular focus on urinary analysis. J. Proteome Res. 15 (2), 360–373.

Emwas, A.H., Roy, R., McKay, R.T., Tenori, L., Saccenti, E., Gowda, G.A.N., et al., 2019. NMR spectroscopy for metabolomics research. Metabolites 9 (7), E123.

Fillet, M., Frédérich, M., 2015. The emergence of metabolomics as a key discipline in the drug discovery process. Drug Discov. Today Technol. 13, 19–24.

Gao, Q.Q., McNally, E.M., 2015. The dystrophin complex: structure, function, and implications for therapy. Comp. Physiol. 5 (3), 1223–1239.

Gillard, E.F., Chamberlain, J.S., Murphy, E.G., Duff, C.L., Smith, B., Burghes, A.H., et al., 1989. Molecular and phenotypic analysis of patients with deletions within the deletion-rich region of the Duchenne muscular dystrophy (DMD) gene. Am. J. Hum. Genet. 45 (4), 507–520.

Gouri-Devi, M., Venkatram, B.S., 1983. Concept of disorders of muscles in 'Charak Samhita', an ancient Indian medical treatise—relevance to modern myology. Neurol. India 31, 13–14.

Griffin, J.L., 2006. The Cinderella story of metabolic profiling: does metabolomics get to go to the functional genomics ball. Philos. Trans. R. Soc. Lond. B Biol. Sci. 361 (1465), 147–161.

Hilton-Jones, D., Turner, M., Turner, M.R., 2014. Oxford Textbook of Neuromuscular Disorders. Oxford University Press.

Hoffman, E.P., Brown, R.H., Kunkel, L.M., 1987. Dystrophin: the protein product of the Duchenne muscular dystrophy locus. Cell 51 (6), 919–928.

Hoffman, E.P., Fischbeck, K.H., Brown, R.H., Johnson, M., Medori, R., Loike, J.D., et al., 1988. Dystrophin characterization in muscle biopsies from Duchenne and Becker muscular dystrophy patients. New Engl. J. Med. 318, 1363–1368.

Hu, X.Y., Burghes, A.H., Ray, P.N., Thompson, M.W., Murphy, E.G., Worton, R.G., 1988. Partial gene duplication in Duchenne and Becker muscular dystrophies. J. Med. Genet. 25 (6), 369–376.

Hu, X.Y., Ray, P.N., Murphy, E.G., Thompson, M.W., Worton, R.G., 1990. Duplicational mutation at the Duchenne muscular dystrophy locus: its frequency, distribution, origin, and phenotypegenotype correlation. Am. J. Hum. Genet. 46 (4), 682–695.

Kakulas, B.A., 1996. The spectrum of dystrophinopathies. In: Lane, R.J. (Ed.), Handbook of Muscle Diseases. Chap. 19. Marcel Dekker Informa Health Care, New York, pp. 235–244.

Kneppers, A.L., Deutz-Terlouw, P.P., den Dunnen, J.T., van Ommen, G.J., Bakker, E., 1995. Point mutation screening for 16 exons of the dystrophin gene by multiplex single-strand conformation polymorphism analysis. Hum. Mutat. 5 (3), 235–242.

Kodaira, M., Hiyama, K., Karakawa, T., Kameo, H., Satoh, C., 1993. Duplication detection in Japanese Duchenne muscular dystrophy patients and identification of carriers with partial gene deletions using pulsed-field gel electrophoresis. Hum. Genet. 92 (3), 237–243.

Koenig, M., Monaco, A.P., Kunkel, L.M., 1988. The complete sequence of dystrophin predicts a rod-shaped cytoskeletal protein. Cell 53 (2), 219–228.

Koenig, M., Beggs, A.H., Moyer, M., Scherpf, S., Heindrich, K., Bettecken, T., et al., 1989. The molecular basis for Duchenne versus Becker muscular dystrophy: correlation of severity with type of deletion. Am. J. Hum. Genet. 45 (4), 498–506.

Kunkel, L.M., Monaco, A.P., Middlesworth, W., Ochs, H.D., Latt, S.A., 1985. Specific cloning of DNA fragments absent from the DNA of a male patient with an X chromosome deletion. Proc. Natl. Acad. Sci. U.S.A. 82 (14), 4778–4782.

Lane, R.J.M., 1996. Handbook of Muscle Disease (Neurological Disease and Therapy), first ed. Informa Health Care.

Le Rumeur, E., 2015. Dystrophin and the two related genetic diseases, Duchenne and Becker muscular dystrophies. Bosn. J. Basic Med. Sci. 15 (3), 14–20.

Lee, B.L., Nam, S.H., Lee, J.H., Ki, C.S., Lee, M., Lee, J., 2012. Genetic analysis of dystrophin gene for affected male and female carriers with Duchenne/Becker muscular dystrophy in Korea. J. Kor. Med. Sci. 27 (3), 274–280.

Lehninger, A.L., Nelson, D.L., Cox, M.M., 2005. Lehninger Principles of Biochemistry, fourth ed. W.H. Freeman, New York.

Malhotra, S.B., Hart, K.A., Klamut, H.J., Thomas, N.S., Bodrug, S.E., Burghes, A.H., et al., 1988. Frame-shift deletions in patients with Duchenne and Becker muscular dystrophy. Science 242 (4879), 755–759.

Meulenbeld, G.J., 1999. A history of Indian medical literature. Groningen:Forsten. 1, 114.

Mishra, U.K., Kalita, J., 2006. Clinical Neurophysiology, second ed. Elsevier.

Mittal, B., Singh, V., Mishra, S., Sinha, S., Mittal, R.D., Chaturvedi, L.S., et al., 1997. Genotype-phenotype correlation in Duchenne/Becker muscular dystrophy patients seen at Lucknow. Indian J. Med. Res. 105, 32–38.

Monaco, A.P., Neve, R.L., Colletti-Feener, C., Bertelson, C.J., Kurnit, D.M., Kunkel, L.M., 1986. Isolation of candidate cDNAs for portions of the Duchenne muscular dystrophy gene. Nature 323 (6089), 646–650.

Murugan, S., Chandramohan, A., Lakshmi, B.R., 2010. Use of multiplex ligation-dependent probe amplification (MLPA) for Duchenne muscular dystrophy (DMD) gene mutation analysis. Indian J. Med. Res. 132, 303–311.

Nigro, V., Piluso, G., 2015. Spectrum of muscular dystrophies associated with sarcolemmal-protein genetic defects. Biochim. Biophys. Acta 1852 (4), 585–593.

Nitahara-Kasahara, Y., Takeda, S., Okada, T., 2016. Inflammatory predisposition predicts disease phenotypes in muscular dystrophy. Inflamm. Regen. 36, 14.

Potnis-Lele, M., 2012. Genetic etiology and diagnostic strategies for Duchenne and Becker muscular dystrophy: a 2012 update. Indian J. Basic Appl. Med. Res. 2, 357–369.

Pozzoli, U., Elgar, G., Cagliani, R., Riva, L., Comi, G.P., Bresolin, N., et al., 2003. Comparative analysis of vertebrate dystrophin loci indicate intron gigantism as a common feature. Genome Res. 13 (5), 764–772.

Pradhan, S., 1994. New clinical sign in Duchenne muscular dystrophy. Pediatr. Neurol. 11 (4), 298–300.

Ray, P.N., Belfall, B., Duff, C., Logan, C., Kean, V., Thompson, M.W., et al., 1985 -1986. Cloning of the breakpoint of an X;21translocation associated with Duchenne muscular dystrophy. Nature 318 (6047), 672–675.

Roberts, R.G., Bobrow, M., Bentley, D.R., 1992. Point mutations in the dystrophin gene. Proc. Natl. Acad. Sci. USA 89 (6), 2331–2335.

Rosenberg, A.S., Puig, M., Nagaraju, K., Hoffman, E.P., Villalta, S.A., Rao, V.A., et al., 2015. Immune-mediated pathology in Duchenne muscular dystrophy. Sci. Transl. Med. 7 (299), 299rv4.

Charak Samhita. Vol. II, III and IV (Original Sanskrit Text with English Translation), 1959. Shree Gulabkunverba Ayurvedic Society, Jamnagar (India).

Savorani, F., Rasmussen, M.A., Mikkelsen, M.S., Engelsen, S.B., 2013. A primer to nutritional metabolomics by NMR spectroscopy and chemometrics. Food Res. Int. 54, 1131–1145.

Schapira, A.H.V., Griggs, R.C., 1999. Muscle Diseases, first ed. Butterworth-Heinemann.

Shin, J., Tajrishi, M.M., Ogura, Y., Kumar, A., 2013. Wasting mechanisms in muscular dystrophy. Int. J. Biochem. Cell Biol. 45 (10), 2266–2279.

Singh, V., Sinha, S., Mishra, S., Chaturvedi, L.S., Pradhan, S., Mittal, R.D., et al., 1997. Proportion and pattern of dystrophin gene deletions in north Indian Duchenne and Becker muscular dystrophy patients. Hum. Genet. 99 (2), 206–208.

Srivastava, N.K., Pradhan, S., Mittal, B., Gowda, G.A., 2010. High resolution NMR based analysis of serum lipids in Duchenne muscular dystrophy patients and its possible diagnostic significance. NMR Biomed. 23 (1), 13–22.

Srivastava, N.K., Srivastava, A.K., Mukherjee, S., Sharma, R., Mahapatra, A.K., Sharma, D., 2015. Determination of oxidative stress factors in patients with hereditary muscle diseases: one possible diagnostic and optional management of the patients. Int. J. Pharma Bio Sci. 6 (3), 315–335.

Srivastava, N.K., Annarao, S., Sinha, N., 2016a. Metabolic status of patients with muscular dystrophy in early phase of the disease: in vitro, high resolution NMR spectroscopy based metabolomics analysis of serum. Life Sci. 151, 122–129.

Srivastava, N.K., Mukherjee, S., Sinha, N., 2016b. Alteration of phospholipids in the blood of patients with Duchenne muscular dystrophy (DMD): in vitro, high resolution ^{31}P NMR-based study. Acta Neurol. Belg. 116 (4), 573–581.

Srivastava, N.K., Yadav, R., Mukherjee, S., Pal, L., Sinha, N., 2017. Abnormal lipid metabolism in skeletal muscle tissue of patients with muscular dystrophy: in vitro, high-resolution NMR spectroscopy based observation in early phase of the disease. Magn. Reson. Imaging 38, 163–173.

Srivastava, N.K., Yadav, R., Mukherjee, S., Sinha, N., 2018. Perturbation of muscle metabolism in patients with muscular dystrophy in early or acute phase of disease: in vitro, high resolution NMR spectroscopy based analysis. Clin. Chim. Acta 478, 171–181.

Stålberg, E., 2003. Clinical Neurophysiology of Disorders of Muscle and Neuromuscular Junction, Including Fatigue: Handbook of Clinical Neurophysiology, first ed., vol. 2. Elsevier Health Sciences.

Swash, M., Schwartz, M.S., 2013. Neuromuscular Diseases: A Practical Approach to Diagnosis and Management. Springer Science & Business Media.

Tiidus, P.M., 2008. Skeletal Muscle Damage and Repair. Human Kinetics.

Villalta, S.A., Rosenberg, A.S., Bluestone, J.A., 2015. The immune system in Duchenne muscular dystrophy: friend or foe. Rare Dis. 3 (1), e1010966.

Vogel, H., Zamecnik, J., 2005. Diagnostic immunohistology of muscle diseases. J. Neuropathol. Exp. Neurol. 64 (3), 181–193.

Woolf, N., Wotherspoon, A.C., Young, M., Path, M.R.C., 2002. Essentials of Pathology, first ed. WB Saunders, Edinburgh ; New York.

Worton, R.G., Thompson, M.W., 1988. Genetics of Duchenne muscular dystrophy. Annu. Rev. Genet. 22, 601–629.

Young, S.P., Wallace, G.R., 2009. Metabolomic analysis of human disease and its application to the eye. J. Ocul. Biol. Dis. Infor. 2 (4), 235–242.

CHAPTER 6

Viral and Host Cellular Factors Used by Neurotropic Viruses

MOHANAN VALIYA VEETTIL, PHD • GAYATHRI KRISHNA, MSC •
VINOD SOMAN PILLAI, MSC

6.1 INTRODUCTION

Viral tropism is the ability of different viruses to infect different cellular types ultimately to produce a successful infection. Being parasitic entities, cellular tropism is one of the major characteristics of viruses to maintain a successful infectious cycle in target cells. Neurotropic viruses can invade the neurons to infect the central nervous system (CNS) preferably through two major routes: infecting the peripheral nerves or by traversing the blood–brain barrier (BBB). Viruses enter the peripheral nervous system (PNS) and travel through the axon of neurons to reach the CNS. During zoonotic viral infection, intruding the CNS from a natural host is often neuroinvasive and damaging though they are noninvasive in their natural host (Mcgavern and Kang, 2011). Zoonotic neurotropic viruses such as Rabies virus, Nipah virus, West Nile virus, Japanese encephalitis virus, and Zika virus cause lethal infections, potentially leading to a dead-end host. These viruses often damage the CNS possessing a challenge of viral clearance by the host immune system. Viral clearance by virus-specific antibodies and interferon gamma secretion from T cells is favorable compared with cytolysis of irreplaceable neurons (Griffin and Metcalf, 2011). However, host immune cells fail to control few human adapted viruses that infect the CNS such as herpes simplex virus, varicella zoster virus, cytomegalovirus, and JC virus. Alpha herpes viruses remain latent in the neurons seldom reactivating to reinfect other hosts. Here, we discuss on how neurotropic viruses gain access and infect the nervous system by giving a general overview on various viral and host factors. We also discuss the transmission, entry, trafficking, replication, and assembly with emphasis on poliovirus, Zika virus, varicella zoster virus, herpes simplex virus-1, Japanese encephalitis virus, and rabies virus.

6.2 POLIOVIRUS

Poliovirus (PV), the etiological agent of poliomyelitis, belongs to the family Picornaviridae. In humans, PV infection is initiated by binding to its cellular receptor protein poliovirus receptor (PVR or CD155) (Fig. 6.1). After receptor binding and entry, a series of events lead to the virus replication in the human intestinal mucosa to produce infectious viral particles and shedding in the feces (Racaniello, 2006). However, a few viral particles enter the blood circulation causing viremia and infect the neuromuscular junction (NMJ) between the muscles and motor neurons and invade CNS causing motor dysfunction poliomyelitis (Racaniello, 2006).

PV shows two routes of entry into the CNS; firstly, PV enters the CNS from blood by traversing the BBB, and secondly, PV can be carried to the CNS by peripheral nerves during a skeletal muscle injury (Ohka and Nomoto, 2001). In a nerve cell, particularly in axons, microtubules are arranged and oriented with their plus end toward axon terminal and minus end toward the soma while dendrites have mixed polarity (Kapitein and Hoogenraad, 2011). Dynein motor proteins mediate axon terminal to soma retrograde transport, whereas kinesin motor proteins mediate soma to axon end anterograde transport (Kapitein and Hoogenraad, 2011). Therefore, viral particle entering a neuronal cell through axon terminus necessitates dynein motor protein for retrograde transport to the cell body. During PV infection, endocytosed PVR's cytoplasmic domain binds to Tctex-1, light chain of dynein motor protein to facilitate retrograde transport from axon terminal to the cell body of motor neuron where uncoating and replication of the virus happens (Ohka et al., 2004).

Membrane vesicles formed from cell organelles in neuronal cytoplasm serve as a site for RNA replication, which occurs in complexes (Caliguiri and Tamm, 1970; Blondel et al., 2005). RNA replication begins with

The Molecular Immunology of Neurological Diseases. https://doi.org/10.1016/B978-0-12-821974-4.00003-0

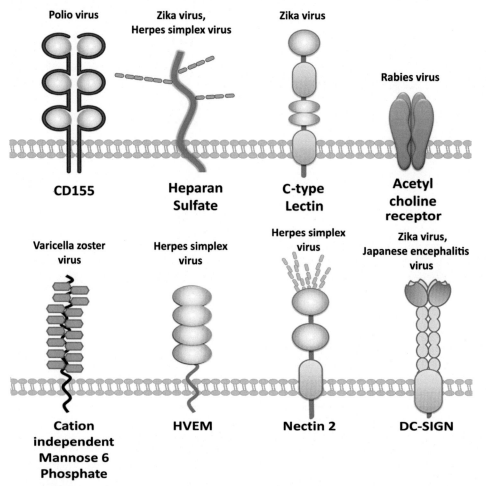

FIG. 6.1 **Cellular Receptors Used by Neurotropic Viruses.** Polioviruses utilize CD155 to infect the host cell. Zika viruses use heparan sulfate as attachment factor to facilitate binding with cellular receptor DC-SIGN, a C-type lectin receptor. Herpes simplex virus-1 uses the attachment factor, heparan sulfate to later interact with HVEM or nectin-2 to mediate entry. Varicella zoster virus utilizes cation independent mannose-6-phosphate for cellular entry. Japanese encephalitis virus use DC-SIGN for entry into the host cell. Rabies viruses utilize acetyl choline receptor to infect host cell.

formation of negative-sense RNA from genomic RNA, which serves as a template for viral RNA-dependent RNA polymerase, 3D^{pol}, while viral proteins 2BC, 2B, and 2C comprise the protein complex essential for virus replication (Flanegan and Van Dyke, 1979; Van Dyke and Flanegan, 1980). Additionally, viral VPg and associated proteins are also required for PV replication (Reuer et al., 1990). Furthermore, translation of PV RNA begins in the cytoplasm of infected neurons with the binding of ribosomes to the internal ribosomal entry site (IRES), a *cis*-acting element required for initiation of PV translation located in the 5′ noncoding sequence of PV RNA (Pelletier and Sonenberg, 1988;

Gromeier et al., 1996; Racaniello, 2006). Host cellular and initiation factors aid translation to produce a polyprotein, which is further processed by viral proteases, 3C^{pro}/3CD^{pro} to release structural and nonstructural proteins (Ypma-Wong et al., 1988; Blondel et al., 2005).

During assembly, virions are accumulated in the cytoplasm, and release of mature virions occurs by cell lysis causing cell to cell spread of PV. During the course of PV infection, cell undergoes various metabolic and morphological changes (Blondel et al., 2005). PV proteases play a major role in shutting off the transcription and translation in neurons. Nonstructural proteins induce membrane vesiculation and inhibit protein trafficking through various pathways. Moreover, infection leads to deregulated nuclear-cytoplasmic protein trafficking causing apoptosis due to CNS cell damage (Griffin and Hardwick, 1999). Collectively, destruction of motor neurons causes paralysis in a PV-infected individual.

6.3 ZIKA VIRUS

Zika virus (ZIKV), a flavivirus of the family Flaviviridae, has been reported to be existing in the nonhuman primates and Aedes mosquito species. ZIKV follows a human–mosquito–human transmission cycle where humans get infected from mosquito bites (White et al., 2016). However, laboratory infections and monkey bites have also reported to cause ZIKV infection. ZIKV can also transmit through sexual contact and blood transfusion. Moreover, in humans, ZIKV crosses the placenta to infect the fetal brain, damaging brain tissue, causing microcephaly (Petersen et al., 2016).

ZIKV is an enveloped positive stranded RNA virus encoding precursor protein, capsid protein, envelop protein, and seven nonstructural proteins (Agrelli et al., 2019). Cellular receptors such as C-type lectin, T-cell immunoglobulin mucin, and tyrosine kinases have been reported to aid entry of ZIKV (Agrelli et al., 2019). Envelop proteins arranged in an icosahedral manner mediate receptor-mediated endocytosis through the host cell receptor. The ZIKV replication begins following entry into cells. In ZIKV replication process, the RNA released from the virion is translated in the cytoplasm and subsequently produces polypeptide chain possessing an endoplasmic reticulum (ER) localization signal, which facilitates a cotranslational insertion into ER. Viral proteases and cellular proteases cleave the polyprotein into individual peptides at the cytoplasmic side and inside the lumen of ER, respectively. The genomic RNA is transcribed into negative-sense RNA by the nonstructural proteins, which later serves as a template for replication (Sager et al., 2018; Garcia-Blanco et al., 2016).

During ZIKV assembly, the assembled virions travel through a secretory pathway and get matured and released into the extracellular space (Sager et al., 2018). ZIKV was detected in infant and fetal brain tissues, amniotic fluid, and placental tissues of ZIKV-infected females. ZIKV passes the placental barrier to infect the placental cells reaching the fetus (Noronha et al., 2016). ZIKV was also detected in maternal trophoblasts and fetal endothelial cells (Miner et al., 2016). Additionally, ZIKV evades immune response by increasing the levels of interleukin-6, interleukin-8, vascular endothelial growth factor (VEGF), monocyte chemoattractive protein 1 (MCP-1), IFN-γ-inducible protein 10 (IP-10), and granulocyte colony-stimulating factor (G-CSF) as detected in amniotic fluid of ZIKV-infected pregnant female with fetal microcephaly (Ornelas et al., 2017). Compared with immature neurons, ZIKV effectively infects human cortical neural progenitor cells (NPCs) (Tang et al., 2016). Cortical thinning and microcephaly was observed in in vivo models where ZIKV replicated in NPCs located in the ventricular zone (VZ) and subventricular zone (SVZ), which show the susceptibility of NPCs to ZIKV infections (Cugola et al., 2016; Li et al., 2016b). Furthermore, astrocytes, neocortical and spinal cord neuroepithelial stem cells, and cranial neural crest cells were also reported to be susceptible to ZIKV infections (Hamel et al., 2017; Bayless et al., 2016). In ZIKV-infected cells, VZs are arrested in S, G1, or G2 phases of the cell cycle, and NPCs in M phase, which impairs cell cycle progression (Li et al., 2016a; Shao et al., 2016). Mitochondrial sequestration of centrosomal phospho-TBK1 (TANK-binding kinase 1) disrupts mitosis, resulting in impaired cell division and chromosomal abnormalities (Dang et al., 2016; Onorati et al., 2016). Collectively, ZIKV infection suppresses NPC proliferation and impairs neurogenesis, leading to cell death and microcephaly.

6.4 HERPES SIMPLEX VIRUS-1

Herpes simplex virus-1 (HSV-1) belongs to the subfamily Alphaherpesvirinae of the family Herpesviridae, which is present in almost 60% of world's human population with humans as the only natural host. HSV-1 is most commonly found in oral mucosa and ocular areas and can reactivate to cause ocular infections, keratitis, and sporadic encephalitis called herpes simplex encephalitis (Kollias et al., 2015; Farooq and Shukla, 2012). HSV-1 is transmitted by oral to oral contact, namely, when the virus is active in mucosal surfaces or sores or saliva; however, it can also be transmitted to the genital areas by oral–genital contact.

HSV-1 is an enveloped, double-stranded DNA virus that establishes latency in host cell after initial infection. Following entry by membrane fusion, the nucleocapsid is released into the cytoplasm where it is trafficked through microtubule transport system toward the nucleus to release the genome. HSV-1 infects the innervating sensory neurons by the interaction of viral glycoproteins with the cell surface glycosaminoglycans and receptor nectin-1 to facilitate pH-independent fusion at the axonal terminal of sensory nerves (Smith, 2012).

Inside a neuronal cell, HSV-1 capsid is associated with motor protein dynein at the axonal terminus to enable retrograde transport toward the neuronal cell body in trigeminal ganglia. HSV-1 outer capsid protein pUL35 was reported to interact with Tctex-1 and rp3 to facilitate movement toward nucleus (Douglas et al., 2004). Once capsid reaches the nuclear pore, the HSV-1 DNA is instilled into the nucleus where a lifelong latent or quiescent phase is established. Various stimuli such as stress, UV light, and immunosuppression can reactivate HSV-1 virus from latency phase (Pires De Mello et al., 2016). Upon reactivation, HSV-1 initiates lytic cycle and newly formed viruses move toward the axon of peripheral sensory neurons by anterograde transport to exit by exocytosis (Smith, 2012). They access the mucosal epithelium where further replication happens and finally enables human to human transmission (Smith, 2012). HSV-1 can mostly be asymptomatic with few cases of ocular diseases. Incidence of encephalitis is comparatively less despite the seroprevalence of HSV-1 (Kennedy and Steiner, 2013).

6.5 RABIES VIRUS

Rabies virus (RABV) is single-stranded negative RNA virus that causes rabies. RABV is a member of Rhabdoviridae family belonging to the genus Lyssavirus. RABV encodes five structural proteins: nucleoprotein (N), phosphoprotein (P), matrix protein (M), glycoprotein (G), and an RNA-dependent RNA polymerase (L). N protein encapsulates the genome to form a ribonucleoprotein with helical capsid.

RABV infection begins with the attachment of virus G protein with host cellular receptors. Several cell surface molecules have been reported as RABV receptors such as nicotinic acetylcholine receptor (nAChR), neural cell adhesion molecule (NCAM), and low-affinity neurotropin receptor p75^{NTR}, while the exact role of some of these molecules is still unknown (Thoulouze et al., 1998; Tuffereau et al., 1998). Muscle cells that express nAchR can get infected by RABV followed by subsequent infection of the neuronal cells. Here, RABV infects the NMJ where NCAM and nAChRs serve as RABV receptors to reach the CNS (Schnell et al., 2010; Thoulouze et al., 1998). Upon infecting the NMJ, RABV along with the putative receptor is internalized by receptor-mediated endocytosis into Rab5-positive early endosome vesicle, which later matures into Rab7-positive early endosome vesicle (Deinhardt et al., 2006). Subsequently, Rab7 recruits dynactin, a part of dynein motor protein to mediate the retrograde transport of RABV-carrying endosomes to the cell body of motor neuron (Deinhardt et al., 2006).

Replication and assembly of RABV occurs in the neuronal cell body. During the early stages of replication, the low pH of endosome releases the capsid into the cytoplasm of cell body by fusion of viral and endosomal membranes (Le Blanc et al., 2005). Replication takes place in the neuronal cytoplasm inclusion bodies called Negri bodies with the help of heat shock protein 70 as functional structures supporting replication, whereas Toll-like receptor 3 helps in the spatial arrangement of RABV replication sites (Lahaye et al., 2009; Menager et al., 2009). Viral polymerase uses the ribonucleoprotein for repetitive rounds of transcription (Ivanov et al., 2011). Transcription of RABV genome starts at the 3′ end, resulting in the formation of short uncapped and unpolyadenylated leader RNA (leRNA), which is followed by the transcription of 5′ end-capped and polyadenylated mRNAs that code for RABV proteins (Tordo et al., 1986; Tordo and Kouknetzoff, 1993). Viral polymerase stops transcription at the intergenic region and later reinitiates at the start signal sequence (Finke et al., 2000). However, expression of nucleoprotein along with phosphoprotein acts as a chaperone to reduce nonspecific RNA binding (Wojczyk et al., 1998). Viral polymerase first produces a full-length antisense RNA genome, which acts as a template for the production of negative-sense RNA. Phosphorylation of nucleoprotein–phosphoprotein complex allows maximizing viral replication while decreasing viral transcription (Yang et al., 1998). Moreover, RABV M protein also plays a significant role in regulating replication and transcription (Schnell et al., 2010).

Viral assembly and budding becomes the last phase of viral life cycle where the capsid moves toward the plasma membrane by an unknown mechanism. M protein and G protein are reported to play an important part in the budding process of RABV, as their absence reduces the efficiency of viral release. Viral G protein gets inserted into the plasma membrane, which initiates the budding process followed by the interaction of M protein with the cytoplasmic end of G protein, ultimately releasing the virus from the cell to infect other neuronal cells (Mebatsion et al., 1996, 1999; Okumura and Harty, 2011). During the course of infection, this can lead to nerve dysfunction and demyelination causing restlessness, confusion and paralysis, unconsciousness, and coma. Moreover, cardiac arrest and respiratory failure can also lead to death.

6.6 VARICELLA ZOSTER VIRUS

Varicella zoster virus (VZV), alpha herpes virus or human herpes virus 3 (HHV-3), is double-stranded DNA causing varicella or chicken pox with only known natural host as humans. On infection, VZV causes chicken pox in cutaneous epithelial cells of humans further infecting T-lymphocytes and eventually evades from immune response by residing within ganglionic neurons. Once inside the ganglionic neurons, they may reappear when the individual's immunity fades or compromised or under immunosuppressant treatment, leading to zoster or shingles, characterized by chronic pain (postherpetic neuralgia [PHN]), ocular, and neurological disorders such as keratitis, retinopathy, vasculopathy, cranial nerve palsies, myelitis, and meningoencephalopathy. VZV is a communicable disease and usually transmitted via airborne cell-free virus from the skin lesions or vesicles. Although a lot of complications associated with VZV are neurological, the virus enters human body via oropharynx through inhalation of infectious respiratory droplets or via direct contact with the cutaneous lesions.

On transmission, VZV proliferates within the oropharynx and particularly in the tonsils, thereby infecting the T cells leading to the circulatory system, and ends up in various organs as well as skin. Further proliferation is usually limited by the innate immunity and eventually by the adaptive immunity, but the shift in immunity time span is exploited by the virus particularly at the cutaneous infection sites, leading to the formation of varicella or pus-filled rashes. In the lytic stage, the virus expresses entire 71 of its proteins including the immediate-early, early, and late genes, while in latency, they express very few genes similar to other herpes viruses. Excessive proliferation at cutaneous sites is further limited by the adaptive immunity; thus the virus subsides by itself gaining latency in ganglionic nerves via neuronal retrograde transport (Fig. 6.2). VZV occurs in latent stage in various ganglia including the cranial nerve ganglia, dorsal root ganglia (DRG), enteric ganglia, and the autonomic ganglia (Gershon and Gershon, 2013). Latent VZV genome is circularized and is nonreplicating in nature, tightly maintained by posttranslationally modified histones selectively maintaining the expression of immediate-early genes of the virus. Immediate-early protein 63 (IE63) is associated with antisilencing function protein 1(ASF1) facilitating more histone binding to the genome, ensuring no late protein genes are being expressed. It is being touted

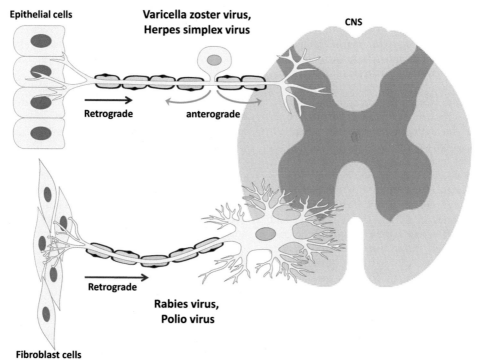

FIG. 6.2 **Transport of Neurotropic Viruses to Central Nervous System.**   Alpha herpes viruses (herpes simplex virus-1 and varicella zoster virus) infect the sensory neurons and get transported to the neuronal soma by retrograde transport; however, lytic reactivation is signified by anterograde neuronal transport. Rabies and poliovirus infect the neuromuscular junction to enter the axon terminal followed by retrograde transport toward the neuronal cell body and central nervous system.

that histone depositioning by ASF1 along with the IE63 may regulate the cellular as well as VZV gene transcription, mediating a lytic reactivation. The lytic reactivation is signified by anterograde neuronal transport to epithelial cells and zoster or shingles disorder of the skin (Ambagala et al., 2009).

Once the virions are fused to the cell membrane via glycoprotein and cellular receptor interactions, uncoating with release of capsid and tegument proteins such as the major viral transcription factor immediate-early protein 62 (IE62) is released into cytoplasm. The IE62 could be transported to the nucleus of the cell resulting in de novo protein synthesis, further anchoring nucleocapsid to the nucleus resulting in nucleopore formation and genome release into the nucleus. Tegument proteins open reading frame 47 (ORF47) and ORF66 are serine/threonine kinases, leading to autophosphorylation of other viral and cellular transcription factors. Similarly, the major transcription factor protein, IE62, activates transcription factor specificity protein (Sp1) having specific binding site along various viral promoters further activating other VZV gene expression.

A primary envelopment and later deenvelopment of nucleocapsid from nucleus to cytoplasm typical of herpes virus happens in VZV also, and the nucleocapsid further translocates to trans-Golgi network (TGN) acquiring the tegument and glycoproteins. The nucleocapsid with the tegument and glycoprotein is transported to the cell membrane via vesicles and appears on cell surface within 9−12 h of a cellular infection. Unlike other herpes viruses, the glycoprotein with the virus on the cell surface mediates cell to cell interaction, thereby further mediating entry to the other cell and cell−cell syncytia formation, further aiding in spread. The viral genetic factors along with cellular factors play in fastening the infectivity, and VZV is understood to extensively employ this mechanism in T-lymphocytes and epithelial cells efficiently. VZV infects and replicates in CD3+, CD4+, CD8+, and CD4+CD8+ T cells, tonsil T cells, and dendritic cells further extending their spread into circulatory system and reaching the skin via CD4+ T cell expressing skin homing markers such as cutaneous leukocyte antigen (CLA) (Ku et al., 2002). STAT3 and Survivin are two important cell factors to have found role in VZV infection in epithelial cells. Autopsy studies have revealed up to 4% of the neurons to have VZV genome in latent state, as well as in ganglia, and also include high expression of ORF63. Similarly, autopsy sample of recent zoster condition reveals expression of VZV proteins and inflammatory proteins even up to 25% of the neurons signifying reactivation extensively in neurons. Dorsal root ganglia—xenografted studies lead to identification of antiapoptotic role of IE63 signifying the latency of VZV in neurons. Neurons and satellite cells could harbor the nucleocapsid via promyelocytic leukemia protein nuclear bodies (PML-NBs), which could also trap aberrant polyglutamine proteins as well as HttQ72, leading to neurological disorders such as Huntington's disease (Janer et al., 2006).

6.7 JAPANESE ENCEPHALITIS VIRUS

Flaviviruses tend to emerge and reemerge as potential pathogens and cause outbreaks now and then, majorly encephalitic infections. One among the family of flaviviruses is Japanese encephalitis virus (JEV), a more common causative of encephalitis and critical neurological illness. JEV causes severe inflammation of the CNS and is often associated with high mortality approximated up to 30%. Like all flaviviruses, JEV replicates in arthropods as well as in all vertebrates. However, an arthropod-borne disease in humans and other vertebrates by infected mosquito bite, pigs, and aquatic birds can also act as natural amplifying host to the virus. Transovarial or vertical transmission among Culex species of mosquito, a major vector carrier with various other species such as Aedes, Anopheles, Mansonia, and Armigeres, is understood as susceptible to the virus under laboratory setting raising concerns over wider spread.

JEV is believed to have neuronal tropism and mostly infects neurons, while certain animal model studies have proven JEV infection of other organs including kidney, liver, and spleen as well as peripheral tissues. Cellular susceptibility for JEV infection has also been observed in immune cells such as monocytes, macrophages, dendritic cells, microglia, and T cells in neuron-associated cells such as astrocytes endothelial cells, pericytes, and fibroblasts. JEV is understood to utilize various receptors facilitating its entry into the cells and further aiding in replication. Receptors including vimentin, laminin, α5β3 integrins, HSP70, and DC-SIGN extensively expressed over various cells could aid in JEV infection. JEV is an enveloped positive-sense single-stranded RNA virus with the genome being encapsulated in a capsid. A 10.5-kb genome is over-flanked with 5′ and 3′ untranslated region or UTR coding a polyprotein, which is further cleaved and post-translationally modified to form 10 different structural and nonstructural proteins. Once JEV establishes receptor interaction on a susceptible cell, the viral capsid is being released into the cytoplasm in a clathrin-dependent or clathrin-independent manner. The envelop protein (E) undergoes a pH-associated conformational change, leading to endocytosed viral capsid to fuse with the endosomal membrane, further facilitating the release of viral genome RNA into the host cell cytoplasm. The viral RNA is further translated into two polyproteins based on the ribosomal frameshift at NS2A coding region of the genome. The translated polyprotein is further cleaved into three structural proteins C, prM (nonglycosylated precursor of matrix protein-M), and E (glycosylated) and seven nonstructural proteins. Once the essential proteins synthesis is attained, the virus genome undergoes replication in an ER-derived organelle carrying the replication complex. The large nonstructural proteins NS-3 and NS-5 mediate negative-sense template RNA synthesis, genomic RNA synthesis, RNA capping, and cap methylation. The structural proteins prM- and E-mediated packaging of viral capsid protein C with the newly replicated genomic RNA enables release of immature virions from ER membrane with prM and E forming multiple heterodimeric spikes on the surface of the capsid. ER release of immature virions is succeeded by viral maturation in the trans-Golgi network by furin-mediated cleavage of prM to M and further rearrangement of the heterodimeric M:E protein complex to form heterotetramers (Zhang et al., 2004). Although maturation happens after the immature release by ER membrane, immature virions are also unusually released from cells. The virions are exocytosed from cell membrane once they get released from ER and mature eventually through the cytoplasm. Various viral factors including the NS2A and NS2B transmembrane domains are reported to interact with host factors such as tripartite motif containing (TRIM) protein 52 as well as signal peptidase complex subunit 1 (SPCS1) respectively. JEV NS5 protein could also interact with mitochondrial trifunctional protein (MTP) disrupting long-chain fatty acid (LCFA) metabolism enhancing viral pathogenesis (Fan et al., 2016; Kao et al., 2015).

Immune evasion and transmigration of JEV are notorious, as they replicate in the skin on mosquito bite later traversing to the adjacent lymph node understated with low viremia. Once in the lymph node they replicate in monocytes, macrophages and dendritic cells, particularly the antigen-presenting cells (APC), pave their way through CNS and further infect the brain. They hide from the immune surveillance to breach the BBB by suppressing the host immune mechanism like in case of myeloid-derived suppressor cells (MDSCs) enhancement, leading to suppression of T cell—mediated immunity. Monocyte-mediated suppression of immunity and CNS breach are also being observed in animal model studies. They utilize mast cells (MCs); a myeloid lineage

cell found in the CNS resident at the BBB gets activated following JEV infection, leading to release of various proteases especially chymases, leading to the breach of BBB and further JEV brain infection. Similarly, they go around the BBB tight junction and infect the brain to further sequester microglia cells at the BBB and recruitment of chemokines and inflammatory cytokines, leading to inflammation, further successfully establishing breach of BBB and cellular infiltration. Microglia may act as a reservoir for JEV infection in the brain, and sustained activation of microglia mediates N-methyl D aspartate receptor (NMDAR) influxing high amounts of Ca^{2+}, leading to aberrant mitochondrial function and neuronal damage (Chen et al., 2018). Apart from the neuronal damage, the neuron to microglia crosstalk in the CNS is mediated via exosome released from the glia. JEV infection of glia could produce distinct exosomes carrying miRNAs and proteins, leading to neuronal apoptosis and degeneration likely resulting in progressive neuronal disorders (Yang et al., 2018).

6.8 CONCLUSION

Viruses are obligate parasites that can infect a variety of cell types to subsequently spread from one cell to another. The interaction of neurotropic viral factors with the host cells is notable and develops intricate signaling pathways and networks in biological systems. Though CNS is protected from neurotropic viruses, they tend to infect neurons and invade the CNS causing lethality. Viral replication and spread to CNS can damage the neurons and also activate host immune response to debilitate the host cell. However, maximizing research with better research methodologies can reveal further new roles of viruses in neuropathogenesis.

REFERENCES

Agrelli, A., De Moura, R.R., Crovella, S., Brandao, L.A.C., 2019. ZIKA virus entry mechanisms in human cells. Infect. Genet. Evol. 69, 22—29.

Ambagala, A.P., Bosma, T., Ali, M.A., Poustovoitov, M., Chen, J.J., Gershon, M.D., Adams, P.D., Cohen, J.I., 2009. Varicella-zoster virus immediate-early 63 protein interacts with human antisilencing function 1 protein and alters its ability to bind histones h3.1 and h3.3. J. Virol. 83, 200—209.

Bayless, N.L., Greenberg, R.S., Swigut, T., Wysocka, J., Blish, C.A., 2016. Zika virus infection induces cranial neural crest cells to produce cytokines at levels detrimental for neurogenesis. Cell Host Microbe 20, 423—428.

Blondel, B., Colbere-Garapin, F., Couderc, T., Wirotius, A., Guivel-Benhassine, F., 2005. Poliovirus, pathogenesis of poliomyelitis, and apoptosis. Curr. Top. Microbiol. Immunol. 289, 25—56.

Caliguiri, L.A., Tamm, I., 1970. Characterization of poliovirus-specific structures associated with cytoplasmic membranes. Virology 42, 112–122.

Chen, Z., Wang, X., Ashraf, U., Zheng, B., Ye, J., Zhou, D., Zhang, H., Song, Y., Chen, H., Zhao, S., Cao, S., 2018. Activation of neuronal N-methyl-D-aspartate receptor plays a pivotal role in Japanese encephalitis virus-induced neuronal cell damage. J. Neuroinflammation 15, 238.

Cugola, F.R., Fernandes, I.R., Russo, F.B., Freitas, B.C., Dias, J.L., Guimaraes, K.P., Benazzato, C., Almeida, N., Pignatari, G.C., Romero, S., Polonio, C.M., Cunha, I., Freitas, C.L., Brandao, W.N., Rossato, C., Andrade, D.G., Faria Dde, P., Garcez, A.T., Buchpigel, C.A., Braconi, C.T., Mendes, E., Sall, A.A., Zanotto, P.M., Peron, J.P., Muotri, A.R., Beltrao-Braga, P.C., 2016. The Brazilian Zika virus strain causes birth defects in experimental models. Nature 534, 267–271.

Dang, J., Tiwari, S.K., Lichinchi, G., Qin, Y., Patil, V.S., Eroshkin, A.M., Rana, T.M., 2016. Zika virus depletes neural progenitors in human cerebral organoids through activation of the innate immune receptor TLR3. Cell Stem Cell 19, 258–265.

Deinhardt, K., Salinas, S., Verastegui, C., Watson, R., Worth, D., Hanrahan, S., Bucci, C., Schiavo, G., 2006. Rab5 and Rab7 control endocytic sorting along the axonal retrograde transport pathway. Neuron 52, 293–305.

Douglas, M.W., Diefenbach, R.J., Homa, F.L., Miranda-Saksena, M., Rixon, F.J., Vittone, V., Byth, K., Cunningham, A.L., 2004. Herpes simplex virus type 1 capsid protein VP26 interacts with dynein light chains RP3 and Tctex1 and plays a role in retrograde cellular transport. J. Biol. Chem. 279, 28522–28530.

Fan, W., Wu, M., Qian, S., Zhou, Y., Chen, H., Li, X., Qian, P., 2016. TRIM52 inhibits Japanese encephalitis virus replication by degrading the viral NS2A. Sci. Rep. 6, 33698.

Farooq, A.V., Shukla, D., 2012. Herpes simplex epithelial and stromal keratitis: an epidemiologic update. Surv. Ophthalmol. 57, 448–462.

Finke, S., Cox, J.H., Conzelmann, K.K., 2000. Differential transcription attenuation of rabies virus genes by intergenic regions: generation of recombinant viruses overexpressing the polymerase gene. J. Virol. 74, 7261–7269.

Flanegan, J.B., Van Dyke, T.A., 1979. Isolation of a soluble and template-dependent poliovirus RNA polymerase that copies virion RNA in vitro. J. Virol. 32, 155–161.

Garcia-Blanco, M.A., Vasudevan, S.G., Bradrick, S.S., Nicchitta, C., 2016. Flavivirus RNA transactions from viral entry to genome replication. Antiviral Res. 134, 244–249.

Gershon, A.A., Gershon, M.D., 2013. Pathogenesis and current approaches to control of varicella-zoster virus infections. Clin. Microbiol. Rev. 26, 728–743.

Griffin, D.E., Hardwick, J.M., 1999. Perspective: virus infections and the death of neurons. Trends Microbiol. 7, 155–160.

Griffin, D.E., Metcalf, T., 2011. Clearance of virus infection from the CNS. Curr. Opin. Virol. 1, 216–221.

Gromeier, M., Alexander, L., Wimmer, E., 1996. Internal ribosomal entry site substitution eliminates neurovirulence in intergeneric poliovirus recombinants. Proc. Natl. Acad. Sci. U.S.A. 93, 2370–2375.

Hamel, R., Ferraris, P., Wichit, S., Diop, F., Talignani, L., Pompon, J., Garcia, D., Liegeois, F., Sall, A.A., Yssel, H., Misse, D., 2017. African and Asian Zika virus strains differentially induce early antiviral responses in primary human astrocytes. Infect. Genet. Evol. 49, 134–137.

Ivanov, I., Yabukarski, F., Ruigrok, R.W., Jamin, M., 2011. Structural insights into the rhabdovirus transcription/replication complex. Virus Res. 162, 126–137.

Janer, A., Martin, E., Muriel, M.P., Latouche, M., Fujigasaki, H., Ruberg, M., Brice, A., Trottier, Y., Sittler, A., 2006. PML clastosomes prevent nuclear accumulation of mutant ataxin-7 and other polyglutamine proteins. J. Cell Biol. 174, 65–76.

Kao, Y.T., Chang, B.L., Liang, J.J., Tsai, H.J., Lee, Y.L., Lin, R.J., Lin, Y.L., 2015. Japanese encephalitis virus nonstructural protein NS5 interacts with mitochondrial trifunctional protein and impairs fatty acid beta-oxidation. PLoS Pathog. 11, e1004750.

Kapitein, L.C., Hoogenraad, C.C., 2011. Which way to go? Cytoskeletal organization and polarized transport in neurons. Mol. Cell. Neurosci. 46, 9–20.

Kennedy, P.G., Steiner, I., 2013. Recent issues in herpes simplex encephalitis. J. Neurovirol. 19, 346–350.

Kollias, C.M., Huneke, R.B., Wigdahl, B., Jennings, S.R., 2015. Animal models of herpes simplex virus immunity and pathogenesis. J. Neurovirol. 21, 8–23.

Ku, C.C., Padilla, J.A., Grose, C., Butcher, E.C., Arvin, A.M., 2002. Tropism of varicella-zoster virus for human tonsillar CD4(+) T lymphocytes that express activation, memory, and skin homing markers. J. Virol. 76, 11425–11433.

Lahaye, X., Vidy, A., Pomier, C., Obiang, L., Harper, F., Gaudin, Y., Blondel, D., 2009. Functional characterization of Negri Bodies (NBs) in rabies virus-infected cells: evidence that NBs are sites of viral transcription and replication. J. Virol. 83, 7948–7958.

Le Blanc, I., Luyet, P.P., Pons, V., Ferguson, C., Emans, N., Petiot, A., Mayran, N., Demaurex, N., Faure, J., Sadoul, R., Parton, R.G., Gruenberg, J., 2005. Endosome-to-cytosol transport of viral nucleocapsids. Nat. Cell Biol. 7, 653–664.

Li, C., Xu, D., Ye, Q., Hong, S., Jiang, Y., Liu, X., Zhang, N., Shi, L., Qin, C.F., Xu, Z., 2016a. Zika virus disrupts neural progenitor development and leads to microcephaly in mice. Cell Stem Cell 19, 120–126.

Li, C., Xu, D., Ye, Q., Hong, S., Jiang, Y., Liu, X., Zhang, N., Shi, L., Qin, C.F., Xu, Z., 2016b. Zika virus disrupts neural progenitor development and leads to microcephaly in mice. Cell Stem Cell 19, 672.

Mcgavern, D.B., Kang, S.S., 2011. Illuminating viral infections in the nervous system. Nat. Rev. Immunol. 11, 318–329.

Mebatsion, T., Konig, M., Conzelmann, K.K., 1996. Budding of rabies virus particles in the absence of the spike glycoprotein. Cell 84, 941–951.

Mebatsion, T., Weiland, F., Conzelmann, K.K., 1999. Matrix protein of rabies virus is responsible for the assembly and budding of bullet-shaped particles and interacts with the transmembrane spike glycoprotein G. J. Virol. 73, 242–250.

Menager, P., Roux, P., Megret, F., Bourgeois, J.P., Le Sourd, A.M., Danckaert, A., Lafage, M., Prehaud, C., Lafon, M., 2009. Toll-like receptor 3 (TLR3) plays a major role in the formation of rabies virus Negri bodies. PLoS Pathog. 5, e1000315.

Miner, J.J., Cao, B., Govero, J., Smith, A.M., Fernandez, E., Cabrera, O.H., Garber, C., Noll, M., Klein, R.S., Noguchi, K.K., Mysorekar, I.U., Diamond, M.S., 2016. Zika virus infection during pregnancy in mice causes placental damage and fetal demise. Cell 165, 1081−1091.

Noronha, L., Zanluca, C., Azevedo, M.L., Luz, K.G., Santos, C.N., 2016. Zika virus damages the human placental barrier and presents marked fetal neurotropism. Mem. Inst. Oswaldo Cruz 111, 287−293.

Ohka, S., Matsuda, N., Tohyama, K., Oda, T., Morikawa, M., Kuge, S., Nomoto, A., 2004. Receptor (CD155)-dependent endocytosis of poliovirus and retrograde axonal transport of the endosome. J. Virol. 78, 7186−7198.

Ohka, S., Nomoto, A., 2001. Recent insights into poliovirus pathogenesis. Trends Microbiol. 9, 501−506.

Okumura, A., Harty, R.N., 2011. Rabies virus assembly and budding. Adv. Virus Res. 79, 23−32.

Onorati, M., Li, Z., Liu, F., Sousa, A.M.M., Nakagawa, N., Li, M., Dell'anno, M.T., Gulden, F.O., Pochareddy, S., Tebbenkamp, A.T.N., Han, W., Pletikos, M., Gao, T., Zhu, Y., Bichsel, C., Varela, L., Szigeti-Buck, K., Lisgo, S., Zhang, Y., Testen, A., Gao, X.B., Mlakar, J., Popovic, M., Flamand, M., Strittmatter, S.M., Kaczmarek, L.K., Anton, E.S., Horvath, T.L., Lindenbach, B.D., Sestan, N., 2016. Zika virus disrupts phospho-TBK1 localization and mitosis in human neuroepithelial stem cells and radial glia. Cell Rep. 16, 2576−2592.

Ornelas, A.M., Pezzuto, P., Silveira, P.P., Melo, F.O., Ferreira, T.A., Oliveira-Szejnfeld, P.S., Leal, J.I., Amorim, M.M., Hamilton, S., Rawlinson, W.D., Cardoso, C.C., Nixon, D.F., Tanuri, A., Melo, A.S., Aguiar, R.S., 2017. Immune activation in amniotic fluid from Zika virus-associated microcephaly. Ann. Neurol. 81, 152−156.

Pelletier, J., Sonenberg, N., 1988. Internal initiation of translation of eukaryotic mRNA directed by a sequence derived from poliovirus RNA. Nature 334, 320−325.

Petersen, L.R., Jamieson, D.J., Powers, A.M., Honein, M.A., 2016. Zika virus. N. Engl. J. Med. 374, 1552−1563.

Pires De Mello, C.P., Bloom, D.C., Paixao, I.C., 2016. Herpes simplex virus type-1: replication, latency, reactivation and its antiviral targets. Antivir. Ther. 21, 277−286.

Racaniello, V.R., 2006. One hundred years of poliovirus pathogenesis. Virology 344, 9−16.

Reuer, Q., Kuhn, R.J., Wimmer, E., 1990. Characterization of poliovirus clones containing lethal and nonlethal mutations in the genome-linked protein VPg. J. Virol. 64, 2967−2975.

Sager, G., Gabaglio, S., Sztul, E., Belov, G.A., 2018. Role of host cell secretory machinery in Zika virus life cycle. Viruses 10.

Schnell, M.J., Mcgettigan, J.P., Wirblich, C., Papaneri, A., 2010. The cell biology of rabies virus: using stealth to reach the brain. Nat. Rev. Microbiol. 8, 51−61.

Shao, Q., Herrlinger, S., Yang, S.L., Lai, F., Moore, J.M., Brindley, M.A., Chen, J.F., 2016. Zika virus infection disrupts neurovascular development and results in postnatal microcephaly with brain damage. Development 143, 4127−4136.

Smith, G., 2012. Herpesvirus transport to the nervous system and back again. Annu. Rev. Microbiol. 66, 153−176.

Tang, H., Hammack, C., Ogden, S.C., Wen, Z., Qian, X., Li, Y., Yao, B., Shin, J., Zhang, F., Lee, E.M., Christian, K.M., Didier, R.A., Jin, P., Song, H., Ming, G.L., 2016. Zika virus infects human cortical neural progenitors and attenuates their growth. Cell Stem Cell 18, 587−590.

Thoulouze, M.I., Lafage, M., Schachner, M., Hartmann, U., Cremer, H., Lafon, M., 1998. The neural cell adhesion molecule is a receptor for rabies virus. J. Virol. 72, 7181−7190.

Tordo, N., Kouknetzoff, A., 1993. The rabies virus genome: an overview. Onderstepoort J. Vet. Res. 60, 263−269.

Tordo, N., Poch, O., Ermine, A., Keith, G., Rougeon, F., 1986. Walking along the rabies genome: is the large G-L intergenic region a remnant gene? Proc. Natl. Acad. Sci. U.S.A. 83, 3914−3918.

Tuffereau, C., Benejean, J., Blondel, D., Kieffer, B., Flamand, A., 1998. Low-affinity nerve-growth factor receptor (P75NTR) can serve as a receptor for rabies virus. EMBO J. 17, 7250−7259.

Van Dyke, T.A., Flanegan, J.B., 1980. Identification of poliovirus polypeptide P63 as a soluble RNA-dependent RNA polymerase. J. Virol. 35, 732−740.

White, M.K., Wollebo, H.S., David Beckham, J., Tyler, K.L., Khalili, K., 2016. Zika virus: an emergent neuropathological agent. Ann. Neurol. 80, 479−489.

Wojczyk, B.S., Stwora-Wojczyk, M., Shakin-Eshleman, S., Wunner, W.H., Spitalnik, S.L., 1998. The role of site-specific N-glycosylation in secretion of soluble forms of rabies virus glycoprotein. Glycobiology 8, 121−130.

Yang, J., Hooper, D.C., Wunner, W.H., Koprowski, H., Dietzschold, B., Fu, Z.F., 1998. The specificity of rabies virus RNA encapsidation by nucleoprotein. Virology 242, 107−117.

Yang, Y., Boza-Serrano, A., Dunning, C.J.R., Clausen, B.H., Lambertsen, K.L., Deierborg, T., 2018. Inflammation leads to distinct populations of extracellular vesicles from microglia. J. Neuroinflammation 15, 168.

Ypma-Wong, M.F., Dewalt, P.G., Johnson, V.H., Lamb, J.G., Semler, B.L., 1988. Protein 3CD is the major poliovirus proteinase responsible for cleavage of the P1 capsid precursor. Virology 166, 265−270.

Zhang, Y., Zhang, W., Ogata, S., Clements, D., Strauss, J.H., Baker, T.S., Kuhn, R.J., Rossmann, M.G., 2004. Conformational changes of the flavivirus E glycoprotein. Structure 12, 1607−1618.

CHAPTER 7

Neurooncogenesis in the Development of Neuroectodermal Cancers

ANJU T. R., PHD • JAYANARAYANAN S, PHD

7.1 INTRODUCTION

Cancer is considered as one of the deadliest diseases in the world. According to the World Health Organization (WHO), cancer is the second leading cause of death globally with 9.6 million deaths, or one in six deaths in 2018. The severity of cancer lies in its capability to affect people of all age groups and almost all body parts. The genetic change in a single cell, when goes uncontrolled, resulted in tumor and can then invade other body parts by metastasis, thereby making it fatal. The risk factors for adult cancer are mainly linked to the altered lifestyles (Irigaray et al., 2007). Broadly, cancer is classified as carcinomas (which affect skin or epidermal tissue lining the internal organs and glands), sarcoma (which affect connective tissues such as bone, muscle, blood vessels, and cartilage), leukemia (affect bone marrow and blood cells), and lymphoma (affect lymphatic system or immune system), of which carcinomas are the major solid tumors which account for majority of all cancer cases.

Cancer is a leading cause of death for children and adolescents around the world. Leukemia, lymphoma, and various solid tumors such as ectodermal tumors, neuroblastoma (NB), and nephroblastoma were the most common childhood cancers reported (Steliarova-Foucher et al., 2017). After leukemia, central nervous system (CNS) cancers were the most prevalent childhood cancer (Ferlay et al.,2010 Siegel et al.,2012); but the etiology of most of these is not well known. Maternal exposure to carcinogen during conception and exposure of children to physical carcinogens such as radiation or biological carcinogenic agents such as Epstein−Barr virus infection (Mawson and Majumdar, 2017) were considered as few risk factors for childhood cancer.

Among the childhood and adolescent cancers, the incidence of a rare carcinoma of neural crest cells called primitive neuroectodermal tumors (PNET), which can affect both CNS and peripheral nervous system (PNS), has been increasing in the recent years (Berthold et al., 2017) and hence needs a better understanding.

7.2 PRIMITIVE NEUROECTODERMAL TUMORS

PNETs are rare malignant tumors, first described by Hart and Earle (1973). These tumors are mostly found in children and young adults and rarely seen in adults (Tong et al., 2015). The WHO classified primitive neuroectodermal tumors as poorly or undifferentiated embryonic tumors of neuroepithelial origin, which can differentiate into various cell lines such as nerve cells, glial cells, ependymal cells, and muscle cells (Patnaik et al., 2012). As the name suggests, primitive neuroectoderm is the origin site of PNETs and may occur both within and outside of the CNS. The neuroectoderm is the region that gives rise to the entire nervous system such as brain and spinal cord (CNS), autonomic nervous system (ANS), dorsal root ganglia, adrenal medulla, neuroendocrine system, and so on during embryonic development (LeDouarin, 1982). Accordingly, in 1996, PNET family of tumors is divided on the basis of the site of origin as: (1) peripheral PNET (pPNET), (2) CNS PNET, and (3) NB. This chapter will identify various types of primitive ectodermal cancers and will discuss the oncogenesis and prognosis of ectodermal cancers.

7.3 CENTRAL NERVOUS SYSTEM PRIMITIVE NEUROECTODERMAL TUMORS

CNS PNETs are rare and aggressive small round cell carcinomas of the brain mostly affecting the childhood population. In the recent classification, the WHO rearranged those CNS embryonic tumors, which was under PNET with typical Homer—Wright rosettes (Votta et al., 2005), as "embryonal tumors with multilayered rosettes (ETMR)" and other ectodermal tumors are classified into four different CNS tumors.

In 1993, the WHO included primitive neuroectodermal tumor under embryonal CNS tumors along with NB/ganglioneuroblastoma (GNB), ependymoblastoma, medulloblastoma, and medulloepithelioma. In 2007, all CNS embryonal tumors were classified under medulloblastoma, atypical teratoid/rhabdoid tumor, and CNS PNET. The supratentorial PNET was renamed as CNS PNET, and NB, GNB, medulloepithelioma, and ependymoblastoma were considered as histologic subtypes/variants of CNS PNET (Louis et al., 2007). In the recent 2016 classification, changes were incorporated in CNS embryonic tumors such as (1) classifying medulloblastoma into four subtypes based on genetics, and not otherwise specified (NOS) group, (2) replacing CNS PNET as CNS embryonal tumor, NOS, (3) removing ependymoblastoma from CNS PNET and replaced as embryonal tumor with multilayered rosettes (ETMR) (Louis et al., 2016). Currently, we can consider CNS NB, CNS GNB, medulloepithelioma, and CNS embryonal tumor, NOS as different neuroectodermal cancers based on the recent classification (Fig. 7. 1).

7.4 ONCOGENESIS OF CENTRAL NERVOUS SYSTEM PRIMITIVE NEUROECTODERMAL TUMOR

7.4.1 Central Nervous System Neuroblastoma and Ganglioneuroblastoma

NB is the most frequent solid tumor found outside the cranium and accounts for 8%—10% of all pediatric cancers (Park et al., 2008). Even though it can be found in adults, it most often occurs within the first 5 years of life. As these tumors are caused from primordial neural crest cells, they can start from anywhere in the sympathetic nervous system with numerous possibilities of tumor locations and their clinical presentations (Maris et al., 2007). The WHO included all the CNS NB into grade IV tumor category (Louis et al., 2007). NBs in the CNS are rarely reported, and recurrent metastatic NB in CNS is difficult to cure (Matthay et al., 2016).

The primary NB of CNS is considered to be derived during the second stage of proliferation of neuronal cytogenesis, and later on, the primitive neuroepithelial cells will differentiate to neuroblast (Rubinstein, 1985). Primary CNS NB can affect a number of locations with varied lesion size and mass, which accounts for its wide symptoms. Adaptability exhibited by brain regions surrounding the tumor site in infants resulted in asymptomatic conditions even when the tumor size is

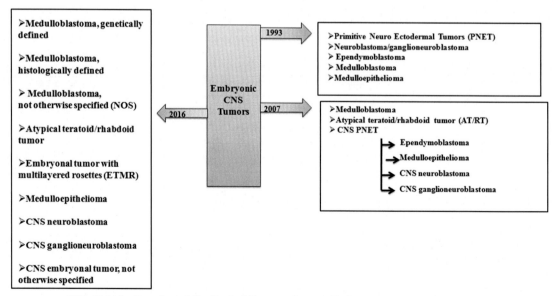

FIG. 7.1 **An Overview of the Central Nervous System Embryonic Tumor Classification.**

high. Another reason for CNS RB to remain subtle is the low metastasis rate (Sillibourne and Bornens, 2010) with only two possible routes identified for metastasis, the cerebrospinal fluid (CSF) pathway, and cervical lymph nodes. CNS NBs are neuroepithelial/neurocytic tumors, which exhibit high cellularity with monomorphic cells having less abundant cytoplasm and irregular or enlarged round nucleus (Dehner et al., 1988). Histologically, these cells are vimentin, GFAP, EMA and PAS negative and MAP2, synaptophysin, Olig2, and Neu-N positive (Bianchi et al., 2018).

CNS NB oncogenesis is a complex process with many transcription factors, oncogenes, and microRNAs involved in it. Anaplastic lymphoma kinase (ALK), v-myc myelocytomatosis viral-related oncogene neuroblastoma-derived (MYCN), and alpha-thalassemia/mental retardation syndrome X-linked (ATRX) are the major genes involved in the tumorigenesis of NB. The paired box (PAX/Pax) transcription factor, Pax3, is one of the central transcription factors implicated in the oncogenesis of CNS NB (Wang et al., 2008). Pax3 is widely expressed in developing nervous system during early neurogenesis, and it regulates the generation of sensory neurons from the neural crest precursor cells (Koblar et al., 1999). Pax3 is expressed in all cells in neural crest during early development and later got restricted to neurons. It is reported that Pax3 expression is abnormally elevated in some NB cell lines and tumors (Koblar et al., 1999).

Pax3 is regulated by the protooncogenes, n-myc, and c-myc at the inverted E box sequence CGCGTG (or CACGCG) of its 5′ promoter region. Thus, expression of n-myc and c-myc results in an abnormal increase in Pax3 expression (Harris et al., 2002). More detailed investigations are needed to identify whether the NB pathogenesis is initiated by the expression of Pax3 or n-myc. Amplification of protooncogene MYCN is a typical feature in both peripheral and CNS NB. This oncogene encodes a pleiotropic nuclear phosphoprotein in the MYC family of helix—loop—helix transcription factors (Schwab et al., 1984). Many important functions such as cell cycle, genomic stability, cellular proliferation/differentiation, cell death, and tumorogenicity are regulated by Myc family of proteins (García-Gutiérrez et al., 2019). Under normal conditions, the expression of MYCN is restricted to embryonic period and B lymphocytes development in adults (Strieder and Lutz, 2003). Amplification of the MYCN gene is observed in 20% of all primary NB tumors. It is present in 50% of all high-risk tumors and is associated with aggressive disease and poor prognosis (Brodeur et al., 1984). Targeted expression of MYCN to the neural crest in transgenic mice causes aggressive NB, and tumorigenesis is positively correlated with MYCN transgene dosage or with the development of additional genetic mutations (Hansford et al., 2004).

Many other reports propose that MYCN is not the only reason of NB progression and aggressiveness; the major cause of familial NB is heritable gain-of-function mutations of ALK. ALK mutations are observed in 10%—12% of cases of sporadic NB (Chen et al., 2008; Mosse et al., 2008). PHOX2B is another important gene mutated in familial NB, and the mutation adversely affects the nerve cell differentiation. ALK gene mutations are more common than PHOX2B gene mutations in familial NB. ALK is a member of the insulin receptor superfamily of tyrosine kinase receptors (Morris et al., 1994). Inactivation of ATRX mutations is observed in a small subgroup of NBs, mainly in adult NB. ATRX is a tumor suppressor gene, and loss-of-function mutations in this gene can trigger the tumor formation by activating ALK pathway (Cheung et al., 2012).

The Polo-like kinase 4 (PLK4) is a member of the polo-like family of serine/threonine protein actively involved in the oncogenesis of CNS NB (Bailey et al., 2018). PLK4 is overexpressed in CNS NB in both primary and metastatic stages. PLK4 is actively involved in cell cycle regulation and mainly localized in the centrosome during cell division. It plays a key role in centriole duplication. Overexpression of PLK4 in the CNS NB indicates centrosome amplification, which is a common factor for genetic instability and spontaneous tumorigenesis (Levine et al., 2017; Ko et al., 2005). A multicenter study on a large sample size found that many CNS NB samples overexpress FOXR2 with the expression of synaptic markers. These samples also showed a gain or loss of function at 1q and 16q, respectively, in chromosome 8. This NB was designated as "CNS NB-FOXR2" (Picard et al. 2012).

CNS GNBs are exceedingly rare tumors composed of both primitive neuroblasts and more mature neoplastic ganglion cells (Gauchan et al., 2017). It mainly occurs in children of ages 2—4 years, and the tumor affects both genders equally. Because GNBs contain two cell types, primitive neuroblasts, and more benign neoplastic ganglion cells, they have the potential to display a highly malignant phenotype.

GNBs may contain differentiated and undifferentiated cells such as ganglion and neuroblasts (Carlos et al., 2008). GNB represents a histological subgroup of neuroblastic tumors with intermediate malignancy potential that is grouped as nodular (GNBn) and intermixed (GNBi) (Okamatsu et al., 2009). GNBi is shown to have a better survival rates than patients with GNBn

(Cohn et al., 2009). Usually, these tumors are less aggressive than NB. Tumorigenesis of the GNB is unknown due to its rarity.

Major treatment method for the CNS NB and CNS GNBs is surgical resection (Gaffney et al., 1985; Bennett and Rubinstein, 1984). In the early stages, therapeutic intervention can be done by the combination of chemo- and radiotherapy, and positive clinical responses to chemotherapy were noticed (Bennett et al., 1984). Main transcription factors and oncogenes involved in the ontogenesis of CNS NB can be used as a useful biomarker/a therapeutic target (Fig. 7.2).

7.4.2 Medulloepitheliomas

Medulloepitheliomas are rare tumors that have embryonic origin and are histologically similar to the neural tube. The original classification of these tumors was done by Bailey and Cushing in the mid-1920s. These tumors, which show cell lines of mesenchymal, neural,

and glial origin, have high proliferative rate and exhibit differ rosettes such as ependymoblastomatous and ependymal. CNS and the structures directly attached to it like optic nerve and neuroepithelial structures of eye are the major sites of these tumors. These tumors are predominant in the early life between 2 and 10 years affecting both sexes (Molloy et al., 1996), but adult cases were also documented (Litricin and Latkovic, 1985).

Intraocular medulloepithelioma is a rare form of this tumor, which originates from the optic cup medullary epithelium (Shields and Shields, 2008). In rare cases, optic nerve or even retina can be its source of origin (Canning et al., 1988). Optic nerve medulloepitheliomas can be observed at the optic disk or the extraocular regions of neurons (Jakobiec et al., 2015). Ocular medulloepithelioma can affect an entire life span of a patient, with nearly 15%–20% prevalence in adults.

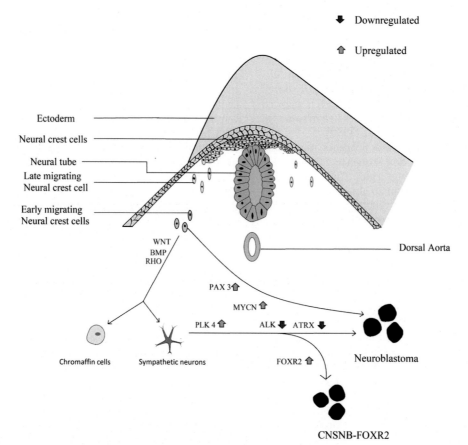

FIG. 7.2 **Oncogenesis of Neuroblastoma.** Increased PAX3 in migrating neural crest cell along with MYCN activation can trigger NB. Altered MYCN, PLK 4, ALK, and ATRX together trigger the tumorigenesis in NB. FroxR2 overexpression along with other factors triggers CNS NB-FOXR2.

Due to their symptomatic nature, even though these tumors are small, with a maximum diameter of 2 cm, they can be detected in the initial stages of tumor formation (Jakobiec et al., 2015). The common sites for CNS medulloepithelioma are temporal, parietal, occipital and frontal lobes, and posterior fossa (Molloy et al., 1996). Medulloepitheliomas located in intraocular regions can be further categorized into "teratoid or nonteratoid" depending on the presence or absence of heteroplastic elements. In the CNS, these tumors did not exhibit such morphological variations (Korshunov et al., 2015).

The first case reported by Pfister et al. has multilayered rosettes and amplification at 19q13.42 with increased microRNAs cluster as distinguishable features of the embryonic tumors. An interesting study by Li et al. (2009) observed same pattern of amplification of 19q13.41 in 25% of CNS-PNET cases analyzed. This chromosomal amplification in the tumors is characterized by high expression of chromosome 19 microRNA cluster (C19MC). The oncogenicity of some of the miRNAs is attributed partially to the enhanced WNT pathway signaling, resulting in the regulation of cell survival pathways and increased growth of untransformed neural stem cells without differentiation (Pfister et al., 2009). One of the key biological factors in medulloepitheliomas is the overexpression of LIN28A, a known marker for malignancy (Jakobiec et al., 2015).

The C19MC molecular signature was not observed in ocular medulloepitheliomas. LIN28A, the RNA-binding protein, when expressed in ocular medulloepitheliomas shows no C19MC amplification (Perry and Brat, 2010). Aggressive medulloepitheliomas showed high intensity LIN28A staining (Grossniklaus et al., 2018). Even though not entirely specific, LIN28A is thought to be a malignancy marker for embryonic tumors of eyes and CNS (Perry and Brat, 2010) because of the high expression of LIN28A protein in certain regions of these tumors (Papagiannakopoulos et al., 2008).

LIN-28 family RNA-binding protein is expressed during embryonic development for temporal regulation of embryonic stem cell development and self-renewal via direct mRNA regulation or inhibiting certain miRNAs. It can regulate genes posttranscriptionally by its interaction with target mRNA or regulation of certain miRNAs maturation. miRNA expression pattern can help in identifying tumor types and predicting tumor biology (Chan et al., 2011). A group of miRNAs named oncomirs have been implicated as regulators of the gene expression involved in several human malignancies (Esquela-Kerscher and Slack, 2006). Oncomirs can indirectly act as oncogenes or tumor suppressors by altering their expression pattern, thereby regulating various oncogenic pathways

(Farzaneh et al., 2017). Lethal-7 (Let-7), a member of a highly conserved homonymous miRNA family, is an example for oncomir miRNA family (Farzaneh et al., 2017). The let-7 micro-RNA family, a well-known tumor suppressor that downregulates the Ras oncogenes involved in cellular growth and differentiation, is found to be mutated in about 15%–30% of human tumors. The RNA-binding protein LIN28A can prevent the terminal processing of the LET7 micro-RNA family.

Intraocular medulloepithelioma is not usually found in patients with germline mutations in DICER 1 (Tadepalli et al., 2019), but it is frequently observed in cases of somatic mutations in DICER1 and KMT2D (Johnson et al., 2005). Dicer is a ribonuclease RNase III-like enzyme that processes long double-stranded RNA or pre-microRNA hairpin precursors into small interfering RNAs or microRNA (MacRae and Doudna, 2007). Germline DICER syndrome was not linked to medulloepitheliomas of the optic nerve (Johnson et al., 2005). After DICER 1, the next commonly mutated gene found in intraocular medulloepithelioma is the histone methyltransferase encoding gene, KMT2D, which exhibits interspersed mutations throughout the genome of the affected children (Johnson et al., 2005). The role of systemic chemotherapy is not well established for medulloepithelioma. However, there have been reports of metastatic medulloepithelioma effectively treated with chemotherapy in combination with radiotherapy and surgery (Hellman et al., 2018).

7.4.3 Central Nervous System Embryonal Tumor, Not Otherwise Specified

When diagnostic criteria specified by the WHO cannot be attained to define a CNS embryonic tumor, it is classified under "not otherwise specified or NOS."

7.5 PERIPHERAL PRIMITIVE NEUROECTODERMAL TUMORS

Based on the histological presentation and level of differentiation, the WHO considers pPNET under "Ewing's sarcoma family of tumors (ESFT)" along with classic Ewing's sarcoma and extraskeletal Ewing's sarcoma (Ambros et al., 1991). Despite the level of neuroectodermal differentiation, from the least differentiated Ewing's sarcoma to the most differentiated pPNET (Fletcher, 1995), the "Ewing's sarcoma family of tumors" are defined by explicit chromosomal abnormalities, mainly the t(11,22) (q24,12) translocation (Ambros et al., 1991). Hence, the exact diagnosis of PNETs relies on histopathology and immunohistochemistry studies (Ellis et al., 2011).

7.5.1 Oncogenesis of Peripheral Primitive neuroectodermal tumors

ESFT with characteristic small, blue, round malignant tumor cells originates from the neural crest progenitor cells of bone and soft tissue, and its occurrence is reported at both osseous and extraosseous sites (Choudhury et al., 2011). The major sites of ESFTs are diaphysis, pelvis, ribs, and vertebrae with rare chance of occurrence in the skull (Kumar et al., 2017). ESFT usually affects children and young adults of the age group 10–20. Most cases of ESFT are related to bone and are found to be one of the most common bone cancers in children (Bernstein et al., 2006), even though in few cases only soft tissues are affected without bone involvement (Arvand and Denny, 2001).

7.5.2 Genetic Changes in Ewing's sarcoma family of tumors

Chromosomal rearrangements leading to the fusion of EWS gene located on chromosome 22 with the genes of E26 transformation-specific (ETS) family of transcription factors located at chromosome 11, 21, 7, 2, and/or 17 are a characteristic feature observed in ESFTs. Among these, the translocation t(11; 22) (q24; q12), resulting in the fusion of EWS gene with FLI1 gene on chromosome 11, is the most commonly observed chromosomal rearrangement in ESFT. This translocation resulted in the synthesis of a chimeric transcript during gene expression. The proteins produced from these chimeric transcripts have the N-terminus of EWS fused to the C-terminal portion of FLI1 (Delattre et al., 1994). Other ESFT translocations at t(21; 22) (q22; q12); t(7; 22) (p22; q12); t(2; 21; 22) (q33; q22; q12); and t(17; 22) (q12; q12) resulted in the fusion of EWS to four other ETS family of transcription factors (Sorensen et al., 1995; Jeon et al., 1995; Peter et al., 1997; Urano et al., 1996). Thus, five different EWS/ETS fusion proteins — EWS/FLI1, EWS/ERG, EWS/ETV1, EWS/FEV, and EWS/E1AF — resulted in ESFTs.

In case of EWS/FLI1 translocation, the point of breakage is tightly clustered within a region of 8 kilobase pairs (Kb) in EWS locus, whereas it is loosely dispersed around 35 kb in the FLI1 locus (Delattre et al., 1992). This accounts for the presence of different EWS/FLI1 fusion proteins in ESFT patients due to the incorporation of various combinations of EWS and FLI1 exons in the fusion genes, which are expressed in ESFTs (Zoubek et al., 1996). These variations in the fusion proteins play a commendable role in the prognosis of the disease (Zoubek et al., 1996).

7.5.3 Impact of EWS/ETS Fusion on Oncogenesis of Ewing's Sarcoma Family of Tumors

The proteins encoded by EWS/ETS fusion genes, which are typical to ESFT pathogenesis, are found to be abnormal transcription factors, which act as dominant oncoproteins. In vitro (Teitell et al., 1999) and in vivo (Thompson et al., 1999) studies reported the role of EWS/ETS fusion genes especially EWS/FLI1 in the genesis and maintenance of ESFTs.

7.5.4 Normal Functions of EWS and ETS

EWS (Ewing sarcoma breakpoint region 1) gene encodes an RNA-binding protein EWS (or EWSR1) that is mainly involved in RNA processing and transcriptional regulation. The N-terminal of EWS protein is a transcriptional activation domain, and a C-terminal is an RNA-binding domain. EWS protein is grouped under the Ten-eleven translocation (TET) family proteins based on their distinct RNA-binding domain (RBD) of 87 amino acids (Bertolotti et al., 1996), a transcriptional-activation domain (EAD), 3 RGG (arginine–glycine–glycine)-rich regions, and a zinc finger domain (Svetoni et al., 2016).

The EWS proteins that are mainly confined to the nucleus can complex with transcription factor II D (TFIID) and RNA polymerase II (Bertolotti et al., 1996, 1998), thereby influencing transcription and alternative splicing. When subjected to different environmental signals, these proteins can translocate to cytoplasm or undergo posttranslational modifications in RBD and RGG domains to modulate their activity (Svetoni et al., 2016). Evidences suggest the critical role of EWS in cell signaling pathways via its interaction with calmodulin (Deloulme et al., 1997) and tyrosine kinases (Powers et al., 1998), thus conferring an important role for EWS proteins in determining genomic stability.

E26 transformation-specific/erythroblast transformation-specific (ETS) family of transcription factors encoded by 28 genes in humans (Sharrocks, 2001) has an 85 amino acid evolutionary conserved DNA-binding domain (DBD), flaked by protein–protein interaction domains, which regulates transcription (Degnan et al., 1993). The Ets proteins are found to play an important role in the cellular differentiation and responses to cell signaling pathways in a tissue-specific manner (Lelièvre et al., 2001) (Fig. 7.3).

7.5.5 Oncogenic Potential of EWS and ETS

The predominant ETS protein involved in the EWS–ETS fusion is FLI1, which is a nuclear protein restricted to hematopoietic cells in adults. Regions such as spleen and thymus show high expression of FLI1 where lymphocytes and precursors of megakaryocyte and erythroid cells are highly expressed (Maroulakou and Bowe, 2000). Even though FLI1 plays a significant role in normal hematopoietic development, its expression is reported in the oncogenesis of viral-induced leukemia and lymphoma (Blair and Athanasiou, 2000). A study on erythroleukemia cell line linked the oncogenic potential of FLI1 to its downregulation of tumor suppressor retinoblastoma (RB) gene (Tamir et al., 1999). Thus, the same transcription factor, FLI1, is linked to both normal and malignant hematopoietic development.

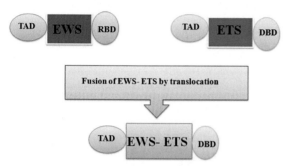

FIG. 7.3 **The Molecular Level Changes During Oncogenenesis of Ewing's Sarcoma Family Tumor.** Fusion of EWS-ETS loci results in encoding a fusion protein with N- terminal transcription activation domain (TAD) of EWS and C-terminal DNA-binding domain (DBD) of ETS. TAD, transcription activation domain; DBD, DNA-binding domain; RB, RNA-binding domain.

The oncogenesis in ESFTs is mainly determined by the EWS/ETS fusion proteins, which regulate transcription by acting as aberrant transcriptional factors. All EWS/ETS fusions exhibit an intact ETS DNA-binding domain (ETS DBD) with an N-terminus EWS transcriptional activation domains. The specificity of the proteins or transcription factors encoded by EWS/ETS fusions is dictated by the specificity of protein–DNA interaction by ETS DBD and protein–protein interaction by EWS transcriptional activation domains (Arvand and Denny, 2001). Thus, the chimeric fusion can transcriptionally regulate a range of target genes, which differ from the normal ETS gene expression in its level of expression or even the target gene expressed.

In addition to the transcriptional regulation through aberrant transcriptional factors, EWS/ETS fusions can promote oncogenesis in ESFT by regulating RNA processing during transcription. EWS binds to TASR proteins, a family of putative RNA splicing factors to regulate splicing and RNA processing. As EWS/ETS fusions resulted in the loss of ETS protein-binding domain, this causes an inference to normal RNA processing (Yang et al., 2000), which also contribute to their oncogenic potential (Fig. 7.4).

7.5.6 Other Pathways in Oncogenesis of Ewing's Sarcoma Family of Tumors

Polycomb group family proteins are the key proteins involved in cellular pluripotency by maintaining the self-renewal of stem cells (Bracken et al., 2006). Among polycomb proteins, BMI-1 has an important role in

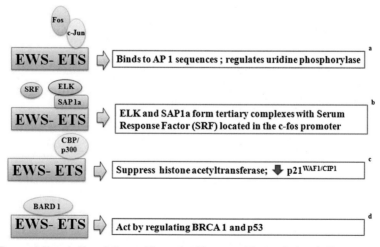

FIG. 7.4 **Transcriptional Regulations Through Aberrant Transcriptional Factors of EWS/ETS Fusions.** The oncogenesis of ESFT is attributed to the interactions of aberrant proteins produced by the intrinsically disordered regions of EWS–ETS fusions. [a] Kim et al. (2006), [b] Watson et al. (1997), [c] Nakatani et al. (2003), [d] Spahn et al. (2002).

maintaining the stemness of hematopoietic and neural crest stem cells by transcriptional repression of cyclin-dependent kinase inhibitor 2A (*CDKN2A*) to prevent cellular senescence (Jacobs et al., 1999). *Cdkn2a* encodes p16 (INK4A) and the p14 (ARF) proteins to regulate cell cycle and apoptosis by modulation of tumor suppressor genes—RB and p53. Deletion of p16 locus is reported as one of the genetic alterations observed in ESFT (Kovar et al., 1997). In ESFT, overexpression of BMI-1 promotes tumorogenicity by *CDKN2A* repression or even independent of p16 locus or *CDKN2A* repression by modulating adhesion pathways (Douglas et al., 2008).

The involvement of NF-κB signaling pathways (Javelaud et al., 2000), FAS-mediated apoptosis (Mitsiades et al., 1999), and interferon-β-induced apoptosis (Sancéau et al., 2000) are other signaling pathways involved in the oncogenesis of ESFT.

7.6 CONCLUSION

Neuroectodermal tumor, a common pediatric tumor of either central or peripheral nervous system, manifests itself with a wide range of histopathologies. Since pediatric tumors are not as extensively studied as adult tumors, there is a lot gap in our understanding of the exact tumorogenesis of PNET. Those embryonic tumors, which were once classified as primitive neuroectodermal tumors, were now reclassified under the broad umbrella of embryonic tumors with peripheral PNET classified under ESFTs and CNS PNET broadly divided into NB, medulloepithelioma, and CNS embryonal tumor, NOS. The oncogenesis of all PNET was found to be regulated by transcription factors, oncogenes, or microRNAs. Our current understanding of the oncogenic pathways regulating these embryonic tumors is too little to delineate the interdependence of different PNETs and the complex mechanisms involved in its tumorogenesis. The oncogenesis of peripheral PNET is mainly linked to the chromosomal translocation and resultant fusion protein expressions, but each type of embryonic tumor in CNS shows its own oncogenic pathways. More studies need to be carried out to understand the tumorogenesis and tumor biology and to devise more effective treatment strategies.

REFERENCES

Ambros, I.M., Ambros, P.F., Strehl, S., Kovar, H., Gadner, H., Salzer-Kuntschik, M., 1991. MIC2 is a specific marker for Ewing's sarcoma and peripheral primitive neuroectodermal tumors. Evidence for a common histogenesis of Ewing's sarcoma and peripheral primitive neuroectodermal tumors from MIC2 expression and specific chromosome aberration. Cancer 67, 1886–1893.

Arvand, A., Denny, C., 2001. Biology of EWS/ETS fusions in Ewing's family tumors. Oncogene 20, 5747–5754.

Bailey, P., Cushing, H.A., 1927. A classification of the tumours of the glioma group on a histogenetic basis with a correlated study of prognosis. Br. J. Surg. 14 (55), 554–555.

Bailey, A.W., Suri, A., Chou, P.M., Pundy, T., Gadd, S., Raimondi, S.L., et al., 2018. Polo-like kinase 4 (PLK4) is overexpressed in central nervous system neuroblastoma (CNS-NB). Bioengineering 5 (4), 96.

Bennett Jr., J.P., Rubinstein, L.J., 1984. The biological behavior of primary cerebral neuroblastoma: a reappraisal of the clinical course in a series of 70 cases. Ann. Neurol. 16 (1), 21–27.

Bernstein, M., Kovar, H., Paulussen, M., Randall, R.L., Schuck, A., et al., 2006. Ewing's sarcoma family of tumors: current management. Oncologist 11, 503–519.

Berthold, F., Spix, C., Kaatsch, P., Lampert, F., 2017. Incidence, survival, and treatment of localized and metastatic neuroblastoma in Germany 1979–2015. Paediatr. Drugs 19 (6), 577–593.

Bertolotti, A., Lutz, Y., Heard, D.J., Chambon, P., Tora, L., 1996. hTAF(II)68, a novel RNA/ssDNA-binding protein with homology to the pro-oncoproteins TLS/FUS and EWS is associated with both TFIID and RNA polymerase II. EMBO J. 15, 5022–5031.

Bertolotti, A., Melot, T., Acker, J., Vigneron, M., Delattre, O., Tora, L., 1998. EWS, but not EWS-FLI-1, is associated with both TFIID and RNA polymerase II: interactions between two members of the TET family, EWS and hTAF$_{II}$68, and subunits of TFIID and RNA polymerase II complexes. Mol. Cell. Biol. 18, 1489–1497.

Bianchi, F., Tamburrini, G., Gessi, M., Frassanito, P., Massimi, L., Caldarelli, M., 2018. Central nervous system (CNS) neuroblastoma. A case-based update. Child's Nerv. Syst. 34 (5), 817–823.

Blair, D.G., Athanasiou, M., 2000. Ets and retroviruses—transduction and activation of members of the Ets oncogene family in viral oncogenesis. Oncogene 19, 6472–6481.

Bracken, A.P., Dietrich, N., Pasini, D., Hansen, K.H., Helin, K., 2006. Genome-wide mapping of Polycomb target genes unravels their roles in cell fate transitions. Genes Dev. 20 (9), 1123–1136.

Brodeur, G.M., Seeger, R.C., Schwab, M., Varmus, H.E., Bishop, J.M., 1984. Amplification of N-myc in untreated human neuroblastomas correlates with advanced disease stage. Science 224 (4653), 1121–1124.

Canning, C.R., McCartney, A.C., Hungerford, J., 1988. Medulloepithelioma (diktyoma). Br. J. Ophthalmol. 72 (10), 764–767.

Carlos, P.A., Edward, C.H., Luther, W.B., 2008. Pediatric Tumors-Neuroblastoma. Principles and Practice of Radiation Oncology, fifth ed. Lippincott Williams & Wilkins, Baltimore, p. 1860.

Chan, E., Prado, D.E., Weidhaas, J.B., 2011. Cancer microRNAs: from subtype profiling to predictors of response to therapy. Trends Mol. Med. 17 (5), 235–243.

Chen, Y., Takita, J., Choi, Y.L., Kato, M., Ohira, M., Sanada, M., et al., 2008. Oncogenic mutations of ALK kinase in neuroblastoma. Nature 455 (7215), 971–974.

Cheung, N.K., Zhang, J., Lu, C., Parker, M., Bahrami, A., Tickoo, S.K., et al., 2012. Association of age at diagnosis and genetic mutations in patients with neuroblastoma. J. Am. Med. Assoc. 307 (10), 1062–1071.

Choudhury, K.B., Sharma, S., Kothari, R., Majumder, A., 2011. Primary extraosseous intracranial Ewing's sarcoma: case report and literature review. Indian J. Med. Paediatr. Oncol. 32, 118–121.

Cohn, S.L., Pearson, A.D., London, W.B., Monclair, T., Ambros, P.F., Brodeur, G.M., et al., 2009. The International Neuroblastoma Risk Group (INRG) classification system: an INRG task force report. J. Clin. Oncol. 27 (2), 289–297.

Degnan, B.M., Degnan, S.M., Naganuma, T., Morse, D.E., 1993. The ets multigene family is conserved throughout the Metazoa. Nucleic Acids Res. 21 (15), 3479–3484.

Dehner, L.P., Abenoza, P., Sibley, R.K., 1988. Primary cerebral neuroectodermal tumors: neuroblastoma, differentiated neuroblastoma, and composite neuroectodermal tumor. Ultrastruct. Pathol. 12 (5), 479–494.

Delattre, O., Zucman, J., Melot, T., Garau, X.S., Zucker, J.M., Lenoir, G.M., et al., 1994. The ewing family of tumors – a subgroup of small-round-cell tumors defined by specific chimeric transcripts. N. Engl. J. Med. 331, 294–299.

Delattre, O., Zucman, J., Plougastel, B., Desmaze, C., Melot, T., Peter, M., et al., 1992. Gene fusion with an ETS DNA-binding domain caused by chromosome translocation in human tumours. Nature 359, 162–165.

Deloulme, J.C., Prichard, L., Delattre, O., Storm, D.R., 1997. The prooncoprotein EWS binds calmodulin and is phosphorylated by protein kinase C through an IQ domain. J. Biol. Chem. 272, 27369–27377.

Douglas, D., Hsu, J.H., Hung, L., Cooper, A., Abdueva, D., van Doorninck, J., et al., 2008. BMI-1 promotes Ewing sarcoma tumorigenicity independent of CDKN2A-repression. Cancer Res. 68 (16), 6507–6515.

Ellis, J.A., Rothrock, R.J., Moise, G., McCormick, P.C.II., Tanji, K., Canoll, P., Kaiser, M.G., McCormick, P.C., 2011. Primitive neuroectodermal tumors of the spine: a comprehensive review with illustrative clinical cases. Neurosurg Focus 30, E1.

Esquela-Kerscher, A., Slack, F.J., 2006. Oncomirs – microRNAs with a role in cancer. Nat. Rev. Cancer 6 (4), 259–269.

Farzaneh, M., Attari, F., Khoshnam, S.E., 2017. Concise review: LIN28/let-7 signaling, a critical double-negative feedback loop during pluripotency, reprogramming, and tumorigenicity. Cell. Reprogr. 19 (5), 289–293.

Ferlay, J., Shin, H.R., Bray, F., Forman, D., Mathers, C., Parkin, D.M., 2010. Estimates of worldwide burden of cancer in 2008: GLOBOCAN 2008. Int. J. Cancer 127 (12), 2893–2917.

Fletcher, C., 1995. Peripheral neural tumors: Diagnostic histopathology of tumors. Churchil Livingstone, Edinburgh, pp. 1239–1241.

Gaffney, C.C., Sloane, J.P., Bradley, N.J., Bloom, H.J., 1985. Primitive neuroectodermal tumours of the cerebrum. Pathology and treatment. J. Neurooncol. 3 (1), 23–33.

García-Gutiérrez, L., Delgado, M.D., León, J., 2019. MYC oncogene contributions to release of cell cycle brakes. Genes (Basel) 10 (3), 244.

Gauchan, E., Sharma, P., Ghartimagar, D., Ghosh, A., 2017. Ganglioneuroblastoma in a newborn with multiple metastases: a case report. J. Med. Case Rep. 11, 239.

Grossniklaus, H.E., Eberhart, C.G., Kivela, T.T., 2018. WHO Classification of Tumours of the Eye, fourth ed.

Hansford, L.M., Thomas, W.D., Keating, J.M., Burkhart, C.A., Peaston, A.E., Norris, M.D., et al., 2004. Mechanisms of embryonal tumor initiation: distinct roles for MycN expression and MYCN amplification. Proc. Natl. Acad. Sci. USA 101 (34), 12664–12669.

Harris, R.G., White, E., Phillips, E.S., Lillycrop, K.A., 2002. The expression of the developmentally regulated proto-oncogene Pax-3 is modulated by N-Myc. J. Biol. Chem. 277 (38), 34815–34825.

Hart, M.N., Earle, K.M., 1973. Primitive neuroectodermal tumors of the brain in children. Cancer 32 (4), 890–897.

Hellman, J.B., Harocopos, G.J., Lin, L.K., 2018. Successful treatment of metastatic congenital intraocular medulloepithelioma with neoadjuvant chemotherapy, enucleation and superficial parotidectomy. Am. J. Ophthalmol. Case Rep. 11, 124–127.

Irigaray, P., Newby, J.A., Clapp, R., Hardell, L., Howard, V., Montagnier, L., etal, 2007. Lifestyle-related factors and environmental agents causing cancer: an overview. Biomed. Pharmacother. 61 (10), 640–658.

Jacobs, J.J., Kieboom, K., Marino, S., DePinho, R.A., van Lohuizen, M., 1999. The oncogene and Polycomb-group gene bmi-1 regulates cell proliferation and senescence through the ink4a locus. Nature 397 (6715), 164–168.

Jakobiec, F.A., Kool, M., Stagner, A.M., Pfister, S.M., Eagle, R.C., Proia, A.D., Korshunov, A., 2015. Intraocular medulloepitheliomas and embryonal tumors with multilayered rosettes of the brain: comparative roles of LIN28A and C19MC. Am. J. Ophthalmol. 159 (6), 1065–1074.e1.

Javelaud, D., Wietzerbin, J., Delattre, O., Besancon, F., 2000. Induction of p21$^{Waf1/Cip1}$ by TNFα requires NF-κB activity and antagonizes apoptosis in Ewing tumor cells. Oncogene 19, 61–68.

Jeon, I.S., Davis, J.N., Braun, B.S., Sublett, J.E., Roussel, M.F., Denny, C.T., Shapiro, D.N., 1995. A variant Ewing's sarcoma translocation (7; 22) fuses the EWS gene to the ETS gene ETV1. Oncogene 10, 1229–1234.

Johnson, S.M., Grosshans, H., Shingara, J., Byrom, M., Jarvis, R., Cheng, A., etal, 2005. RAS is regulated by the let-7 microRNA family. Cell 120 (5), 635–647.

Kim, S., Denny, C.T., Wisdom, R., 2006. Cooperative DNA binding with AP-1 proteins is required for transformation by EWS-Ets fusion proteins. Mol. Cell Biol. 26, 2467–2478.

Ko, M.A., Rosario, C.O., Hudson, J.W., Kulkarni, S., Pollett, A., Dennis, J.W., Swallow, C.J., 2005. Plk4 haploinsufficiency causes mitotic infidelity and carcinogenesis. Nat. Genet. 37 (8), 883–888.

Koblar, S.A., Murphy, M., Barrett, G.L., Underhill, A., Gros, P., Bartlett, P.F., 1999. Pax-3 regulates neurogenesis in neural crest-derived precursor cells. J. Neurosci. Res. 56 (5), 518–530.

Korshunov, A., Jakobiec, F.A., Eberhart, C.G., Hovestadt, V., Capper, D., Jones, D.T., et al., 2015. Comparative integrated molecular analysis of intraocular medulloepitheliomas and central nervous system embryonal tumors with multilayered rosettes confirms that they are distinct nosologic entities. Neuropathology 35 (6), 538−544.

Kovar, H., Jug, G., Aryee, D.N., et al., 1997. Among genes involved in the RB dependent cell cycle regulatory cascade, the p16 tumor suppressor gene is frequently lost in the Ewing family of tumors. Oncogene 15, 2225−2232.

Kumar, V., Singh, A., Sharma, V., Kumar, M., 2017. Primary intracranial dural-based Ewing sarcoma/peripheral primitive neuroectodermal tumor mimicking a meningioma: a rare tumor with review of literature. Asian J. Neurosurg. 12, 351−357.

LeDouarin, 1982. The Neural Crest. Cambridge University Press, NewYork.

Lelièvre, E., Lionneton, F., Soncin, F., Vandenbunder, B., 2001. The Ets family contains transcriptional activators and repressors involved in angiogenesis. Int. J. Biochem. Cell Biol. 33 (4), 391−407.

Levine, M.S., Bakker, B., Boeckx, B., Moyett, J., Lu, J., Vitre, B., et al., 2017. Centrosome amplification is sufficient to promote spontaneous tumorigenesis in mammals. Dev. Cell 40 (3), 313−322.e5.

Li, M., Lee, K.F., Lu, Y., Clarke, I., Shih, D., Eberhart, C., et al., 2009. Frequent amplification of a chr19q13.41 microRNA polycistron in aggressive primitive neuroectodermal brain tumors. Cancer Cell 16 (6), 533−546.

Litricin, O., Latkovic, Z., 1985. Malignant teratoid medulloepithelioma in an adult. Ophthalmologica 191 (1), 17−21.

Louis, D.N., Ohgaki, H., Wiestler, O.D., Cavenee, W.K., Burger, P.C., Jouvet, A., et al., 2007. The 2007 WHO classification of tumours of the central nervous system. Acta Neuropathol. 114 (2), 97−109.

Louis, D.N., Perry, A., Reifenberger, G., von Deimling, A., Figarella-Branger, D., Cavenee, W.K., et al., 2016. The 2016 World Health Organization classification of tumors of the central nervous system: a summary. Acta Neuropathol. 131 (6), 803−820.

Maris, J.M., Hogarty, M.D., Bagatell, R., Cohn, S.L., 2007. Neuroblastoma. Lancet 369 (9579), 2106−2120.

Maroulakou, I.G., Bowe, D.B., 2000. Expression and function of Ets transcription factors in mammalian development: a regulatory network. Oncogene 19, 6432−6442.

Matthay, K.K., Maris, J.M., Schleiermacher, G., Nakagawara, A., Mackall, C.L., Diller, L., Weiss, W.A., 2016. Neuroblastoma. Nat. Rev. Dis. Primers 2, 16078.

Mawson, A.R., Majumdar, S., 2017. Malaria, Epstein-Barr virus infection and the pathogenesis of Burkitt's lymphoma. Int. J. Cancer 141 (9), 1849−1855.

MacRae, I.J., Doudna, J.A., 2007. Ribonuclease revisited: structural insights into ribonuclease III family enzymes. Curr. Opin. Struct. Biol. 17 (1), 138−145.

Mitsiades, N., Poulaki, V., Leone, A., Tsokos, M., 1999. Fas-mediated apoptosis in Ewing's sarcoma cell lines by metalloproteinase inhibitors. J. Natl. Cancer Inst. 91, 1678−1684.

Molloy, P.T., Yachnis, A.T., Rorke, L.B., et al., 1996. Central nervous system medulloepithelioma a series of eight cases including two arising in the pons. J. Neurosurg. 84, 430−436.

Morris, S.W., Kirstein, M.N., Valentine, M.B., Dittmer, K.G., Shapiro, D.N., Saltman, D.L., Look, A.T., 1994. Fusion of a kinase gene, ALK, to a nucleolar protein gene, NPM, in non-Hodgkin's lymphoma. Science 263 (5151), 1281−1284.

Mossé, Y.P., Laudenslager, M., Longo, L., Cole, K.A., Wood, A., Attiyeh, E.F., et al., 2008. Identification of ALK as a major familial neuroblastoma predisposition gene. Nature 455 (7215), 930−935.

Nakatani, F., Tanaka, K., Sakimura, R., et al., 2003. Identification of p21WAF1/CIP1 as a direct target of EWS-Fli1 oncogenic fusion protein. J. Biol. Chem. 278, 15105−15115.

Okamatsu, C., London, W.B., Naranjo, A., Hogarty, M.D., Gastier-Foster, J.M., Look, A.T., et al., 2009. Clinicopathological characteristics of ganglioneuroma and ganglioneuroblastoma: a report from the CCG and COG. Pediatr. Blood Cancer 53 (4), 563−569.

Papagiannakopoulos, T., Shapiro, A., Kosik, K.S., 2008. MicroRNA-21 targets a network of key tumor-suppressive pathways in glioblastoma cells. Cancer Res. 68 (19), 8164−8172.

Park, J.R., Eggert, A., Caron, H., 2008. Neuroblastoma: biology, prognosis, and treatment. Pediatr. Clin. 55 (1), 97.

Patnaik, A., Mishra, S., Mishra, S., Deo, R., 2012. Primary spinal primitive neuroectodermal tumour: report of two cases mimicking neurofibroma and review of the literature. Neurol. Neurochir. Pol. 46 (5), 480−488.

Perry, A., Brat, D.J., 2010. Practical Surgical Neuropathology A Diagnostic Approach, first ed. Elseiver.

Peter, M., Couturier, J., Pacquement, H., Michon, J., Thomas, G., Magdelenat, H., Delattre, O., 1997. A new member of the ETS family fused to EWS in Ewing tumors. Oncogene 14, 1159−1164.

Pfister, S., Remke, M., Castoldi, M., Bai, A.H., Muckenthaler, M.U., Kulozik, A., et al., 2009. Novel genomic amplification targeting the microRNA cluster at 19q13.42 in a pediatric embryonal tumor with abundant neuropil and true rosettes. Acta Neuropathol. 117 (4), 457−464.

Picard, D., Miller, S., Hawkins, C.E., Bouffet, E., Rogers, H.A., Chan, T.S., et al., 2012. Markers of survival and metastatic potential in childhood CNS primitive neuro-ectodermal brain tumours: an integrative genomic analysis. Lancet Oncol. 13, 838−848.

Powers, C.A., Mathur, M., Raaka, B.M., Ron, D., Samuels, H.H., 1998. TLS (translocated-in-liposarcoma) is a high-affinity interactor for steroid, thyroid hormone, and retinoid receptors. Mol. Endocrinol. 12, 4−18.

Rubinstein, L.J., 1985. Embryonal central neuroepithelial tumors and their differentiating potential. A cytogenetic view of a complex neuro-oncological problem. J. Neurosurg. 62 (6), 795−805.

Sancéau, J., Hiscott, J., Delattre, O., Wietzerbin, J., 2000. IFN-beta induces serine phosphorylation of Stat-1 in Ewing's sarcoma cells and mediates apoptosis via induction of IRF-1 and activation of caspase-7. Oncogene 19, 3372−3383.

Schwab, M., Ellison, J., Busch, M., Rosenau, W., Varmus, H.E., Bishop, J.M., 1984. Enhanced expression of the human gene N-myc consequent to amplification of DNA may contribute to malignant progression of neuroblastoma. Proc. Natl. Acad. Sci. USA 81 (15), 4940–4944.

Sharrocks, A.D., 2001. The ETS-domain transcription factor family. Nat. Rev. Mol. Cell Biol. 2 (11), 827–837.

Shields, J.A., Shields, C.L., 2008. Congenital neoplasms (medulloepithelioma). In: Shields, J.A., Shields, C.L. (Eds.), Intraocular Tumors. An Atlas and Textbook, second ed. Lippincott Williams and Wilkins, Philadelphia PA, pp. 482–490.

Siegel, R., Naishadham, D., Jemal, A., 2012. Cancer statistics, 2012. Ca – Cancer J. Clin. 62 (1), 10–29.

Sillibourne, J.E., Bornens, M., 2010. Polo-like kinase 4: the odd one out of the family. Cell Div. 5, 25.

Sorensen, P.H., Shimada, H., Liu, X.F., Lim, J.F., Thomas, G., Triche, T.J., 1995. Biphenotypic sarcomas with myogenic and neural differentiation express the Ewing's sarcoma EWS/FLI1 fusion gene. Cancer Res. 55, 1385–1392.

Spahn, L., Petermann, R., Siligan, C., Schmid, J.A., Aryee, D.N., Kovar, H., 2002. Interaction of the EWS NH2 terminus with BARD1 links the Ewing's sarcoma gene to a common tumor suppressor pathway. Cancer Res. 62, 4583–4587.

Steliarova-Foucher, E., Colombet, M., Ries, L., Moreno, F., Dolya, A., Bray, F., et al., 2017. International incidence of childhood cancer, 2001–10: a population-based registry study. Lancet Oncol. 18 (6), 719–731.

Strieder, V., Lutz, W., 2003. E2F proteins regulate MYCN expression in neuroblastomas. J. Biol. Chem. 278, 2983–2989.

Svetoni, F., Frisone, P., Paronetto, M.P., 2016. Role of FET proteins in neurodegenerative disorders. RNA Biol. 13, 1089–1102.

Tadepalli, S.H., Shields, C.L., Shields, J.A., Honavar, S.G., 2019. Intraocular medulloepithelioma – a review of clinical features, DICER 1 mutation, and management. Indian J. Ophthalmol. 67 (6), 755–762.

Tamir, A., Howard, J., Higgins, R.R., Li, Y.J., Berger, L., Zacksenhaus, E., et al., 1999. Fli-1, an ets-related transcription factor, regulates erythropoietin-induced erythroid proliferation and differentiation: evidence for direct transcriptional repression of the Rb gene during differentiation. Mol. Cell. Biol. 19, 4452–4464.

Teitell, M.A., Thompson, A.D., Sorensen, P.H., Shimada, H., Triche, T.J., Denny, C.T., 1999. EWS/ETS fusion genes induce epithelial and neuroectodermal differentiation in NIH 3T3 fibroblasts. Lab. Invest. 79, 1535–1543.

Thompson, A.D., Teitell, M.A., Arvand, A., Denny, C.T., 1999. Divergent Ewing's sarcoma EWS/ETS fusions confer a common tumorigenic phenotype on NIH3T3 cells. Oncogene 18, 5506–5513.

Tong, X., Deng, X., Yang, T., Yang, C., Wu, L., Wu, J., et al., 2015. Clinical presentation and long-term outcome of primary spinal peripheral primitive neuroectodermal tumors. J. Neurooncol. 124 (3), 455–463.

Urano, F., Umezawa, A., Hong, W., Kikuchi, H., Hata, J., 1996. A novel chimera gene betweenEWSandE1A-F, encoding the adenovirus E1A enhancer-binding protein, in extraosseous Ewing's sarcoma. Biochem. Biophys. Res. Commun. 219, 608–612.

Votta, T.J., Fantuzzo, J.J., Boyd, B.C., 2005. Peripheral primitive neuroectodermal tumor associated with the anterior mandible: a case report and review of the literature. Oral Surg., Oral Med., Oral Pathol., Oral Radiol., Endod. 100 (5), 592–597.

Wang, Q., Fang, W.H., Krupinski, J., Kumar, S., Slevin, M., Kumar, P., 2008. Pax genes in embryogenesis and oncogenesis. J. Cell Mol. Med. 12 (6A), 2281–2294.

Watson, D.K., Robinson, L., Hodge, D.R., Kola, I., Papas, T.S., Seth, A., 1997. FLI1 and EWS-FLI1 function as ternary complex factors and ELK1 and SAP1a function as ternary and quaternary complex factors on the Egr1 promoter serum response elements. Oncogene 14, 213–221.

Yang, L., Chansky, H.A., Hickstein, D.D., 2000. EWS·Fli-1 fusion protein interacts with hyperphosphorylated RNA polymerase II and interferes with serine-arginine protein-mediated RNA splicing. J. Biol. Chem. 275, 37612–37618.

Zoubek, A., Dockhorn-Dworniczak, B., Delattre, O., Christiansen, H., Niggli, F., Gatterer-Menz, I., et al., 1996. Does expression of different EWS chimeric transcripts define clinically distinct risk groups of Ewing tumor patients? J. Clin. Oncol. 14, 1245–1251.

CHAPTER 8

Neurological Diseases and Mitochondrial Genes

AYSWARIA DEEPTI, MSC, PHD • BABY CHAKRAPANI P.S., MSC, PHD

8.1 INTRODUCTION

Mitochondria are the semi-autonomous, membrane-bound, energy-converting organelles present in almost all eukaryotic cells. The size of mitochondria varies considerably, ranging from 0.75 to 3 μm^2 in area, but unless specifically stained is not visible under light microscopes (Wiemerslage and Lee, 2016). These are one of the most dynamic organelles that perform many important roles in the cells, including the most vital functions like the supply of energy to the cell through the process of oxidative respiration. The first observation of these intracellular structures was reported in the 1840s (Ernster and Schatz, 1981). In 1890, Richard Altmann established them as cell organelles calling them "bioblasts" and Carl Benda in 1898 coined the term "Mitochondria" from the Greek words "mitos" meaning thread and "chondros" meaning grain (Ernster and Schatz, 1981; Westermann, 2012). Benjamin F. Kingsbury first suggested mitochondria's function to be involved in cellular respiration in 1912, and its role in the respiratory chain was confirmed by the discovery of cytochromes by David Keilin in 1925 (Ernster and Schatz, 1981). Philip Siekevitz in 1957 first used the popular term "powerhouse of the cell" to describe mitochondria since it is the major source of adenosine triphosphate (ATP), which is a source of chemical energy for the cell (Willis, 1992; Ernster and Schatz, 1981).

Currently, two hypotheses exist regarding the origin of mitochondria: endosymbiosis and autogenous. The endosymbiotic hypothesis states that mitochondria originated as aerobic bacteria or prokaryotic cells capable of oxidative respiration, which were subsequently engulfed by eukaryotic cells forming an endosymbiotic relationship, where the prokaryotes generated the energy for the cells efficiently through oxidative respiration. On the other hand, the autogenous hypothesis suggests that mitochondria originated by the splitting off of nuclear DNA and later got enclosed by a membrane during cellular development. Since the mitochondria show characters similar to the bacteria, the endosymbiotic hypothesis is most widely accepted (Gray, 2012; Emelyanov, 2003; McBride et al., 2006).

Mitochondria play a vital role in supplying energy to the brain, and its deterioration has detrimental consequences in the optimal functioning of the organ leading to disruption of neural plasticity and affecting most of the neuronal functions. Although the human brain represents only 2% of total body weight, it accounts for around 20%−25% of oxygen consumption of the body indicating high metabolic activities. Hence, disruption of mitochondrial dynamics is thought to have a pivotal role in the pathology of several neurological diseases and aging (De Castro et al., 2010).

Three major neurodegenerative diseases correlated with mitochondrial dysfunction are Alzheimer's disease (AD), Parkinson's disease (PD), and Huntington's disease (HD). AD is characterized by loss of cognitive abilities, memory, and changes in behavior. Pathologically, AD is characterized by the accumulation of β-amyloid proteins (amyloid plaques), and the formation of neurofibrillary tangles composed of hyperphosphorylated Tau proteins (tau tangles) (Swomley et al., 2014). PD is characterized by resting tremors, bradykinesia, rigidity, and postural instability due to the loss of pigmented dopaminergic neurons in substantia nigra pars compacta of the brain affecting the motor system (Kalia and Lang, 2015). PD is also referred as synucleinopathy due to the formation of Lewy bodies, caused by aggregation of proteins where alpha-synuclein is the major component, in the brain (Galpern and Lang, 2006; Lesage and Brice, 2009). HD or Huntington's chorea is a neurodegenerative disease characterized by a gradual decline in cognitive and physical capacities caused due to the autosomal dominant mutation of the huntingtin (HTT) gene (Dayalu and Albin, 2015).

The Molecular Immunology of Neurological Diseases. https://doi.org/10.1016/B978-0-12-821974-4.00009-1

Mutations in mitochondrial genes are implicated in the above neurodegenerative conditions, which will be discussed later in this chapter.

8.2 MITOCHONDRIAL DYNAMICS AND BIOLOGICAL CONSEQUENCES OF MITOCHONDRIAL DISFUNCTION

Mitochondria are present in the cell as either independent morphologically distinct organelles or as large interconnected networks, depending on the cell type and morphological conditions (Kuznetsov et al., 2009; Bereiter-Hahn, 1990). Mitochondria constantly undergo fission and fusion, which is crucial for its dynamics maintenance, which in turn regulates its morphology, size, and number (Chen and Chan, 2009; Zhu et al., 2018). Mitochondrial dynamics apart from functional maintenance, has a variety of functions in the cell-like mitochondrial inheritance, distribution, remodeling of mitochondria during development and coordination of cell death programs (Westermann, 2010, 2012; Ishihara et al., 2009; Chen et al., 2007; Hales and Fuller, 1997; Youle and Karbowski, 2005). In addition to the quality check of mtDNA, the fusion of mitochondria is essential for the activities of the cells as it mediates the spread of enzymes, enabling protein complementation and distribution of metabolites effectively in the mitochondrial compartment. These processes are also vital for preventing the aging of the cells (Chen and Chan, 2009). Fission on the other hand enables equal division of mitochondria to daughter cells during cell division and plays a vital role in clearing the damaged organelle by autophagy (Westermann, 2010, 2012). Dysfunction of mitochondrial dynamics leads to the accumulation of defective mitochondria resulting in cellular damage and apoptosis (Zhu et al., 2018).

The central machinery of mitochondrial dynamics, which is conserved during evolution, consists of three large GTPases that fuse and divide the mitochondrial membrane. Mitofusins (Mfn1 and Mfn2) are critical for the outer membrane fusion (Santel and Fuller, 2001) while optic dominant atrophy 1 (Opa1), which is a dynamin-related protein, of the intermembrane space (IMS) is required for fusion of inner membrane (Cipolat et al., 2004). Assembly of another dynamin-related protein Drp1 on the mitochondrial membrane is needed to aid mitochondrial fission and distribution of mitochondria (Smirnova et al., 1998) through its interaction with mitochondrial elongation factor 1 (MIEF1), an outer membrane protein (Zhao et al., 2011). MIEF1 overexpression results in excessive

mitochondrial fusion, while underexpression leads to excessive fragmentation of the mitochondria (Zhao et al., 2011).

Mitochondria move along the cytoskeletal tracks consisting of actin filaments with the help of myosin motors for a short distance and through microtubules aided by dynein or dynactin and kinesin for a longer distance (Perier and Vila, 2012). Defects in fission and fusion mechanisms are shown to hamper mitochondrial motility. Mouse knockouts of each of the three genes involved in fission and fusion of mitochondria resulted in embryonic lethality due to the dysfunction of mitochondria (Chen et al., 2003; Alavi et al., 2007; Davies et al., 2007). Mfn2 plays an important role in the maintenance of the mitochondrial structure and hence, defect in the protein affects mitochondrial movement (Chen et al., 2003). In addition, embryonic fibroblasts lacking either Mfn1 or Mfn2 display fragmented mitochondria due to a severe reduction in mitochondrial fusion (Chen et al., 2003).

For synaptic transmission, neurons require high levels of ATP and hence, optimal functioning of mitochondria is necessary for the continuous supply of energy. Hence, a perturbation in mitochondrial dynamics can lead to distinctive defects in the neurons (Chen and Chan, 2009). In highly polarised cells like neurons, mitochondria are needed in sites distant from the cell body and hence, disruption in mitochondrial motility has shown to be affecting neuronal function (Campello et al., 2006). Defects in both fission and fusion are shown to disrupt mitochondrial movement. In neuronal cells where mitochondrial fusion was disrupted, increase in mitochondrial diameter and aggregations of mitochondria were observed, which blocked their efficient entry into neurites resulting in neurodegeneration (Chen et al., 2007). Since Mfn2 is believed to play a role in the maintenance of microtubule structure, defects in this protein can affect mitochondrial motility (Chen et al., 2003). Another set of proteins that plays a vital role in mitochondrial motility is mitochondrial Rho GTPase 1 and 2 (Miro1 and Miro2). Damage to these proteins or its unavailability is shown to pivot fission–fusion ratio depleting the number of mitochondria in the neurons resulting in defective neurotransmission.

Loss of axonal mitochondria due to the knockdown of mitochondrial genes Milton and Miro enhanced microtubule disassembly due to tau-induced neurodegeneration (Iijima-Ando et al., 2012). Reduction in mitochondrial membrane potential (MMP) is shown to induce PINK1 accumulation in the outer mitochondrial membrane (OMM) thereby recruiting Parkin, an

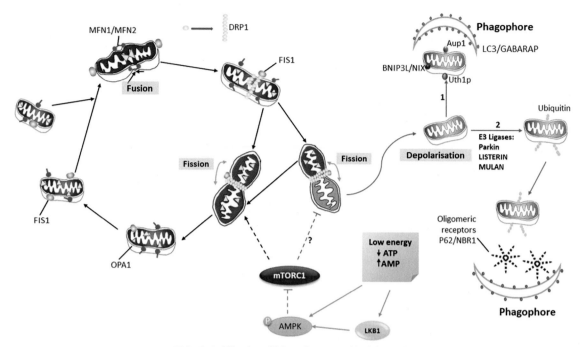

FIG. 8.1 Mitochondrial quality control inside a cell.

E3 ubiquitin ligase, from the cytosol, which in turn activates mitophagy (Ashrafi et al., 2014). Mutation in PINK1 is shown to change the mechanism of fission and fusion, which in turn induces neurodegeneration in PD.

Mitochondria have multiple quality control mechanisms to eliminate dysfunctional or damaged organelles through various degradation processes like intramitochondrial proteases, mitochondrial-derived vesicles, mitophagy (autophagy-mediated degradation of mitochondria), and proteasomal degradation (Itoh et al., 2013). Parkin, as mentioned above, plays an important role in mitophagy where the protein is recruited to dysfunctional mitochondria, ubiquitinating mitochondrial proteins, which in turn aid in proteasomal degradation and promoting engulfment by autophagosomes. During mitophagy, the mitochondrial dynamics are shifted from fusion to fission due to proteasomal degradation of MFNs. Enhanced mitochondrial division aids in mitophagy of dysfunctional mitochondria, whereas mitochondrial fission is downregulated during autophagy induced by starvation resulting in elongated mitochondrial tubules. Elongated mitochondria are protected from degradation as they cannot be efficiently engulfed by autophagosomes, thereby enabling the cells to continuously produce energy (Itoh et al., 2013). The mechanism of

mitochondrial fission, fusion, and quality control is shown in Fig. 8.1.

8.3 MITOCHONDRIA AND AGING

Mitochondria have been recognized as an important player in the aging of an organism. A decline in mitochondrial function is believed to contribute to the age-dependent deterioration in the function of an organ and also in age-related diseases. Accumulation of mitochondrial DNA (mtDNA) mutations and augmented production of reactive oxygen species (ROS) due to aging is considered as the greatest risk factor associated with a plethora of neurodegenerative diseases like AD, PD, and amyotrophic lateral sclerosis (ALS). Since mitochondria have their own protein synthesis mechanism and genetic system, disease and aging could easily disrupt the organelle's pathways. In addition, these organelles are considered to be easily impacted by environmental factors and toxins. In addition, mitochondria's role in regulating oxidative stress and apoptosis makes it a key element in the process of aging. The organelle's 10-fold greater mutation rate compared with nuclear DNA and reduced repair capacity is also shown to have a role in aging and diseases like cancer. Since mitochondria play a crucial role in the regulation of metabolism in all the organs, disruption of this organelle's homeostasis could easily impact them by diseases and accelerate aging process (Haas, 2019).

Although the nuclear genome is responsible for the majority of the protein-coding of mitochondrial proteins, it also possesses several copies of their own DNA called mtDNA. Human mtDNA is circular with 16,569 base pairs encoding 13 polypeptide components of the respiratory chain. In addition, mtDNA codes for the rRNAs and tRNAs essential for intramitochondrial protein synthesis. Inherited mutations in mtDNA are known to be the reason for many of the diseases affecting tissues, which need high energy like the brain and muscles (Lin and Beal, 2006). Mitochondrial mutations are noted to increase in incidence as the organism grows older in a tissue dependant manner where it is hypothesized that during aging and aging-related neurodegeneration, somatic mtDNA mutations are acquired which in turn contributes to the physiological decline (Lin and Beal, 2006; Sun et al., 2016). It is a well-established fact that with the aging of an organism, mtDNA accumulates mutations especially point mutations and large-scale deletions. The accumulation of these point mutations and deletions occurring with aging correlates with deterioration in mitochondrial function. Increased point mutation levels are observed in cytochrome oxidase gene (CO1) during aging, which in turn reduces cytochrome oxidase activity, and hence affecting electron transport chain (ETC). High levels of these mutations are observed in *substantia nigra* of the brain correlating with cytochrome oxidase deficiency in this region. This may contribute to the age dependence of PD.

mtDNA replication is carried out by mtDNA polymerase, POLGγ, which is the only mtDNA polymerase. In addition to its 5'-to-3' polymerase activity, these polymerases have 3'-to-5' exonuclease (proofreading) activity as well. MtDNA mutations are accumulated in scenarios where the proofreading activity of POLG is eliminated, still preserving its polymerase activity. This increase in mtDNA mutations results in decreased ATP production and respiratory enzyme activity. Mice were knocked in with mutated D257A a proofreading-deficient form of mutation at position 257 results in the enzyme, which retains the polymerase activity but have impaired proofreading function, and these mice exhibited accelerated aging because of the mitochondrial mutations (Kujoth et al., 2005; Trifunovic et al., 2004). These animals exhibited several symptoms of aging like weight loss, osteoporosis, kyphosis, alopecia, cardiomyopathy, anemia, sarcopenia, etc. The mutations also reduced the life expectancy of such animals. In humans, mutations in POLG exhibits parkinsonism and myopathy (Lin and Beal, 2006).

It is widely being accepted that when organism ages it is generally accompanied by a decline in mitochondrial enzyme activities, a decrease in the mitochondrial respiratory capacity indicated by reduced phosphocreatine (PCr) recovery time and increased ROS. Coordination of mitochondrial biogenesis is largely carried out by the transcriptional coactivator peroxisome proliferator-activated receptor γ coactivator-1 α (PGC-1α), which regulates the activity of transcriptional factors which are involved in mitochondriogenesis. Increasing the levels of PGC-1α in skeletal muscle cells of mice could prevent age-related sarcopenia, inflammation, and metabolic diseases emphasizing the role of this protein in aging (Sun et al., 2016).

Production of ROS is another significant mechanism where mitochondria are believed to contribute to aging. Oxidative damage initiated by ROS is believed to be the major contributor to functional decline, which is characteristic of aging. Mitochondria contain a wide-ranging network of antioxidant defences and multiple electron carriers that are capable of producing ROS. Damage to mitochondria, including oxidative damages, can cause an imbalance in the production and removal of ROS, resulting in ROS overload. Overexpression of mitochondrial antioxidant enzymes reduced the levels of mitochondrial deletions. This in turn showed reduced hydrogen peroxide production indicative of ROS biosynthesis and prolonged the life span of the organism and reduces age-related disorders like arteriosclerosis, cataract which shows the importance of mitochondrial ROS production in the process of aging (Lin and Beal, 2006). Gene expression studies in human brain also suggest a major role of oxidative damage in age-related cognitive decline. In human brain, as the age of the individual advances, a decrease in expression of genes involved in mitochondrial functions, synaptic plasticity, and vesicular transport is observed. On the other hand, aged individuals show increased expression of stress proteins indicative of higher ROS levels. Age increased oxidative DNA damage, mainly thought to be mediated by affecting the promoter regions, perhaps due to the G/C-rich sequences, which are sensitive to oxidation. Since vulnerable-gene promoters are deficient in repair and are more sensitive to oxidative stress, thus mitochondrial dysfunction could aggravate both these functions by increasing ROS or decreasing the ATP availability, which is necessary for repair of the cell. Hence, it is safe to assume that aging induces mitochondrial dysfunction, which in turn contributes to the damage of vulnerable genes in the brain.

8.4 CELL DEATH RELATED TO MITOCHONDRIAL DISFUNCTION

The first report on a highly conserved form of programmed cell death, involved in embryonic development, which is genetically regulated, was done by Kerr et al. in 1972. The mechanism, later named as apoptosis is an intricate mechanism of cell suicide that can be triggered by a variety of mechanisms involving cellular homeostasis and development in multicellular organisms (Wyllie et al., 1980). Apoptosis and its subsequent clearing mechanism are necessary for the system to defend against infected or hyperproliferative cells. It is involved in a wide variety of biological systems involved in host mechanisms like embryonic development, surveillance, and cellular differentiation (Rupinder et al., 2007). The balance between cell proliferation and cell death is vital for the physiological process to function properly. Disruption of this balance can attribute to neurodegenerative diseases and cancer (Wang and Youle, 2009; Arun et al., 2016).

A series of distinct morphological changes characterize apoptotic cell death. In its initial phase, the cell destined for apoptosis loses its contact with neighboring cells of the tissue and shrinks due to loss of cytoplasmic volume, condensation of cytoplasmic proteins, and nuclear chromatin. In the second phase, the cell membrane blebs, the membrane is ruffled due to flipping of phosphatidylserine, cells are fragmented, and apoptotic bodies are formed (Arends and Wyllie, 1991). The induction of apoptosis is characterized by intranucleosomal cleavage of DNA and is observed as a classic marker of apoptosis (Wyllie et al., 1980). Finally, the dying cells are engulfed by macrophages and surrounding healthy cells. Apoptosis is generally considered as a caspase-dependant form of cell death.

For the full expression of apoptosis, highly specific proteolytic cleavage, and signaling cascade mediated by caspases, a family of cysteinyl aspartate-specific proteases are necessary. In mammals, 14 caspases are identified and are functionally divided into three families. The initiator caspases (caspase-2, 8, 9, 10) get activated on early apoptotic signals and activates the executioner caspases (caspase-3, 6, 7) which implements apoptosis by cleavage of substrate proteins. The third set of caspases are cytokine processors (caspase-1, 4, 5). Caspases are highly substrate-specific.

Apoptosis is regulated by mitochondria, which are characterized by the condensation and breakage of DNA and cell shrinkage. In addition, mitochondrial fractionation and release of several proteins from mitochondria to the cytoplasm is necessary for apoptosis. Bcl-2 family of proteins regulated mitochondrial homeostasis during apoptosis. Mitochondria, as we have previously mentioned, have two well-defined compartments, namely the matrix and the IMS. The IMS in normal physiological conditions is impermeable. The protons pumped into the intermembrane during electron transport forms mitochondrial membrane potential (MMP (ΨM)). It is assumed that permeability of the mitochondrial membrane is originated from an increase in permeability of the inner mitochondrial membrane (IMM) due to the opening of multiprotein pore called permiability transition pore (PTP). The opening of PTP leads to swelling of the mitochondrial matrix, depolarization of membrane potential ($\Delta\Psi$M), subsequent rupture of the outer membrane, and nonselective release of intermembrane space proteins (IMS proteins). Maintenance of ΨM is essential for the integrity and the oxidative phosphorylation of mitochondria. Many factors inducing apoptosis disrupt ΨM and also form permeability transition pores (PTPs). PTP formation disrupts the organelle's proton gradient and the expansion of the matrix space. The matrix expansion can ultimately rupture the outer membrane of mitochondrial-releasing caspases activating proteins such as cytochrome c to the cytosol. One of the molecules that aids in the formation of mitochondrial membrane channels is Bax, a pro-apoptotic Bcl-2 protein.

The Bcl-2 family of proteins is one of the key classes of proteins regulating mitochondria-mediated apoptosis. This family of proteins includes both pro- and anti-apoptotic proteins, which are localized in the nucleus, endoplasmic reticulum (ER), and outer membrane of mitochondria. The cytosolic Bcl-2 proteins translocate to mitochondria during apoptosis, regulating cell death in part, by affecting the release of mitochondrial proteins to the cytosol. Bcl-2 family of proteins gets involved in the apoptotic pathway by either activating anti-apoptotic Bcl-2 proteins or by inactivating the pro-apoptotic Bcl-2 members (Shimizu and Tsujimoto, 2000). The anti-apoptotic proteins Bcl-2 and Bcl-xL stabilizes ΨM, whereas the pro-apoptotic Bax and Bak insert themselves to mitochondrial membranes promoting the release of cytochrome c through the mitochondrial PTP-triggering caspase cascade. Bax is localized in the cytosol as a monomer in healthy cells, which during apoptosis translocates to mitochondria undergoing conformational changes to form oligomers.

In contrast, Bak resides in the mitochondria, and apoptotic signals oligomerize the molecule bringing out conformational changes. Translocation of Bax and release of Cyt c are regarded as two significant upstream molecular events of apoptosis. Another Bcl-2 family member, which is involved in regulating cytochrome c release is Bid, where the truncated Bid (tBid) is

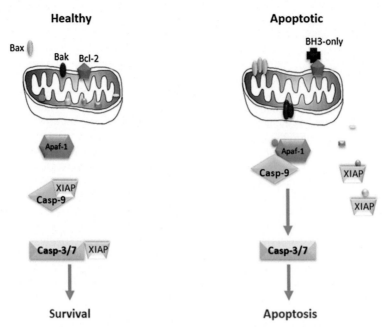

FIG. 8.2 Mitochondrial-mediated apoptosis and survival machinery regulated by Bcl-2 family proteins.

translocated to the mitochondria thereby triggering the release. The mitochondria-specific lipid called cardiolipin provides specificity for the targeting of tBid to the organelle (Yang et al., 1997; Wang and Youle, 2009; Tait and Green, 2010; Peña-Blanco and García-Sáez, 2018) (Fig. 8.2).

Mitochondrial dysfunction leads to elevated levels of ROS, leading to oxidative damage affecting apoptosis signaling pathways. In addition, excessive production of ROS can lead to oxidation of macromolecules like cardiolipin, which is the anchor of cytochrome c. This in turn facilitates the release of cytochrome c triggering the caspase cascade. ROS also affects the ubiquitin-proteasome pathway, which mediates the elimination of damaged and misfolded proteins. The increased levels of polyubiquitinated proteins result in cytosolic oxidative stress and subsequent loss of MMP. The introduction of antioxidants and an increase in its concentrations can prevent or reverse ΨM loss.

Changes in Ca^{2+} ion levels can enable the release of pro-apoptotic factors from mitochondria, thereby promoting apoptosis. Uptake of Ca^{2+} by the mitochondria can trigger the opening of mitochondrial PTP, thereby altering the organelle's morphology and functional activity. Even though the opening of the mitochondrial PTP is controlled by Ca^{2+}, ATP depletion and oxidative stress enhance the pore formation by accelerating apoptosis.

Several elements that are constituents of mitochondria play an important role in the induction of apoptosis. Cytochrome c is a crucial element of the mitochondrial ETC. This is one of the three apoptotic proteases activating factors (Apafs) that activate caspases. Release of Cyt c from the mitochondria requires two steps, namely mobilization and translocation. Mobilization of Cyt c involves its dissociation from the membrane phospholipid cardiolipin and detachment from the IMM. After translocation, cytochrome c, along with Apaf-1 forms the apoptosome complex, which is necessary for the recruitment and cleavage of caspase 9.

Smac/DIABLO, another mitochondrial protein is an enhancer of Cyt c-mediated activation of caspase − 3. This protein facilitates activation and release of caspases by binding to XIAP (X-linked inhibitor of apoptosis *protein*). Omi/HtrA2, through its protease activity, can irreversibly degrade inhibitor of apoptosis protein (IAP). This mitochondrial protein enhances apoptosis through its protease activity and IAP binding activity via both caspase dependant and independent pathways.

EndoG/endonuclease G belongs to DNA/RNA nonspecific $\beta\beta\alpha$-Me-finger nucleases family. This protein is released from the mitochondria enabling the degradation of nuclear chromatin and once released from the mitochondria, the function of EndoG is independent of caspase activity. The release of EndoG

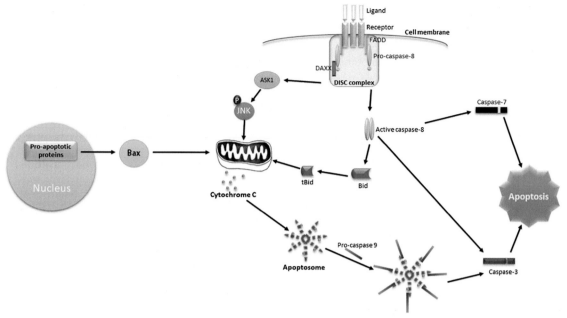

FIG. 8.3 Apoptotic pathway in the cell—intrinsic and extrinsic pathways with mitochondria as the central execution mechanism.

is facilitated by tBid and requires additional inner membrane remodeling during apoptosis, unlike Cyt *c* (Wang and Youle, 2009).

AIF (apoptosis-inducing factor) is a flavin adenine dinucleotide–binding oxidoreductase, which is another protein released from mitochondria during apoptosis. AIF undergoes proteolysis during apoptosis producing a soluble AIF protein, which is necessary for its release in a caspase-independent manner. After cleavage, AIF is released from the mitochondria where it translocates to the nucleus triggering chromatin condensation and DNA degradation (Fig. 8.3).

8.5 GENETICS OF NEURODEGENERATIVE DISEASES

The formation of protein aggregates in the brain is considered to be an important mechanism in the progression of many neurodegenerative diseases. AD is one of the most common neurodegenerative disease characterized by the loss of memory and cognitive function due to the loss of neurons. AD accounts for around 80% of dementia cases. AD is attracting both clinical and government attention due to the increasing social impact and cost associated with it. It is estimated that by the year 2050, around 50% of people above the age of 85 would be affected by this condition. AD is

classified into two, namely sporadic and familial, and among that only a small fraction of the population (10%) are familial. Sporadic AD is caused by many external factors like stress, reduced sleep, other environmental and lifestyle factors, and the greatest risk factor identified is age. Pathologically, all forms of AD are characterized by the progressive loss of neurons in regions of the brain associated with memory formation. AD is characterized by neurofibrillary tangles consisting of hyperphosphorylated tau protein aggregates and β-amyloid proteins. Aβ aggregation is considered as the most relevant phenomenon implicated in the cause of AD.

Most of the autosomal dominant familial AD is caused due to the mutation in three genes, amyloid precursor protein (APP) and two presenilins (PSEN1 and PSEN2). Unfolding and tangling of β-amyloid protein forms plaques, which are toxic to the neurons leading to neurodegeneration. The presenilin genes cleave APP by contributing components to the gamma-secretase complex. The other two genes associated with autosomal-dominant AD are ABCA7 and SORL1. However, in most cases, AD is contributed due to the changes in environmental and genetic factors. The best-known genetic risk factor in AD is the inheritance of ε4 allele of the apolipoprotein E (APOE). Recently, variants of clusterin (CLU) and phosphatidylinositol-binding clathrin assembly (PIC-ALM) proteins have also been identified to be associated with AD (De Castro et al., 2010; Kim, 2018).

PD is the most common neurodegenerative disease affecting the motor system showing symptoms like tremors, slow movements, and difficulty in walking, which affects around 1% of the population above 60 years of age. The most common clinical feature of the disease is the critical loss of neuromelanin pigmented neurons in the *substantia nigra*. Loss of neurons in brain regions controlling autonomic functions, mood, and cognition is also implicated in PD. PD is generally considered a nongenetic disorder due to a large number of sporadic cases, but a small proportion of familial cases are also identified. Neurons degenerated in PD contains intracellular inclusions of α-synuclein (SNCA), a major component of Lewy body. Mutations of 17 genes so far have been implicated in the pathogenesis of PD. Mutation of genes encoding α-synuclein (SNCA), parkin (encoded by PARK2) leucine-rich repeat kinase 2—LRRK2 (encoded by PARK8) and PINK1 results in protein aggregates, mitochondrial dysfunction and oxidative stress in the *substantia nigra* region of the midbrain. Mutations in the mitochondrial serine protease HTRA2 are also described in familial PD. Around 5% of people affected with PD have a mutation in the GBA1 gene. Other genes reported in the pathogenesis of PD are POLG, PRKN, PINK1, DJ1/PARK7, VPS35, EIF4G1, DNAJC13, CHCHD2, and UCHL1 (Samii et al., 2004; Dexter and Jenner, 2013; Kalia and Lang, 2015).

HD or Huntington's chorea is a fatal, dominantly inherited genetic disorder resulting in the death of brain cells characterized by a lack of coordination and uncontrolled muscle movements. The symptoms of HD arise due to the selective loss of long-projection neurons called medium spiny neurons, which release γ-aminobutyric acid (GABA) in the striatal region of the brain, which controls movement, emotions, and memory. HD is a slowly progressing condition where the patient survives for around 15—20 years from the onset of the disease. Generally, HD is an inherited condition, but around 10% of cases are seen to be due to new mutations. HTT, an extremely large protein of 350 kDa size, acts as a scaffold protein regulating signaling pathways and organelle trafficking. The aberrant repetition of the CAG trinucleotide sequence in exon 1 of the HTT gene is implemented as the reason for the disease. An increase in the repeat count is considered as the reason for the pathogenesis of the disease. Less than 35 polyglutamine stretch does not seem to cause the disease, but an abnormal stretch of more than 35

glutamine residues leads to the pathogenesis of the disease (Walker, 2007; Dayalu and Albin, 2015; De Castro et al., 2010).

ALS otherwise known as motor neuron disease or Lou Gehrig's disease is caused by the death of neurons controlling voluntary muscles characterized by stiff muscles, a decrease in muscle size and muscle twitching. Worldwide, ALS has a prevalence of 4—6 cases per 100,000 individuals. Familial cases of ALS account to 10% where four genes (C9orf72, SOD1, FUS, and TARDBP) are accounted for the majority of cases. Genetics of sporadic ALS is poorly understood, but around 16 genes are identified to be involved in the disease (van Es et al., 2017; Corcia et al., 2017; Smith et al., 2019).

The list of genes identified to be involved in various neurological conditions are tabulated in Table 8.1.

8.6 NEUROLOGICAL DISEASES ASSOCIATED WITH MITOCHONDRIA

As discussed earlier in this chapter, mitochondria possess their own genome and replicate through binary fission similar to bacterial reproduction. mtDNA is ~16.6 kb long, comprised of 37 genes. Thirteen genes of mtDNA encode for the proteins involved in ATP production, while twenty-four are involved in mtDNA translation. Being circular in nature, mtDNA is different from nuclear DNA, and their mutations have been implicated in neuronal diseases like PD, AD, HD, and ALS (De Castro et al., 2010).

8.6.1 Alzheimer's Disease

In the brains of AD patients, a higher concentration of damaged mitochondria is observed corresponding to the presence of mtDNA and mitochondrial resident proteins in the cytosol. This may be due to various reasons, and a few are mentioned below. As previously mentioned, AD is characterized by the presence of amyloid plaques, which are comprised of Aβ proteins (β-amyloid peptides). Aβ proteins are formed by the proteolytic cleavage of transmembrane APP.

Aβ species have been found to be accumulating intracellularly in mitochondria, and this phenomenon has been identified to be occurring before the extracellular accumulation providing a direct link between the protein and pathological dysfunction of the organelle in AD. The existence of Aβ aggregates in mitochondria impairs the cytochrome *c* oxidase enzymatic activity and inhibits the mitochondrial amyloid-β-binding alcohol dehydrogenase (ABAD) activity. Aβ has a strong effect on mitochondrial enzymatic activities and

TABLE 8.1
Interaction Between Mitochondria and Proteins Involved in Neurodegenerative Diseases.

Gene	Gene Product	Link to Mitochondria
ALZHEIMER'S DISEASE		
Amyloid precursor protein (APP)	β-Amyloid	APP accumulates in the mitochondrial protein import channels where it stably associates with translocates of the outer mitochondrial membrane and inner mitochondrial membrane
PS1 and PS2	Presinilin 1 and 2	PS1 and PS2 are enriched in mitochondria-associated membranes, increases Ca^{2+} overload
MAPT	Tau	Hyperphosphorylation and overexpression of tau impair axonal transport of mitochondria
PARKINSON'S DISEASE		
PARK1/4	α-Synuclein	Oxidative stress
PARK2	Parkin	Oxidative stress and mitochondrial damage
PARK6	PINK1	Deficiency sensitizes mitochondria to rotenone and induces dopaminergic neuron degeneration
PARK7	DJ1	Oxidative stress if dysfunctional
PARK8	LRRK2	Bind with parkin and associates with the outer mitochondrial membrane.
PARK13	OMI/HTRA2	Increased production of ROS and accumulation of unfolded proteins in mitochondria
HUNTINGTON'S DISEASE		
HTT	Huntingtin	Regulation of lactate levels, mitochondrial membrane potential, respiratory function through complex II, mitochondrial calcium uptake, mitochondrial mobility and ultrastructural changes in mitochondria.
AMYOTROPHIC LATERAL SCLEROSIS (ALS)		
ALS1	Cu/Zn superoxide dismutase	Redox signaling
ALS2	Alsin	Oxidative stress, mitochondrial quality control
ALS4	Senataxin	Mitochondrial biogenesis
ALS6	FUS	Interacts with the mitochondrial ATP synthase β-subunit (ATP5B), suppresses mitochondrial ATP synthase activity, disrupts ATP synthase complex assembly and activates the mitochondrial unfolded protein response
ALS8	VAPB	Part of endoplasmic reticulum—mitochondria tethering complex
ALS10	TAR DNA-binding protein	Mitochondria-dependent cell death
ALS11	FIG4	Mitochondrial transport
ANG	Angiogenin	Inhibitor of Cyt c release from mitochondria

Continued

TABLE 8.1
Interaction Between Mitochondria and Proteins Involved in Neurodegenerative Diseases.—cont'd

Gene	Gene Product	Link to Mitochondria
DCTN1	Dynactin	Mitochondrial fission and localization
DPP6	Dipeptidyl-peptidase 6	
NEFH	Neurofilament	Aerobic glycolysis and mitochondrial stability
PRPH	Peripherin	Maintenance of mitochondrial homeostasis
SMN1 and 2	Survival motor neuron	Maintenance of mitochondrial homeostasis

respiratory dysfunctions. Toxic Aβ proteins decrease the enzymatic activity of complexes III and IV, thereby inhibiting mitochondrial respiration and accelerating apoptosis and promoting neurodegeneration. In the brains of AD patients, selective decrease in complex IV is observed, resulting in an imbalance in the ETC compromising the general metabolism of neurons (De Castro et al., 2010).

Aggregation of Aβ is accompanied by the generation of ROS and lipid peroxidation. In mitochondria, this affects the components of oxidative phosphorylation by depleting ATP levels affecting mitochondrial ATPase. Aβ protein aggregation also activates the NF-κB pathway, which regulates inflammation, cell death, and division. Activation of NF-κB increases ROS production leading to mitochondrial damage-inducing neurodegeneration. Aβ levels are significantly increased during oxidative stress encouraging phosphorylation and polymerization of tau (Zhao and Zhao, 2013).

In human AD brains, Aβ peptides interact with a mitochondrial matrix protein cyclophilin D (CYPD), which is involved in mitochondrial PTP formation. This interaction is particularly observed during oxidative stress—inducing neurodegeneration.

The mitochondrial dysfunction in AD is not only associated with Aβ but also with mechanisms independent of amyloid aggregation. A majority of familial cases of AD carry mutations in presenilin genes (PSEN1 and PSEN2), a γ-secretase subcomponent, which is responsible for cleavage of APP to produce Aβ. Mutations in presenilins has been associated with early onset of AD (Arun et al., 2016). One hundred eighty-five different mutations of PSEN1 have been identified associated with AD early onset. These mutations could destabilize mitochondrial functions through two possible mechanisms: (1) disturbing ER calcium homeostasis, thereby promoting mitochondrial Ca^{2+} uptake and (2) impairing the transport of

the organelle through axons affecting synaptic activities of the neurons (De Castro et al., 2010). Both presenilins are major components of the catalytic center of the γ-secretase complex, which is known to produce toxic Aβ species.

Defects in mitochondrial transmission by small GTPase - Miro may be a contributing factor to the pathogenesis of AD. Neurons lacking Miro and Milton, another mitochondrial protein, are shown to possess axons lacking mitochondria, which increases phosphorylation of the mitochondrial protein tau and induces neurodegeneration.

Specific mtDNA mutations have resulted in elevated oxidative damage and have been associated with AD. The excision of mtDNAdelta4977 has been associated with late-onset AD.

Altogether, causes of AD are attributed to two interrelated pathogenic events: misfolded protein aggregation and mitochondrial dysfunction. Despite the general agreement that both processes have pivotal roles in the progression of AD, the evidence supporting the concept that mitochondrial alterations are the primary cause of AD is insufficient and is being assumed as a secondary event in the disease process.

8.6.2 Parkinson's Disease

Mitochondrial dysfunction has been associated with the selective loss of dopaminergic neurons in PD. Even though the majority of PD cases occur sporadically due to unknown causes, a reduction in mitochondrial complex I activity is commonly associated with the disease, in addition, exposure to the toxins rotenone and MPTP are shown to cause an acute and irreversible parkinsonian syndrome, which may involve the specific inhibition of complex I suggesting that one of the key causes of motor impairment and neuron loss in PD is due to mitochondrial stress in dopaminergic neurons (De Castro et al., 2010).

Many genes related to mitochondria are shown to be associated with the pathogenesis of PD. The major common functional effects of the proteins encoded by these genes relate to mitochondria: PINK1, a kinase that is regulated by a canonical N-terminal mitochondrial-targeting sequence, PARK7 (DJ-1) the molecular chaperone that translocates to the mitochondria during oxidative stress and stabilize complex I. Parkin and PINK1 are involved in the regulation of mitochondrial dynamics, turnover, and morphology. HTRA2/OMI and PINK1 are the components of the same mitochondrial stress-sensing pathway. In addition, α-synuclein and LRRK2 are partly localized to mitochondria. Since the majority of genes related to PD are either directly or indirectly related to mitochondria, the involvement of mitochondria in the pathogenesis of PD is obvious.

PINK1 and Parkin regulate the mitochondrial dynamics by their communication with the fusion–fission machinery. In addition, these proteins affect mitochondrial dynamics by disturbing calcium and ROS homeostasis. PINK1 and Parkin are suggested to have roles in mitochondrial fission, as loss of these proteins lead to mitochondrial enlargement.

Parkin, which acts as an E3 ubiquitin ligase, is mostly localized to the cytoplasm. However, on loss of ΔψM, Parkin is recruited to the mitochondria leading to the elimination of dysfunctional organelles through selective autophagic engulfment. This recruitment is due to the direct phosphorylation of Parkin on Thr175 by PINK1. PINK1 causes loss of MMP in mammalian systems, thereby flagging dysfunctional mitochondria for degradation. In addition, PINK1 acts as a substrate for two IMS proteins TRAP1/Hsp75 and the serine protease HTRA2/OMI, which are crucial for mitochondrial quality control. These proteins interact with PINK1 under stress conditions preventing the accumulation of misfolded proteins.

In addition to the dysfunction of mitochondria, protein aggregation also has an important role in the pathogenesis of PD. As we discussed earlier, α-synuclein, the main component of Lewy bodies is a fibrillar aggregation-prone protein. This protein, even though cytosolic in localization, has a direct role in regulating mitochondria, as mutations in the gene cause mitochondrial abnormalities. As mentioned above, PINK1 prevents the accumulation of misfolded proteins, and α-synuclein is one among them. In addition, the molecular chaperone DJ-1, which redistributes to mitochondria during oxidative stress prevents aggregation and subsequent toxicity of α-synuclein. In addition, Parkin, PINK1, and DJ-1 form a functional ubiquitin E3 ligase complex promoting ubiquitination and degradation of misfolded proteins protecting the cells from degeneration. Hence, damage to mitochondria can lead to unfolded protein response further leading to apoptosis and neurodegeneration, especially PD.

Another PD-associated protein LRRK2, a serine/threonine kinase accountable for the phosphorylation of actin cross-linkers interacts with PINK1, DJ-1, and Parkin. LRRK2 is a key mediator of stress responses binding to moesin, ezin, radixin, and eukaryotic initiation factor 4E (eIF4E)-binding protein (4E-BP). LRRK2 is responsible for the activation of PINK1 and Parkin. Mutation of the LRRK2 gene can lead to the accumulation of misfolded proteins leading to neurodegeneration (De Castro et al., 2010; Arun et al., 2016).

8.6.3 Huntington's Disease

Evidence suggests that HD pathogenesis is strongly linked with mitochondrial impairments. Postmortem brain samples of HD patients showed reduced activity of mitochondrial respiratory complexes II, III, and IV. Mutations in the HTT gene might cause neuronal toxicity by evoking mitochondrial dysfunctions. HTT binds to Htt-associated protein 1 (HAP1), which may act as a docking platform interacting with the molecular motor dynein–dynactin and kinesin, which is known to regulate mitochondrial transport. Neuronal functions depend on the effective trafficking of organelles and molecules, and as previously mentioned, defect in mitochondrial transport impairs neuronal function and induces neurodegeneration. In addition to blocking the movement of mitochondria, mutant HTT also induces mitochondrial fission–fusion imbalance and increases mitochondrial fragmentation (De Castro et al., 2010). The mitochondrial fusion proteins like Mfn2 are noted to be reduced in expression in grade III and IV HD. In addition, the increase in mitochondrial fission observed in later stages of HD may contribute to mitochondrial swelling.

These two mechanisms may be linked to the increase of polyQ (a series of glutamine residues) repeats, which are characteristics of HD (Arun et al., 2016).

8.6.4 Amyotrophic Lateral Sclerosis

The dysfunction of mitochondria is considered as a core ALS disease component. Mitochondrial morphological changes like giant mitochondria and spiny or stubby organelle are frequently seen in ALS. In addition, abnormal accumulation of mitochondria in the initial segment of axons and axonal hillock is also observed in ALS. Changes in ETC activities are also observed in biopsies from ALS patients.

ALS associated with mitochondrial dysfunction comes in many pretexts, including defects in oxidative phosphorylation, defective mitochondrial dynamics, ROS production, and impaired calcium buffering capacity. One of the first changes detected in ALS patient motor neurons are structurally altered and aggregated mitochondria. In addition, sporadic ALS cases display axonal swellings consisting of swollen mitochondria and fragmented mitochondrial populations.

A number of proteins that have been associated with sporadic and familial ALS have been shown to interact with mitochondria. These proteins include SOD1, C9orf72, TDP-43, FUS, and the C9orf72 GGGGCC repeat expansion-associated glycine/arginine (GR) dipeptide repeat protein (DPR). The interaction of these proteins with the mitochondria appears to be the reason for the induction of mitochondrial damage associated with ALS.

ALS-associated mutant protein SOD1 is shown to localize to the IMS of mitochondria, where it aggregates and reduces the ETC complex activity. Besides, SOD1 aggregates interfere with the voltage-dependent anion channel 1 (VDAC1) activity, which is responsible for the exchange of respiratory substrates across the outer mitochondrial membrane. ALS mutant SOD1's direct interaction with VDAC1 inhibits ion channel conductance and reduces its permeability to ADP in both presymptomatic and symptomatic disease conditions.

ALS linked with mutations in TAR-DNA-binding protein 43 or TDP-43 is associated with mitochondria, where it preferentially binds the mtDNA-encoded complex I subunits ND3 and ND6 mRNAs, causing complex I weakening and affecting their transcription. Cytoplasmic TDP-43 accumulation is a hallmark pathology in most ALS, which explains mitochondrial defects observed in ALS.

An increase in levels of ROS and oxidative damage has been reported in ALS where increased markers of oxidative damage are found in biofluids of sporadic ALS patients and postmortem tissues. In addition, an increase in ROS levels is reported in lymphoblasts of SOD1-mutated familial ALS cases. Besides, SOD1 itself is a target of oxidative damage and leading to its misfolding and aggregation. The misfolded and aggregated SOD1 has been shown to increase superoxide production and disrupt mitochondrial function. Hence, a vicious cycle mechanism is proposed in which misfolded SOD1 causes mitochondrial damage and oxidative stress, which leads to the aggravation of SOD1 misfolding and downstream mitochondrial damage. Oxidative damage can also promote TDP-43 aggregation.

Substantial evidence suggests that mitochondrial dysfunction plays an important role in ALS pathogenesis (Smith et al., 2019; Lezi and Swerdlow, 2012).

8.7 CONCLUSION

As discussed, mitochondrial dysfunction and abnormalities contribute to various neurodegenerative diseases. The extend of mitochondrial damage increases as an individual age and is proportional to the progression of diseases. In addition, a remarkable number of proteins that are involved in neurodegeneration and their mutations are shown to have a direct link to this organelle. Hence, it has been considered that mitochondria play an important role in sporadic neurodegeneration. Henceforth, drugs designed to eliminate misfolded proteins through autophagy, scavenge ROS, and improve mitochondrial functions have the potential to eliminate or halt various neurodegenerative conditions.

REFERENCES

Alavi, M.V., Bette, S., Schimpf, S., Schuettauf, F., Schraermeyer, U., Wehrl, H.F., Ruttiger, L., Beck, S.C., Tonagel, F., Pichler, B.J., Knipper, M., Peters, T., Laufs, J., Wissinger, B., 2007. A splice site mutation in the murine Opa1 gene features pathology of autosomal dominant optic atrophy. Brain 130, 1029–1042.

Arends, M.J., Wyllie, A.H., 1991. Apoptosis: mechanisms and roles in pathology. Int. Rev. Exp. Pathol. 32, 223–254.

Arun, S., Liu, L., Donmez, G., 2016. Mitochondrial biology and neurological diseases. Curr. Neuropharmacol. 14, 143–154.

Ashrafi, G., Schlehe, J.S., Lavoie, M.J., Schwarz, T.L., 2014. Mitophagy of damaged mitochondria occurs locally in distal neuronal axons and requires PINK1 and Parkin. J. Cell Biol. 206, 655–670.

Bereiter-Hahn, J., 1990. Behavior of mitochondria in the living cell. Int. Rev. Cytol. 122, 1–63.

Campello, S., Lacalle, R.A., Bettella, M., Mañes, S., Scorrano, L., Viola, A., 2006. Orchestration of lymphocyte chemotaxis by mitochondrial dynamics. J. Exp. Med. 203, 2879–2886.

Chen, H., Chan, D.C., 2009. Mitochondrial dynamics–fusion, fission, movement, and mitophagy–in neurodegenerative diseases. Hum. Mol. Genet. 18, R169–R176.

Chen, H., Detmer, S.A., Ewald, A.J., Griffin, E.E., Fraser, S.E., Chan, D.C., 2003. Mitofusins Mfn1 and Mfn2 coordinately regulate mitochondrial fusion and are essential for embryonic development. J. Cell Biol. 160, 189–200.

Chen, H., Mccaffery, J.M., Chan, D.C., 2007. Mitochondrial fusion protects against neurodegeneration in the cerebellum. Cell 130, 548–562.

Cipolat, S., Martins De Brito, O., Dal Zilio, B., Scorrano, L., 2004. OPA1 requires mitofusin 1 to promote mitochondrial fusion. Proc. Natl. Acad. Sci. U.S.A. 101, 15927–15932.

Corcia, P., Couratier, P., Blasco, H., Andres, C.R., Beltran, S., Meininger, V., Vourc'h, P., 2017. Genetics of amyotrophic lateral sclerosis. Rev. Neurol. (Paris) 173, 254–262.

Davies, V.J., Hollins, A.J., Piechota, M.J., Yip, W., Davies, J.R., White, K.E., Nicols, P.P., Boulton, M.E., Votruba, M., 2007. Opa1 deficiency in a mouse model of autosomal dominant optic atrophy impairs mitochondrial morphology, optic nerve structure and visual function. Hum. Mol. Genet. 16, 1307–1318.

Dayalu, P., Albin, R.L., 2015. Huntington disease: pathogenesis and treatment. Neurol. Clin. 33, 101–114.

De Castro, I.P., Martins, L.M., Tufi, R., 2010. Mitochondrial quality control and neurological disease: an emerging connection. Expet Rev. Mol. Med. 12, e12.

Dexter, D.T., Jenner, P., 2013. Parkinson disease: from pathology to molecular disease mechanisms. Free Radic. Biol. Med. 62, 132–144.

Emelyanov, V.V., 2003. Mitochondrial connection to the origin of the eukaryotic cell. Eur. J. Biochem. 270, 1599–1618.

Ernster, L., Schatz, G., 1981. Mitochondria: a historical review. J. Cell Biol. 91, 227s–255s.

Galpern, W.R., Lang, A.E., 2006. Interface between tauopathies and synucleinopathies: a tale of two proteins. Ann. Neurol. 59, 449–458.

Gray, M.W., 2012. Mitochondrial evolution. Cold Spring Harb. Perspect. Biol. 4, a011403.

Haas, R.H., 2019. Mitochondrial dysfunction in aging and diseases of aging. Biology 8.

Hales, K.G., Fuller, M.T., 1997. Developmentally regulated mitochondrial fusion mediated by a conserved, novel, predicted GTPase. Cell 90, 121–129.

Iijima-Ando, K., Sekiya, M., Maruko-Otake, A., Ohtake, Y., Suzuki, E., Lu, B., Iijima, K.M., 2012. Loss of axonal mitochondria promotes tau-mediated neurodegeneration and Alzheimer's disease-related tau phosphorylation via PAR-1. PLoS Genet. 8, e1002918.

Ishihara, N., Nomura, M., Jofuku, A., Kato, H., Suzuki, S.O., Masuda, K., Otera, H., Nakanishi, Y., Nonaka, I., Goto, Y., Taguchi, N., Morinaga, H., Maeda, M., Takayanagi, R., Yokota, S., Mihara, K., 2009. Mitochondrial fission factor Drp1 is essential for embryonic development and synapse formation in mice. Nat. Cell Biol. 11, 958–966.

Itoh, K., Nakamura, K., Iijima, M., Sesaki, H., 2013. Mitochondrial dynamics in neurodegeneration. Trends Cell Biol. 23, 64–71.

Kalia, L.V., Lang, A.E., 2015. Parkinson's disease. Lancet 386, 896–912.

Kim, J.H., 2018. Genetics of Alzheimer's disease. Dement. Neurocogn. Disord. 17, 131–136.

Kujoth, G.C., Hiona, A., Pugh, T.D., Someya, S., Panzer, K., Wohlgemuth, S.E., Hofer, T., Seo, A.Y., Sullivan, R., Jobling, W.A., Morrow, J.D., Van Remmen, H., Sedivy, J.M., Yamasoba, T., Tanokura, M., Weindruch, R., Leeuwenburgh, C., Prolla, T.A., 2005. Mitochondrial DNA mutations, oxidative stress, and apoptosis in mammalian aging. Science 309, 481–484.

Kuznetsov, A.V., Hermann, M., Saks, V., Hengster, P., Margreiter, R., 2009. The cell-type specificity of mitochondrial dynamics. Int. J. Biochem. Cell Biol. 41, 1928–1939.

Lesage, S., Brice, A., 2009. Parkinson's disease: from monogenic forms to genetic susceptibility factors. Hum. Mol. Genet. 18, R48–R59.

Lezi, E., Swerdlow, R.H., 2012. Mitochondria in neurodegeneration. Adv. Exp. Med. Biol. 942, 269–286.

Lin, M.T., Beal, M.F., 2006. Mitochondrial dysfunction and oxidative stress in neurodegenerative diseases. Nature 443, 787–795.

Mcbride, H.M., Neuspiel, M., Wasiak, S., 2006. Mitochondria: more than just a powerhouse. Curr. Biol. 16, R551–R560.

Perier, C., Vila, M., 2012. Mitochondrial biology and Parkinson's disease. Cold Spring Harb. Perspect. Med. 2, a009332.

Peña-Blanco, A., García-Sáez, A.J., 2018. Bax, Bak and beyond - mitochondrial performance in apoptosis. FEBS J. 285, 416–431.

Rupinder, S.K., Gurpreet, A.K., Manjeet, S., 2007. Cell suicide and caspases. Vasc. Pharmacol. 46, 383–393.

Samii, A., Nutt, J.G., Ransom, B.R., 2004. Parkinson's disease. Lancet 363, 1783–1793.

Santel, A., Fuller, M.T., 2001. Control of mitochondrial morphology by a human mitofusin. J. Cell Sci. 114, 867–874.

Shimizu, S., Tsujimoto, Y., 2000. Proapoptotic BH3-only Bcl-2 family members induce cytochrome c release, but not mitochondrial membrane potential loss, and do not directly modulate voltage-dependent anion channel activity. Proc. Natl. Acad. Sci. U.S.A. 97, 577–582.

Smirnova, E., Shurland, D.L., Ryazantsev, S.N., Van Der Bliek, A.M., 1998. A human dynamin-related protein controls the distribution of mitochondria. J. Cell Biol. 143, 351–358.

Smith, E.F., Shaw, P.J., De Vos, K.J., 2019. The role of mitochondria in amyotrophic lateral sclerosis. Neurosci. Lett. 710, 132933.

Sun, N., Youle, R.J., Finkel, T., 2016. The mitochondrial basis of aging. Mol. Cell 61, 654–666.

Swomley, A.M., Förster, S., Keeney, J.T., Triplett, J., Zhang, Z., Sultana, R., Butterfield, D.A., 2014. Abeta, oxidative stress in Alzheimer disease: evidence based on proteomics studies. Biochim. Biophys. Acta 1842, 1248–1257.

Tait, S.W., Green, D.R., 2010. Mitochondria and cell death: outer membrane permeabilization and beyond. Nat. Rev. Mol. Cell Biol. 11, 621–632.

Trifunovic, A., Wredenberg, A., Falkenberg, M., Spelbrink, J.N., Rovio, A.T., Bruder, C.E., Bohlooly-Y, M., Gidlöf, S., Oldfors, A., Wibom, R., Törnell, J., Jacobs, H.T., Larsson, N.G., 2004. Premature ageing in mice expressing defective mitochondrial DNA polymerase. Nature 429, 417–423.

Van Es, M.A., Hardiman, O., Chio, A., Al-Chalabi, A., Pasterkamp, R.J., Veldink, J.H., Van Den Berg, L.H., 2017. Amyotrophic lateral sclerosis. Lancet 390, 2084–2098.

Walker, F.O., 2007. Huntington's disease. Lancet 369, 218–228.

Wang, C., Youle, R.J., 2009. The role of mitochondria in apoptosis*. Annu. Rev. Genet. 43, 95–118.

Westermann, B., 2010. Mitochondrial fusion and fission in cell life and death. Nat. Rev. Mol. Cell Biol. 11, 872–884.

Westermann, B., 2012. Bioenergetic role of mitochondrial fusion and fission. Biochim. Biophys. Acta 1817, 1833–1838.

Wiemerslage, L., Lee, D., 2016. Quantification of mitochondrial morphology in neurites of dopaminergic neurons using multiple parameters. J. Neurosci. Methods 262, 56–65.

Willis, E.J., 1992. The powerhouse of the cell. Ultrastruct. Pathol. 16, iii–vi.

Wyllie, A.H., Kerr, J.F., Currie, A.R., 1980. Cell death: the significance of apoptosis. Int. Rev. Cytol. 68, 251–306.

Yang, J., Liu, X., Bhalla, K., Kim, C.N., Ibrado, A.M., Cai, J., Peng, T.I., Jones, D.P., Wang, X., 1997. Prevention of apoptosis by Bcl-2: release of cytochrome c from mitochondria blocked. Science 275, 1129–1132.

Youle, R.J., Karbowski, M., 2005. Mitochondrial fission in apoptosis. Nat. Rev. Mol. Cell Biol. 6, 657–663.

Zhao, J., Liu, T., Jin, S., Wang, X., Qu, M., Uhlén, P., Tomilin, N., Shupliakov, O., Lendahl, U., Nistér, M., 2011. Human MIEF1 recruits Drp1 to mitochondrial outer membranes and promotes mitochondrial fusion rather than fission. EMBO J. 30, 2762–2778.

Zhao, Y., Zhao, B., 2013. Oxidative stress and the pathogenesis of Alzheimer's disease. Oxid. Med. Cell Longev. 2013, 316523.

Zhu, T., Chen, J.L., Wang, Q., Shao, W., Qi, B., 2018. Modulation of mitochondrial dynamics in neurodegenerative diseases: an insight into prion diseases. Front. Aging Neurosci. 10, 336.

CHAPTER 9

Viral Genes in Neurological Disorders

AMRESH KUMAR SINGH, MD • VIVEK GAUR, MSC • ANAND KUMAR MAURYA, PHD • URMILA GUPTA, MSC

9.1 INTRODUCTION

Viral genes play an important role in the effective treatment of rare neurological disorders. The primary neurological symptoms are related with damage to the white matter tracts within the central nervous system (CNS). The duration and treatment of the disease can vary from the most common relapsing–remitting form to primary progressive cases(Fogdell-Hahn et al., 2002). The neurological complications include meningitis, encephalitis, myelitis, vasculopathies, lysosomal storage diseases, spinal muscular atrophy, acute and chronic radiculoneuritis, various inflammatory diseases of the eye and dominant toxic mutations (amyotrophic lateral sclerosis) and conditions of mixed etiology (Alzheimer's disease [AD], Parkinson's disease, and brain tumors) (Holz et al., 2017; Bowers et al., 2011). Many neurological disorders are among the most crucial to treat with established and recommended pharmacological approaches because it is very difficult for the drugs used in neurological diseases to reach target cells at a sufficient concentration due to the blood–brain barrier (BBB), blood–cerebrospinal fluid barrier, and meningeal–cerebrospinal fluid barrier, which may lead to side effects in other organs and treatment also affected due to the complexity of the nervous system (Zhang et al., 2012). Few significant surgical and clinical advancements have been made, including delivery of genes, sequence targeted regulatory molecules, genetically modified cells, using oligonucleotides to treat a variety of genetic diseases, and the use of either naked DNA or vector (via nanoparticle) delivery of DNA. Vector-mediated gene transfer is a modified and promising strategy to treat genetic CNS disorders, especially among monogenic diseases (Kalburgi et al., 2013). The selection of the viral vector depends on the tropism of the virus and its ability to allow sustained therapeutic gene expression in the target cells (Lukashchuk et al., 2016).

Supply of gene products during genetic interventions helps to restore function permanently and even induce replacement and repairment of lost or affected cells.

Such kinds of approaches are defined as gene therapy, in which DNA or RNA is used as the pharmacological agent to treat genetic or neurological diseases (Simonato et al., 2013). The major viral vector systems used for neurological diseases are recombinant lentivirus (LV), adenovirus (Ad), herpes simplex virus (HSV) (Fogdell-Hahn et al., 2002), and serotype 9 (AAV9). Adeno-associated virus (AAV) vector has become a choice of preference for CNS delivery, because of its increased capability to cross the BBB compared with other AAV serotypes. By using self-complementary (SC), AAV9 ensures a rapid onset of gene transcription due to its double-stranded conformation of the genome in comparison with conventional recombinant AAVs (Lukashchuk et al., 2016).

This chapter aims to provide an updated overview of the role and mechanisms of viral gene involved in the disorder and pathogenesis of CNS infections, rather than by providing an exhaustive overview of the knowledge of all neurological disorders.

9.2 VIRAL INFECTIONS AND ASSOCIATED NEUROLOGICAL DISORDERS

The CNS is protected by a highly complex barrier system (CBS), but large group of viruses are still manage to get access and induce diseases and may cause immediate or delayed neurological manifestations in humans as shown in Table 9.1. There are more than 600 neurological disorders that can occur throughout a person's lifetime. In several studies, the rate of CNS viral infections increasing each year is greater in comparison with bacterial, fungal, and protozoa infections (Swanson and McGavern, 2015). A mostly acute and chronic viral infection starts from the periphery, frequently affecting epithelial or endothelial cell surfaces. However, these viruses are ubiquitous among general population and typically take decades to acquire before disease presentation (Leibovitch and Jacobson, 2018). Infections by neurotropic viruses are usually associated with autoimmune

The Molecular Immunology of Neurological Diseases. https://doi.org/10.1016/B978-0-12-821974-4.00012-1

TABLE 9.1
Virus-Induced Neurological Disorders (Koyuncu et al., 2013).

S.NO	Viruses	Genome	Route of Entry	Clinical Presentation
1.	Alphavirus	ssRNA	BBB and ORN	Meningitis, Encephalitis
2.	Flavivirus	ssRNA	BBB and BMVECs	Meningitis, Encephalitis
3.	Herpesvirus	Large enveloped dsDNA	Sensory nerve ending, ORN, BBB, and BMVECs	Meningitis, Encephalitis, Myelitis, Polyradiculitis, and neuropathies
4.	Picornavirus	ssRNA	BBB and NMJs	Meningitis, Encephalitis, and poliomyelitis
5.	Paramyxoviruses	Nonsegmented ssRNA	BBB, ORN, and NMJs	Meningitis, Encephalitis
6.	Influenza virus	Segmented ssRNA	Peripheral nerve and ORN	Seizure, encephalitis, and encephalopathy
7.	Boronavirus	ssRNA	ORN	Myoclonus, Ocular paresis, and Encephalitis
8.	Rhabdovirus	Nonsegmented ssRNA	NMJs and ORN	Encephalitis and dysautonomia

BBB, blood−brain barrier; *BMVECs*, Bovine brain microvascular endothelial cells.

response, which cause few neurological disorders that may consider being long-term and delayed virus-induced disorders such as multiple sclerosis, Guillain–Barré syndrome, and encephalitis (Li et al., 2019). These virus-induced infections and their association with neurological disorders will be discussed in more detail in the Fig. 9.1 (Koyuncu et al., 2013).

9.2.1 Multiple Sclerosis
It is an inflammatory disorder affecting the CNS and resulting in tissue damage that may lead to neurologic dysfunction (Leibovitch and Jacobson, 2018). Viruses that today remain the most strongly association with multiple sclerosis (MS) are the human herpes virus 6 (HHV-6), Epstein−Barr virus (EBV), and Varicella zoster virus (VZV). These viruses lead to an infection in immune-compromised patients and may induce multi-focal encephalitis, cerebral infarcts, and macrophage-rich demyelinating "MS-like" periventricular cerebral and spinal cord lesions (Li et al., 2019).

9.2.2 Multifocal leukoencephalopathy
Human polyomavirus, picornavirus, and enterovirus 71 (EV71) are able to infiltrate into the CNS by the Trojan horse mode of entry. Infiltration of infected B cells into the CNS leads to infection of oligodendrocytes and astrocytes in immune-suppressed patients can result in the fatal inflammatory disease of the brain (Koyuncu et al., 2013).

9.2.3 Meningitis
Viral meningitis can occur at any age, but it is most common in children. Majority of aseptic meningitis in adult and pediatric cases are due to human enteroviruses (HEVs). Many other enterovirus types such as mumps, HSV-1 and 2, and St. Louis encephalitis viruses having the same potential can give rise to neurological manifestations ranging from aseptic meningitis to meningoencephalitis and paralytic poliomyelitis. Coxsackie B viruses and echoviruses account for most of the cases of enterovirus meningitis (Logan and MacMahon, 2008). The viral meningitis cases among childrens; are especially at risk for developing severe morbidity and mortality as shown in Fig. 9.2 (Swanson and McGavern, 2015).

9.2.4 Encephalitis
The intense inflammation of the parenchyma is known as encephalitis, while meningoencephalitis is due to the inflammation of meninges, leading to headache, fever, and neurological discomfort. Viruses that cause these abnormalities are arboviruses, HEVs, and alphaviruses, such as Eastern, Western, and Venezuelan equine encephalitis viruses (EEEV, WEEV, VEEV), which primarily infect neurons within the CNS and later has a cytopathic effect (Swanson and McGavern, 2015). In a study conducted by Wong et al., postmortem analysis of EV71-infected brains revealed virus-infected neurons, heavy immune cell infiltration, and destruction of nerve cells by phagocytes along the medulla, thalamus,

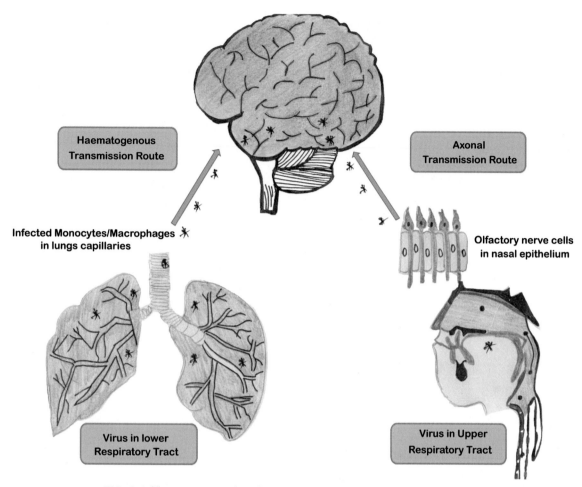

FIG. 9.1 Virus entry routes into the central nervous system (Koyuncu et al., 2013).

hypothalamus, dentate nucleus, and cerebral cortex (Wong et al., 2008).

9.2.5 Alzheimer's Disease

It is a neurodegenerative disease in aged people, which is defined as the progressive loss of memory and cognitive dysfunction. HSV-1, cytomegalovirus, and *Chlamydophila pneumoniae* and some other pathogens are indicated to contribute to the pathogenesis of AD. These pathogens usually cause infection that may lead to chronic inflammation after the period of time, in which level of proinflammatory factors by activating glial cells in the CNS increases by DNA and RNA of pathogen and destroys neuron directly or indirectly (Li et al., 2019).

9.2.6 Acute Encephalopathy

It is an interesting example of communicable disease, which leads to neuropathogenesis caused due to

paramyxoviruses, measles virus (MV), and mumps virus (MuV). Both of them are the causative agents of highly communicable diseases and can also influence serious CNS infections. Primarily, infections from MV and MuV start in the upper respiratory tract; after that, infection of lymphoid tissue of lymphatic system causes viremia and then gradually spreads into other parts of the body. MuV is highly neurotropic kind of virus and can cause acute encephalopathy in children with high incidence of dementia, seizures, coma, and death. Diagnosis of MuV-associated acute encephalopathy is made with the detection of elevated levels of multiple cytokines in cerebrospinal fluids (CSF) of children (Koyuncu et al., 2013).

9.2.7 Microcephaly

Similar to the Flaviviridae family, such as dengue virus, yellow fever, Japanese encephalitis, and West Nile viruses, risk of ZIKA virus–associated viral infection

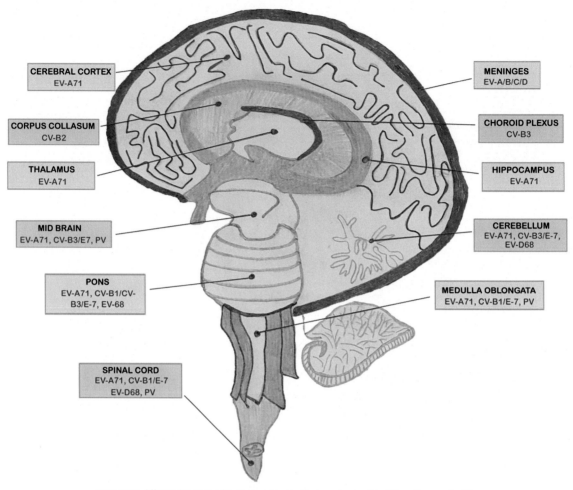

FIG. 9.2 Affected region of brain by viruses that cause meningitis and encephalitis.

markedly increases and is responsible for congenital microcephaly and serious neurologic complications. A hypothesis according to *Zhexing* et al. was that the alteration in amino acid sequence due to nucleotide mutations found in ZIKA virus strains from the current Brazilian outbreak might be the major cause of contributing the increased incidence of microcephaly. This virus is a key mediator to infiltrate transplacental barrier, which infects fetal brain during pregnancy. Changes in transplacental nutrient transport and inflammatory responses may influence fetal development deficits as shown in Fig. 9.3 (Wen et al., 2017).

9.2.8 Bipolar Disease or Schizophrenia

Borna disease virus (BDV) is a slowly replicating virus that can enter the brain, infect neurons, most

commonly in the limbic system, and remain significantly for long periods of time in the CNS without developing neuronal lysis. It is very difficult to imagine that BDV can cause neurological dysfunction in so many species including humans. According to Van den Pol et al., an increased incidence of BDV has been reported in patients in some psychiatric hospitals, where it has been associated with psychiatric and brain diseases (Van den Pol, 2009).

9.2.9 Myelitis

Some neurotropic viruses cause disease in the spinal cord commonly known as poliomyelitis. Many viruses such as WNV, EV71, poliovirus, EV71, WNV, TBEV, and JEV viruses induced myelitis and have proven their association with acute flaccid paralysis. In each and every case, viral protein was found in neuronal cell bodies dwelling in

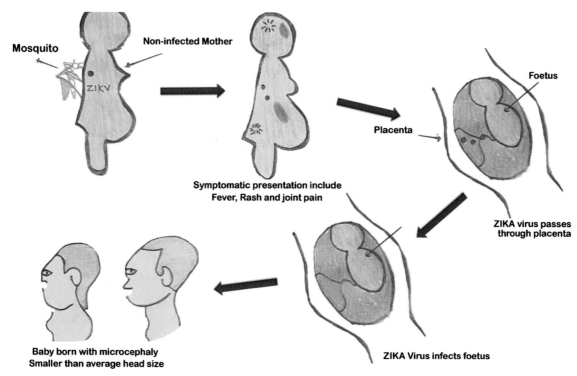

FIG. 9.3 Vertical transmission of ZIKA virus.

the front horn of the spinal rope and the cytopathology of viral infection contributed to the destruction of motor neurons (Swanson and McGavern, 2015).

9.3 GENE THERAPY AND ITS TYPES

Dr. Harry L. Malech from the National Institute of Health (NIH) started the work of human gene therapy. The Recombinant DNA Advisory Committee of the NIH in 1988 reviewed and approved the first guidelines for clinical trials for gene therapy. The first clinical trials were done in 1989 with introduction of gene markers into melanoma patients. In 1990, the first therapeutic trials began with the introduction of the adenosine deaminase (ADA) gene into the T cells of a patient with immunodeficiency disease (Dickler and Collier, 1994). Gene therapy for CNS disorder is an experimental concept. This method is potential for therapies, targeting DNA or RNA, and for clinical utilization which helps in improvement of sign and symptoms, slowing progression of disease, and correction of genetic diseases (Axelsen and Woldbye, 2018; Costantini et al., 2000). Lack of cure in multiple neurological disorders is because of limited regenerative capacity of the brain tissue and the difficulty in conveying standard drugs to the organ due to the BBB

(Lykken et al., 2018; Ingusci et al., 2019). Gene therapy is emerging as an incredible treatment method with great potential to treat and even fix some of the most common neurological diseases. It has been made possible through progress in understanding the fundamental components of illness, especially those involving sensory neurons, also by enhancement of therapeutic gene selection, vector, and methods of delivery in CNS disorders (Simonato et al., 2013).

It helps in introducing correct copy of altered gene into the cells whose malfunction might be responsible to cause a monogenic disease. However, it is a complex process that starts with the choice of a delivery system of functional gene, called as transgene (gene therapy by vector) (Ingusci et al., 2019).

9.3.1 Type of Gene Therapy

There are many methods of gene therapy, and it can be applied in two ways: somatic cell gene therapy and germline gene therapy. In somatic cell gene therapy, the gene is inserted into a vector that will carry the desired gene into the nucleus of bone marrow pluripotent cells, peripheral blood, or even directly intracerebrally into brain matter. But in germline gene therapy, the desired gene is transferred into either of gametes (sperms or ovum), consequently making them heritable to next generation (less efficient). Due to high safety and institutional ethical concerns, now it is limited to somatic cell gene therapy only. The other classification is based on whether genetic correction occurs inside (in vivo) or outside (in vitro) the body as shown in Table 9.2 (Axelsen and Woldbye, 2018).

9.3.1.1 In vivo gene therapy
In this method, vector first entered into the target cell, then the therapeutic gene will be transferred into the target cell's nucleus, and then functional protein is expressed, thereby returning the cell to normal/original state. The therapeutic targeted gene, which is carried by the vector, remains free in the nucleus as extrachromosomal material, or later it can be integrated into the genome. The in vivo strategy involves the direct delivery system of DNA (via a viral vector) to resident cells of the target tissue (Fig. 9.4; Bora, 2014). This strategy is based on two requirements; firstly, that target cells be easily accessible for infusion or injection of virus, and secondly, that the transfer vector readily and specifically infects, integrates, and then expresses the therapeutic gene (transgene) in target cells and not surrounding cells at effective levels for extended time periods. The characteristics of the most commonly used vectors are listed in Table 9.3. Retroviral vectors derived from the mouse Moloney leukemia virus and HSV-derived vectors are of limited value in most in vivo clinical situations (Miller et al., 1990; Marconi et al., 1996; Chiocca et al., 1990). However, in vivo application in several diseases has been established, such as HSV in peripheral neuropathies and retroviral vectors for malignancies (Pulkkanen et al., 2001; Solly et al., 2003). The use of nonviral

and lipophilic molecules to transfer genetic material into the cells in vivo has been restricted by low transfer efficiency (Uchida et al., 2002).

9.3.1.2 In vitro gene therapy
The inserted DNA is incorporated into the specific cell and produces the needed protein encoded by the inserted gene. Therapeutic genes are inserted into viral DNA, liposome, or plasmid DNA.

In in vitro gene therapy, the viral vector is altered genetically, so that it cannot reproduce. In the first step, the targeted genetic material is inserted into the genome of the vector. After insertion of targeted genetic material, the viral vector is mixed with the cells collected from the patient and reinfused after transduction. Then the multiplication of cells takes place in vivo and produces the functional enzyme/protein (Panda and Panda, 2019; Kaufmann et al., 2013). The in vitro approach involves the transfer of a therapeutic gene to cells in vitro (in culture) followed by transplantation of these modified cells to the target tissue as shown in Fig. 9.5 (Bora, 2014). The modified, transplanted cells act as an engineered secretory tissue, synthesizing and releasing desired proteins to the local environment. To exploit the method successfully, an appropriate surrogate cell population must be identified. This cell population should be endowed with specific characteristics that fulfill several criteria. The cells should (1) be readily available and relatively easily obtained; (2) be able to survive for long periods of time in vivo; (3) be able to express a transgene at high levels for extended durations; and (4) not elicit a host-mediated immune reaction (Selkirk, 2004).

9.4 METHODS FOR GENE THERAPY
Gene therapy is basically used to fix the altered genes responsible for genetic neurological disorders by following different approaches such that abnormal gene could be swapped or repaired for a normal homologous gene recombination through selective reverse mutation. Most commonly, nonfunctional gene is replaced by a normal gene by inserting it into a nonspecific location within the genome and also regulates the extent to which a gene is allowed to turned on or off (Panda and Panda, 2019). Most of the vectors used in gene therapy trials are nonimmunogenic and nonpathogenic small stable, which does not interfere with native genome (Axelsen and Woldbye, 2018). The characteristics of an ideal vector system are the ability to infect a wide range of host cells, efficient delivery of the therapeutic gene, and postinfection gene

TABLE 9.2
Difference between in vivo and in vitro Gene Delivery System.

S. No.	In Vivo	In Vitro
1.	Technically simple	Complicated system
2.	Vectors introduced directly	Vector did not introduce directly
3.	Less control over target cells	High control
4.	Less invasive	More invasive
5.	Safety check not possible	Safety check possible

Viral vector   Liposome   Plasmid DNA

1. THERAPEUTIC GENES ARE INSERTED INTO VIRAL DNA, LIPOSOME OR AS PLASMID DNA

2. GENETICALLY ALTERED DNA IS INSERTED INTO PATIENTS BODY BY DIRECT TISSUE INJECTION

3. THE INSERTED DNA IS INCORPORATED INTO THE SPECIFIC CELL AND PRODUCES THE NEEDED PROTEIN ENCODED BY THE INSERTED GENE

FIG. 9.4 Therapeutic gene transfer into the target cells (Bora, 2014).

TABLE 9.3
Features of the Viral Vectors (Simonato et al., 2013).

Characteristic	Adenoviral Vector	Adeno-Associated Viral Vector	RETROVIRAL VECTOR		Herpes Viral Vector
			Murine Vector	Lentiviral Vector	
Wild-type virus	Double-stranded linear DNA (36 kb)	Single-stranded DNA (4.7 kb)	Diploid positive strand RNA (9.2 kb)	Diploid positive strand RNA (9.2 kb)	Double-stranded linear DNA (152 kb)
Derived from pathogenic virus	Yes	No	Yes	Yes	Yes
Maximum size of gene insert	~7.5 kb[a]	~4.5 kb	~8 kb	~8 kb	~20–40 kb
Integration into host-cell genome	No	No[b]	Yes	Yes	No
Achievable titer	High	High	Low	Low	High
Duration of transgene expression	Transient (days to weeks)	Long-lasting (months to years)	Long-lasting (months to years)	Long-lasting (months to years)	Transient (days to weeks)[c]
Target cells	Dividing and nondividing	Dividing and nondividing	Dividing	Dividing and nondividing	Dividing and nondividing
Safety issues	Immune and inflammatory response	Insertional mutagenesis[d]	Insertional mutagenesis	Insertional mutagenesis	Immune and inflammatory response

[a] 35 kb for gutless vectors.
[b] Some integration at a very low frequency occurs.
[c] Years in the PNS.
[d] Minimal compared with other vectors.

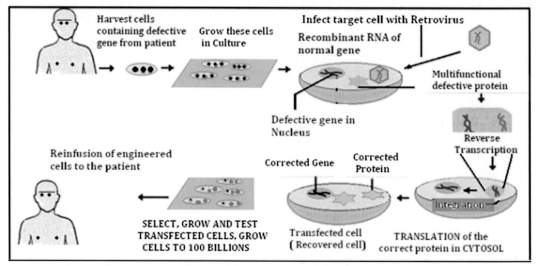

FIG. 9.5 Therapeutic gene transfer into the target cells (Bora, 2014).

expression. Such a vector should evoke little immune response, offer tissue specificity, and be able to maintain high-titer preparations as shown in Table 9.3 (Tang and Chiocca, 1997). A good vector should fulfill many requirements; some of essential characters are mentioned in the following:

- **Manipulation:** The vector should be easily manipulated for recombination and propagation in suitable susceptible hosts.
- **High cloning capacity:** The vector should allow the introduction of one or more desired genes and the regulatory sequences that assure the desired spatial and temporal restriction of the transgene expression.
- **Minimal invasiveness:** The vector should not cause uncontrolled or undesired alterations of the host genome sequences. The integration of a vector into the cellular genome can induce insertional mutagenesis.
- **Selectivity for the cellular target:** The transgene should be expressed specifically in the target cells.
- **Absence of immunogenicity:** The vector should not contain genes that induce immune responses or other immune/offending factors that may be harmful to the body.
- **Stability over time:** The vector should be transferred in an unaltered manner in the cell progeny and/or must allow a correct and prolonged expression of the transgene(s).

Fig. 9.6 represents the genome of the virus along with that of the corresponding viral vector, showing viral structural genes, viral genes involved in replication, and the genes that are essential or nonessential (accessory) for replication or growth of virus.

9.5 VECTORS FOR GENE THERAPY

There are many vectors used for gene therapy. These vectors can be classified into two types as shown in Table 9.4.
1. Nonviral
2. Viral vectors

9.5.1 Nonviral Vectors

This type of vectors offers some advantages low cost, simple technique for the production, and less pathogenicity. DNA carried through nonviral vectors must cross the extracellular and intracellular barriers, limiting the efficiency to reduce the insertion of bacterial plasmid with foreign genetic materials.

The most commonly used tools to deliver nonviral vectors are cationic polymers and cationic lipids. Their efficiency is dependent on cationic charge, saturation, and linker stability. The polyethylene glycol (PEG) is used to increase the stability of DNA in the circulation. The other method is bioresponsive polymers that endeavor the physical–chemical properties of the microbiological environment (pH, presence of reducing agents, etc.). For the promotion and release of the genetic material intracellularly, and another tested trials are over acetyl bonds, which degrade at the pH of the endosomal environment, and for disulfide bridges which are reduced in the cytoplasmic environment (Ingusci et al., 2019; Kay et al., 1997).

9.5.2 Viral Vectors

Retroviruses, adenoviruses, and AAV are some commonly used viral vectors, whereas some less commonly used viral vectors are derived from the HSV-1, the Baculovirus, and so on. We will discuss these vectors in detail.

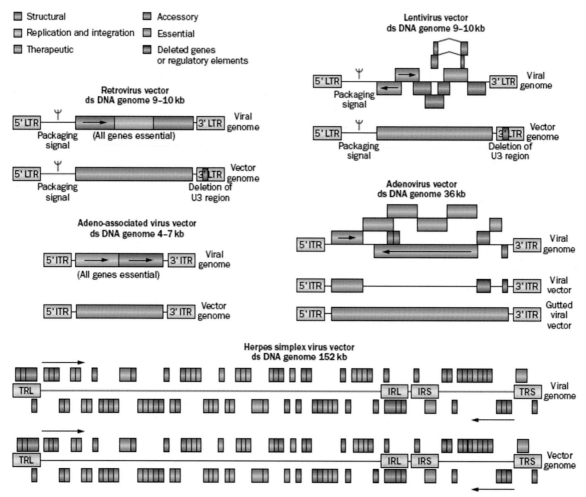

FIG. 9.6 Various viral vectors genomes used in gene therapy (Simonato et al., 2013).

9.5.2.1 Adeno-associated virus vectors

Among vector systems, these viruses are the new members; it consists of small nonpathogenic virion which is 20–24 nm in diameter and carries single-stranded DNA genome that belongs to the *Parvoviridae* family. AAVs have the capability of accommodating <4.5 kb transgene capacity and inverted terminal repeats (ITRs) that enhance extrachromosomal replication process and genomic integration of the transgene. The advantages of AAVs are that they are resistant to many physical and chemical factors such as detergents and may survive in variable pH, freezing and thawing cycles, and lyophilizing and during denaturation process upto 60°C (Selkirk, 2004). In phenotypic infections of AAV, replicative amplification of the impact of ITR-flanked genome facilitates integration mediated by a set of *rep* proteins

TABLE 9.4
Types of Vector Used in Gene Therapy.

S. No.	Nonviral Vectors	Viral Vectors
1.	Lipid complex	Adenovirus
2.	Liposomes	Retrovirus
3.	Peptide/protein	Adeno-associated virus
4.	Polymers	Lentivirus
5.	—	HSV
6.	—	Vaccinia virus

encoded by the *rep* gene, into the host cell genome. AAV-based vectors are facilitating long-term expression of transgene both by incorporation and maintenance as an episomal component within the host cell nucleus (Costantini et al., 2000; Lykken et al., 2018).

9.5.2.2 Retrovirus vectors and ex vivo delivery

A broad range of small RNA viruses belong to the *Retroviridae* family and are classified as the most common vector for non-CNS use and in vitro approaches. The other viruses from the same family are γ-retrovirus, Moloney murine leukemia virus, lentivirus, HIV, and Rous sarcoma virus. Retroviral vectors express the retroviral *gag–pol–env* genes and are able to release vectors continuously. The retroviral vectors are bounded by long terminal repeat (LTR) regions, a virion packaging signal (psi), and a primer binding site for reverse transcription (Ingusci et al., 2019; Choudhury et al., 2017).

The older group of viruses are more suitable for in vitro gene therapy applications because they are unable to infect nondividing cells and also very difficult to reach in high viral counts. But others infect both proliferating and nonproliferating cells, because long-term gene expression is not possible in the presence of agent–host inflammatory responses. The disadvantages of using retrovirus as a vectors include the possibility of insertional mutation in endogenous cellular genes and rapid inhibition of the viral genome, resulting in low transduction capacity (Tang and Chiocca, 1997).

9.5.2.3 Adenovirus vectors

It is the first generation of replication-defective Ad vectors without viral genes, constructed by erasing E1a, E1b, and E3 genes, and because of a high host immune response to the viral antigens, this vector is confined to the constrained use in gene therapy. These are classified into seven subgroups, A to G, and are contrast in cell tropism (Ingusci et al., 2019). New changes in the vector reduced the expression of viral proteins by deleting E2a and E4 genes, as well as explicit ORF in the E4 gene. Maintenance of E3 also diminishes the antigenicity of the AV (Costantini et al., 2000). The benefits of this vector incorporate high transfection viability and the ability to accommodate large genes. Further advantages include the ability to be concentrated to high-titer preparations, and the disadvantages of this vector are lytic infection and intense neuronal toxicity (Tang and Chiocca, 1997; Smith, 1998).

9.5.2.4 HSV-1 and other HSV-1-associated vectors

Study led on HSV inertness by Berges et al. determines the condition of the genome, gene expression, and the role of immunity in maintaining the latent state. A few attributes make herpes virus more efficient as a viral vector and suitable for the treatment of neurological disorders; HSV shows high infectivity for both dividing and nondividing cells, and it expresses over 80 distinct quality of genes, out of which many of them are not required for its replication cycle (Berges et al., 2007). This virus vector can possibly carry a substantial payload; this advantage allows the insertion of multiple or very large transgenes in highly defective vectors (Artusi et al., 2018). It is extremely difficult to create HSV vectors for CNS disorders to demonstrate durable therapeutic gene expression without the activation of viral genes. The other HSV-associated vectors generated to accomplish these goals are HSV/AAV hybrid amplicon vectors, HSV-1 amplicon vectors, and replication-defective HSV-1 vectors.

9.5.2.5 Alphavirus vectors

Alphavirus vectors designed an attractive method for gene delivery and expression of heterologous proteins because of their rapid and basic recombinant virus particle manufacturing capability and their broad range of mammalian host cell transduction (Lundstrom, 2012). Vectors mostly designed for gene delivery are Semliki Forest virus (SFV), Sindbis virus (SIN), and Venezuelan equine encephalitis virus (VEE). The nonstructural genes of alphavirus generate the replicase complex, which is responsible for highly efficient RNA replication susceptibility for neuronal cells and can give outrageous degrees of heterologous gene expression (DiCiommo and Bremner, 1998). These vectors generally generate short-term transient expression, which may limit their therapeutic use in most of the neurological disorders. However, plasmid DNA can be directly transfected into mammalian cells for recombinant protein expression (Leitner et al., 2003).

9.5.2.6 Challenges and future prospects of gene therapy

The current ongoing advancement in gene silencing applying both RNA interference and micro-RNA approaches will help in the expansion of the application range. Moreover, alphaviruses spread out interesting models that lead to dominate research into potential gene therapy solutions for neurological disorders such as demyelinating, monogenetic disorders, and spinal motor diseases where AAV vectors emerge as the most promising treatment method (Lundstrom, 2015).

In a study directed by Ming et al., it was examined that vaccinia-related kinase 2 *(VRK2)* was a significant factor in the beginning and development of schizophrenia (SCZ) and major depressive disorder (MDD). There is an urgent need for additional exploration of its role in the genetic engineering of psychiatric disorders, biological impact on neurodevelopment, brain mechanism and function, and the roles of associated molecules in neurological diseases (Li and Yue, 2018).

Much advancement has been made till date; however, challenges remain same with respect to the long-term efficacy of such treatments because successful clinical trials are very recent, and long-term safety, efficacy, and toxic effects from gene replacement therapy are currently impossible/unknown to determine. In addition, the most important limitation to clinical translation of gene therapy is the requirement of higher doses for gene transfer into the adult CNS for the systemic administration of CNS diseases in older children and adults. The relative unavailability of validated biomarkers and the ability to perform biopsy CNS tissue makes it more difficult to monitor the efficacy of treatments (Mingozzi and High, 2013).

The entry route of viral vectors and genes can impact the immune response to both capsid and transgene, and it can potentially impact safety if the viral vector delivery is not directed into the target site i.e., CNS, so there is a scope for exposure in the periphery. Nowadays, this is a great matter of concern to scientist working in this area. Despite these drawbacks, gene therapy represents a new hope for the treatment of rare neurological diseases. "Gene therapies have the potential to offer life-changing benefits to patients with rare genetic diseases, as acknowledged by the competent regulatory authorities across the world by giving priority and accelerated reviews. We believe that gene therapies have their own role in clinically important treatment option in many neurological disorders and rare monogenic genetic disorders" (Davies, 2019).

ACKNOWLEDGMENT

Authors would like to thank the faculties and all technical staff of department of Microbiology, Baba Raghav Das Medical College, Gorakhpur, Uttar Pradesh, India, for their support during the completion of this chapter.

REFERENCES

Artusi, S., Miyagawa, Y., Goins, W.F., Cohen, J.B., Glorioso, J.C., 2018. Herpes simplex virus vectors for gene transfer to the central nervous system. Diseases 6 (3), 74.

Axelsen, T.M., Woldbye, D.P.D., 2018. Gene therapy for Parkinson's disease, an update. J. Parkinsons Dis. 8 (2), 195−215.

Berges, B.K., Wolfe, J.H., Fraser, N.W., 2007. Transduction of brain by herpes simplex virus vectors. Mol. Ther. 15, 20−29.

Bora, U., 2014. NPTEL − Bio Technology − Genetic Engineering & Applications. https://nptel.ac.in/content/storage2/courses/102103013/pdf/mod8.pdf. IITs and IISc.

Bowers, W.J., Breakefield, X.O., Sena-Esteves, M., 2011. Genetic therapy for the nervous system. Hum. Mol. Genet. 20 (R1), R28−R41.

Chiocca, E.A., Choi, B.B., Cai, W.Z., DeLuca, N.A., Schaffer, P.A., DiFiglia, M., et al., 1990. Transfer and expression of the lacZ gene in rat brain neurons mediated by herpes simplex virus mutants. N. Biol. 2 (8), 739−746.

Choudhury, S.R., Hudry, E., Maguire, C.A., Sena-Esteves, M., Breakefield, X.O., Grandi, P., 2017. Viral vectors for therapy of neurologic diseases. Neuropharmacology 120, 63−80.

Costantini, L.C., Bakowska, J.C., Breakefield, X.O., Isacson, O., 2000. Gene therapy in the CNS. Gene Ther. 7 (2), 93−109.

Davies, N., 2019. Gene therapy Ushers in an era of hope for rare disorders. Neurol. Live. https://www.neurologylive.com/journals/neurologylive/2019/june-2019/gene-therapy-ushers-in-an-era-of-hope-for-rare-neurological-disorders.

DiCiommo, D.P., Bremner, R., 1998. Rapid, high level protein production using DNA-based Semliki Forest virus vectors. J. Biol. Chem. 273 (29), 18060−18066.

Dickler, H.B., Collier, E., 1994. Gene therapy in the treatment of disease. J. Allergy Clin. Immunol. 96, 942−951.

Fogdell-Hahn, A., Soldan, S.S., Jacobson, S., 2002. Association of chronic progressive neurological disease and ubiquitous viral agents: lessons from human herpesvirus 6 and multiple sclerosis. Mol. Psychiatr. 7 (Suppl. 2), S29−S31.

Holz, C.L., Nelli, R.K., Wilson, M.E., Zarski, L.M., Azab, W., Baumgardner, R., et al., 2017. Viral genes and cellular markers associated with neurological complications during herpesvirus infections. J. Gen. Virol. 98 (6), 1439−1454.

Ingusci, S., Verlengia, G., Soukupova, M., Zucchini, S., Simonato, M., 2019. Gene therapy tools for brain diseases. Front. Pharmacol. 10, 724.

Kalburgi, S.N., Khan, N.N., Gray, S.J., 2013. Recent gene therapy advancements for neurological diseases. Discov. Med. 15 (81), 111−119.

Kaufmann, K.B., Büning, H., Galy, A., Schambach, A., Grez, M., 2013. Gene therapy on the move. EMBO Mol. Med. 5 (11), 1642−1661.

Kay, M.A., Liu, D., Hoogerbrugge, P.M., 1997. Gene therapy. Proc. Natl. Acad. Sci. USA 94 (24), 12744−12746.

Koyuncu, O.O., Hogue, I.B., Enquist, L.W., 2013. Virus infections in the nervous system. Cell Host Microbe 13 (4), 379−393.

Leibovitch, E.C., Jacobson, S., 2018. Viruses in chronic progressive neurologic disease. Mult. Scler. 24 (1), 48−52.

Leitner, W.W., Hwang, L.N., deVeer, M.J., Zhou, A., Silverman, R.H., Williams, B.R., et al., 2003. Alphavirus-based DNA vaccine breaks immunological tolerance by activating innate antiviral pathways. Nat. Med. 9 (1), 33−39.

Li, M., Yue, W., 2018. *VRK2*, a candidate gene for psychiatric and neurological disorders. Mol. Neuropsychiatr. 4, 119−133.

Li, L., Mao, S., Wang, J., Ding, X., Zen, J.Y., 2019. Viral infection and neurological disorders—potential role of extracellular nucleotides in neuro-inflammation. ExRNA 1, 26.

Logan, S.A.E., MacMahon, E., 2008. Viral meningitis. Br. Med. J. 336 (7634), 36−40.

Lukashchuk, V., Lewis, K.E., Coldicott, I., Grierson, A.J., Azzouz, M., 2016. AAV9-mediated central nervous system-targeted gene delivery via cisterna magna route in mice. Mol. Ther. Methods Clin. Dev. 3, 15055.

Lundstrom, K., 2012. Alphavirus vectors for therapy of neurological disorders. J. Stem Cell Res. Ther. S4.

Lundstrom, K., 2015. Alphaviruses in gene therapy. Viruses 7 (5), 2321–2333.

Lykken, E.A., Shyng, C., Edwards, R.J., Rozenberg, A., Gray, S.J., 2018. Recent progress and considerations for AAV gene therapies targeting the central nervous system. J. Neurodev. Disord. 10, 16.

Marconi, P., Krisky, D., Oligino, T., Poliani, P.L., Ramakrishnan, R., Goins, W.F., 1996. Replication-defective herpes simplex virus vectors for gene transfer in vivo. Proc. Natl. Acad. Sci. USA 93 (21), 11319–11320.

Miller, D.G., Adam, M.A., Miller, A.D., 1990. Gene transfer by retrovirus vectors occurs only in cells that are actively replicating at the time of infection. Mol. Cell Biol. 10 (8), 4239–4242.

Mingozzi, F., High, K.A., 2013. Immune response to AAV vectors: overcoming barriers to successful gene therapy. Blood 122 (1), 23–36.

Panda, P.K., Panda, P., 2019. A narrative review on gene therapy in pediatric neurological disorders-A promising new avenue in twenty first century. EC Clin. Exp. Anat. 2 (3), 90–96.

Pulkkanen, K.J., Parkkinen, J.J., Laukkanen, J.M., Kettunen, M.I., Tyynela, K., Kauppinen, R.A., et al., 2001. HSV-tk gene therapy for human renal cell carcinoma in nude mice. Canc. Gene Ther. 8 (7), 529–536.

Selkirk, S.M., 2004. Gene therapy in clinical medicine. Postgrad. Med. 80 (948), 560–570.

Simonato, M., Bennett, J., Boulis, N.M., Castro, M.G., Fink, D.J., Goins, W.F., et al., 2013. Progress in gene therapy for neurological disorders. Nat. Rev. Neurol. 9 (5), 277–291.

Smith, G.M., 1998. Adenovirus-mediated gene transfer to treat neurologic disease. Arch. Neurol. 55 (8), 1061–1064.

Solly, S.K., Trajcevski, S., Frisén, C., Holzer, G.W., Nelson, E., Clerc, B., et al., 2003. Replicative retroviral vectors for cancer gene therapy. Cancer Gene Ther. 10 (1), 30–39.

Swanson 2nd, P.A., McGavern, D.B., 2015. Viral diseases of the central nervous system. Curr. Opin. Virol. 11, 44–54.

Tang, G., Chiocca, A., 1997. Gene transfer and delivery in central nervous system disease. Neurosurg. Focus 3 (3), e2.

Uchida, E., Mizuguchi, H., Ishii-Watabe, A., Hayakawa, T., 2002. Comparison of the efficiency and safety of non-viral vector-mediated gene transfer into a wide range of human cells. Biol. Pharm. Bull. 25 (7), 891–897.

Van den Pol, A.N., 2009. Viral infection leading to brain dysfunction: more prevalent than appreciated? Neuron 64 (1), 17–20.

Wen, Z., Song, H., Ming, G.L., 2017. How does Zika virus cause microcephaly? Genes Dev. 31 (9), 849–861.

Wong, K.T., Munisamy, B., Ong, K.C., Kojima, H., Noriyo, N., Chua, K.B., et al., 2008. The distribution of inflammation and virus in human enterovirus 71 encephalomyelitis suggests possible viral spread by neural pathways. J. Neuropathol. Exp. Neurol. 67 (2), 162–169.

Zhang, Y., Zhu, G., Huang, Y., He, X., Liu, W., 2012. Development of foamy virus vectors for gene therapy for neurological disorders and other diseases. In: Chen, K.-S. (Ed.), Advanced Topics in Neurological Disorders. ISBN: 978-953-51-0303-5.

CHAPTER 10

A Gene Map of Brain Injury Disorders

MOHIND C. MOHAN, MSC, PHD • LAKSHMI KESAVAN, MSC •
BABY CHAKRAPANI P.S, MSC, PHD

10.1 TRAUMATIC BRAIN DAMAGE

Traumatic brain injury (TBI), also referred as "silent epidemic" is the cause of death and disability than any other known traumatic insults (Vaishnavi et al., 2009; Rusnak, 2013; Dewan, 2019). Each year, 69 million people are estimated to suffer from TBI worldwide, of which mild TBI (mTBI) affects 55.9 million people and the rest 5.48 million suffers from severe TBI(Dewan, 2019). TBI is one of the leading causes of morbidity and mortality in children with approximately 2685 deaths, 37,000 hospitalizations and around 435,000 emergency visit in the United States alone (Kurowski et al., 2017). According to the report published by the center for disease control and prevention, USA, the number of people hospitalized and death has increased by 53% since 2006 (National Center for Injury Prevention and Control, CDC, 2019).

While it is a known fact that most cases of TBI are left unrecognized (Rusnak, 2013). The pathophysiology of TBI is complex and multifactorial. The brain damage associated with TBI can be classified into primary brain damage and secondary brain damage. Primary brain damage results from a direct insult to the brain parenchyma leading to tissue damage and hemorrhage followed by cerebral edema, increased intracranial pressure, tissue hypoxia, and disruption of the blood–brain barrier. Secondary brain damage may be developed immediately after TBI or may take hours or days. Secondary brain damage is characterized by cellular, neurochemical, and molecular changes. These include neuronal cell death, apoptosis, inflammation, excitotoxicity, Aβ-peptide deposition, altered calcium homeostasis, oxidative stress, and cytoskeletal and mitochondrial dysfunction (Dardiotis et al., 2010). As a result of TBI, there occurs large-scale disruption of brain connections like diffuse axonal injury, changes in white matter integrity, decrease in gray matter volume, neural cellular damage, concussion, hematomas (epidural hematomas, subdural hematomas, subarachnoid hemorrhage, and intracerebral hematoma), contusions (bruising or swelling of brain as a result of bleeding from small blood vessels in the brain), skull fractures, chronic traumatic encephalopathy (CTE), and increased intracranial pressure (NIH, 2020; Raizman et al., 2020).

TBI arises mainly as an outcome of penetrating/open brain injury (object penetrates the skull and enters the brain) and nonpenetrating/closed/blunt head injury (strong jolt on the brain within the skull caused by an external force) (NIH, 2020). Based on the clinical features like duration of injury, changes in consciousness, neurological symptoms like amnesia, observation from imaging techniques like CT and MRI, TBI can be clinically classified as mild, moderate, and severe. Among the three, moderate and severe TBI require neurosurgery and intensive care interventions (Blennow et al., 2016).

10.2 NEED FOR GENE MAPPING OF BRAIN INJURY DISORDERS

mTBI (or concussion) is the least severe form and represents 80%−90% cases of TBI. mTBI is challenging to diagnose due to its vast pathology. Due to this limitation, mTBI can lead to a significant risk of neurodegenerative diseases like Alzheimer's disease (AD) and other neurological disorders like seizures, epilepsy, dementia, headaches and migraines, CTE, posttraumatic stress disorder, emotional, cognitive, and sleeping problems (Arneson et al., 2018). However, no single tests are available, considering the importance of the assessment of mTBI (Blennow et al., 2016). There is a lack of understanding in molecular changes that occur during acute, subacute, and chronic phases of brain injury. Most of the study relates to transcriptome changes that occur at the acute post-TBI phase (Lipponen et al., 2016). Various gene expression studies have shown the role of genes in the progression of brain damage at the secondary stage of TBI (Lei et al., 2009; Dardiotis et al., 2010).

The Molecular Immunology of Neurological Diseases. https://doi.org/10.1016/B978-0-12-821974-4.00002-9

Gene mapping and transcriptome profiling can help in developing a better therapeutic strategy for brain injury disorders and other neurological disorders. Investigating the wide changes in the gene expression can identify the molecular basis of mTBI pathogenesis, which includes problems associated with responding to treatment and injury recovery. Overall dysregulation of cellular function and other effects due to mTBI may trigger the initiation of various neurodegenerative conditions (Diaz-Arrastia et al., 2006; Arneson et al., 2018). The cell-type-specific markers have future potential as molecular diagnostic markers having higher reliability than conventional neuroimaging techniques in the diagnosis of concussion brain injury and other mTBI (Arneson et al., 2018). The genetic factors underlying other neurological disorders are well understood, and detailed gene maps are established (Kayani, 2020), the identification of genes responsible for pathogenesis following brain injury are being investigated and construction of gene maps for brain injury disorders are carried out by various research groups. The emergence of various international collaborations and data sharing among researchers across the globe have paved the way for a better understanding of the underlying genetic factors in various post-TBI cases. One such international collaboration in the field of neurogenetics is ENIGMA (Enhancing Neuroimaging Genetics through Meta-Analysis) consortium consisting of more than 50 working groups spanning across the globe. The workgroups are in the process of developing methodology and tools for analyzing the influence of genetic factors on brain disorders by merging neural genetics, epigenetics, imaging, and electroencephalography data (Thompson et al., 2020). ENIGMA brain injury workgroup is dedicated in developing a reliable biomarker for the assessment of TBI across the injury spectrum (Dennis, 2020).

10.3 GENE MAP OF BRAIN INJURY DISORDERS

As in the case of other genetic disorders, the gene defects for neurological and psychiatric disorders resides in the noncoding regions of the genome. In a study by A. Nott et al., genome-specific regulatory elements of human brain cells like astrocytes, microglia, neurons, and oligodendrocytes have been mapped, which helps elucidate the transcriptional mechanism behind the development of neurological functions in health and disease. This study provides a brain cell–type-specific enhancer–promoter interactome map to identify the genes involved in neurological and psychiatric disorders and also the types of cell involved (Nott et al., 2019;

Koch, 2020). The identification of molecular pathways in various traits and diseases can be answered by genome-wide association studies (GWASs), which define the association between genetic variants and phenotypes of interest. By GWASs, hundreds of single nucleotide polymorphisms (SNPs) have been identified with neuropsychiatric disorders (Maurano et al., 2012; Nott et al., 2019). Thus, GWASs have become a standard tool for identification of genetic variation underlying complex diseases and traits (Tang, 2020). As most of the disease-causing SNPs identified by GWAS resides in the noncoding regions of the genome, it is challenging to elucidate the underlying biological mechanism. Multimarker analysis of genome annotation (MAGMA), a computational tool, is used to answer this problem by aggregating SNP association to the nearest genes. MAGMA does not take into consideration that noncoding SNP can regulate long distal genes by bringing the distal enhancers in contact with gene promoter and also the tissue-specific relationships. This is a major drawback as most of the disease-risk SNPs is present in the regulatory elements of disease-specific tissues. Won and colleagues have devised an improved platform H-MAGMA to overcome these issues where they have assigned the noncoding SNPs to their cognate genes based on the long-range interactions in the disease-relevant tissues. By H-MAGMA, they have developed an improved computational tool for the prediction of brain-disorder risk genes by incorporating brain chromatin interaction profiles. Using H-MAGMA, they have identified genes and cell-type specificities for nine brain disorders (Sey et al., 2020).

While looking at the published literature, we can see that most of these publications focus on the genetic influence of the APOE gene (Zeiler, 2019). A. Nott and his team have shown that the gene variants in transcriptional enhancers for AD are mainly confined to microglial enhancers and present in the noncoding regulatory regions. With the help of interactome maps, which connects disease-risk variants in cell type–specific enhancers to promoters, a vast microglia gene network in the case of AD has been identified (Nott et al., 2019). One of the major epidemiological reason contributing to AD is a head injury. Aβ deposition can be seen in one-third of the mortality occurring due to severe head injury. Various studies have reported a higher frequency of *ApoE4* allele in TBI and Aβ accumulation (Nicoll et al., 1995; Dardiotis et al., 2010). *ApoE4* isoforms bind with Aβ with high affinity and promote rapid aggregation into amyloid fibrils (Dardiotis et al., 2010). In a meta-analysis study carried out by Zhou et al., on 14 cohort studies, which included 2527 participants indicates that *ApoE4* allele increased the risk of poor long-term outcome of TBI arising after 6 months of injury (Diaz-Arrastia et al., 2006; Zhou et al., 2008; Dardiotis et al., 2010).

Zieler and his group (Zeiler et al., 2019a), in their paper, have described genes and SNPs associated with them that may play an important role in the cerebral dysfunction after TBI. Impaired cerebral autoregulation after moderate or severe trauma is associated with mortality and morbidity 6 months post TBI (Czosnyka et al., 1997; Zeiler et al., 2018, 2019a). The genes that are subjected to SNPs that might lead to posttraumatic complications are given in Table 10.1.

Arneson and colleagues have reported the mTBI-induced gene expression in hippocampal cells using single-cell sequencing method. They tried to provide gene markers that can act as a signature of mTBI pathogenesis. While examining the global transcriptome sequence of mTBI and Sham models, Arneson et al., found the significant shifts in the transcriptome of cell types present in dentate gyrus like the granule cells, astrocytes, microglia, ependymal cells, oligodendrocytes, CA1 neurons, and GABAergic interneurons. A large-scale reorganization in the gene coexpression was found to occur among cells like increased interaction between astrocytes and ependymal cells with neurons and microglia to oligodendrocytes. The interaction decreased in case of oligodendrocytes to neurons. These interactions were also reflected in the gene expression patterns. During mTBI, the correlation between neuron *bdnf* and cell metabolism genes in microglia was lost. Likewise, a strong association was seen by *ApoE* (AD risk gene) in astrocytes and ependymal cells with mitochondrial metabolism genes in neurons post-mTBI. The group has also identified DEGs among various cell types in mTBI and Sham model, which could act as potential gene markers. For example, *Id2* gene is specifically upregulated in posttraumatic epilepsy. Similarly, a brain resident microglia marker *P2ry12* is downregulated in post-mTBI and can be potentially used as inflammatory marker post-mTBI. *Klhl2*, a CA1 pyramidal cell-specific DEG, which codes for an actin-binding protein that overexpresses post-mTBI can potentially be used as a marker for blast TBI-associated neuroticism. In CA type 2 cells *Arhgap32*, which encodes for neuron-associated GTPase activating protein, is upregulated and may be important in the transmission of signals across hippocampal cells. The gene *Fxyd5* coding for glycoprotein that stimulates chemokine production and also in downregulating E−cadherin, thus inhibiting cell adhesion is found to be upregulated post-mTBI endothelial cells, suggesting

its potential role in neuroinflammation and disruption of blood−brain barrier (Arneson et al., 2018).

The astrocytic expression of the genes that are linked with metabolic depression such as *MDH1*, *ATP5B*, *COX4I1*, *ATP5A1*, and *NDUFS7* are downregulated in TBI. *NDUFA4*, *ATP5D*, *ATP5G1*, *NDUFV3*, and *COX8A* are downregulated in CA1 neurons following TBI and are associated with metabolic depression. The genes of calcium/calmodulin pathways such as *S100A11* and *SYT1* are upregulated and *CALM1*, *CALM2*, and *CAMK4* are downregulated in astocytes following TBI. The genes of inflammatory pathways are upregulated (*CEBPB*, *IL1B*, and *CXCL*) and downregulated (*SELPLG* and *CX3CR1*) in microglia during TBI. The oligodendrocytes are essential for myelination of the neurons. The *KLK6* gene is upregulated, and *MBP*, *MAL*, *SIRT2*, and *TSPAN2* are downregulated in TBI, which may affect the process of myelination. Oligodendrocyte differentiation is also affected in TBI by the upregulation of *SOX10* and downregulation of *GSTP1*, *CNP*, *TSPAN2*, and *OLIG1* genes (Arneson et al., 2018).

Myelination is also affected in TBI by the downregulation of *MBP*, *PLP1*, *SIRT2*, and *MAG* genes in oligodendrocyte precursor cells. These cells also provide immune response. The genes that are upregulated in TBI are *PRKX* and downregulated are *EGR1* and *FYN*. The genes of the amyloid pathway are affected in TBI. There is endothelial downregulation of *ITM2A* and *ITM2B* and ependymal downregulation of *ITM2C*, *APOE*, *BACE2*, and *APBB1* genes. The genes of the amyloid pathway that are upregulated in TBI are *TTR*, *B2M* (endothelial), and *CST3* (ependymal). Ependymal downregulation of cilia-related pathway is found in TBI. The downregulated genes are *TMEM107*, *IFT43*, *DYNLL1*, *SPAG17*, and *SPEF2*. Pathways of platelet degranulation are dysregulated in TBI as evident with the upregulation of *IGF2* and downregulation of *RARRES2*, *SCG3*, *CLU*, and *PROS1* genes. Genes of dendritic morphogenesis such as *EPHA4*, *STK11*, *BAIAP2*, and *BHLHB9* are downregulated in TBI. Genes that are involved in neurogenesis like *NPY*, *INHBA*, and *ADGRL3* (upregulated) and *NDN* and *CCK* (downregulated) in TBI. Energy-related pathways of neurons are affected in TBI such as *ATP1A1*, *ATP1A2*, *ATP2B2* (upregulation), and *ATP1B1* (downregulation) (Arneson et al., 2018).

Genes of synaptic signaling are upregulated (*SCN1B*, *CPLX2*, *SLC17A7*, *GRIK4*, *GRIN2B*) in neurons and glutamate transport genes are upregulated (*SLC17A7*, *GRIN2B*, *GRIA1*, *GRIA2*) specifically in CA1 neurons. The genes of biosynthetic pathways are upregulated (*NCAN*) and downregulated (*MRPL57*, *EEF1A2*, *FARSB*,

TABLE 10.1
List of Genes and SNPs Associated with TBI.

Gene	Gene Product and major role	TBI associated pathology	References
ABCB1	ATP binding cassette sub family B member1, shows Metabolic effects	Associated with global outcome in TBI	Cousar et al. (2013), Jha et al. (2017), Zeiler et al. (2019a,b)
ABCC1	ATP binding cassette sub family C member1, having metabolic effects	Associated with global outcome in TBI	Cousar et al. (2013), Jha et al. (2017), Zeiler et al. (2019a,b)
ABCC8	ATP binding cassette sub family C member 8, have metabolic effects	Associated with Cerebral edema in TBI	Cousar et al. (2013), Jha et al. (2017), Zeiler et al. (2019a,b)
ACE	Angiotensin Converting Enzyme, have cerebrovascular myogenic response	Subarachnoid hemorrhage related vasospasm, Hemorrhagic stroke, Poor outcome after TBI, Poor neuropsychiatric performance after TBI	(Griessenauer et al. (2017, 2018), Zeiler et al. (2019a))
ADK	Adenosine kinase, maintains the cellular balance of nucleotides	Medical sequelae	(Kurowski et al., 2017)
ADORA1	Adenosin receptor A1, excerts endothelial mediated effects	Associated with development of early and late post traumatic epilepsy and vasospasm in subarachnoid hemorrhage	(Lin et al. (2006, 2007), Zeiler et al. (2019a))
ADORA2A	Adenosin receptor A2A, excerts endothelial mediated effects	vasospasm in subarachnoid hemorrhage	(Lin et al. (2006, 2007), Zeiler et al. (2019a))
AGTR2	Type II angiotensin 2 receptor, have cerebrovascular myogenic response	Subarachnoid hemorrhage related vasospasm	(Griessenauer et al. (2017, 2018), Zeiler et al. (2019a))
ANKK1	Ankyrin repeat and kinase domain contacting 1, important for cognitive functions	The SNP, rs1800497, T-allele mTBI carriers is associated with poor cognitive and slower response latencies when compared with T-allele positive controls.	(McAllister, 2008)
APOE	Apolipoprotein, have roles in lipid and neuronal homeostasis	Survival/global functioning; cognitive; behavioural/emotional; medical sequelae	(Kurowski et al., 2017)
AQP4	Aquaporin 4, shows Metabolic effects	Associated with Cerebral edema, global outcome in TBI	Dardiotis et al. (2014), Zeiler et al. (2019a,b)
BCL2	B-cell lymphoma 2, inhibits apoptosis	Survival/global functioning; behavioural/emotional	(Kurowski et al., 2017)
BDNF	Brain derived neurotropic factor, have neurotransmitter-based effects	Associated with global outcome and neuropsychiatric outcomes in TBI	(Zeiler et al., 2019b; Zeiler, et al., 2019a)
BMX	BMX non-receptor tyrosine kinase, plays important role in cell differentiaton and survival	Behavioural/emotional; medical sequelae	(Kurowski et al., 2017)

TABLE 10.1
List of Genes and SNPs Associated with TBI.—cont'd

C1 or KIBRA	Kidney and brain expressed protein, is important for episodic memory and synaptic plasticity in the temporal lobe of hippocampus.	The SNP, rs17070145 T-allele is associated with poor cognitive performanvein TBI	(Bennett, 2016)
CACNA1A	Voltage dependent P/Q type calcium channel subunit α1A, have neurotransmitter-based effects.	Cerebral edema after TBI	Ferrari (2009), Zeiler et al. (2019a)
COMT	Catechol- O-methyltransferase, have neurotransmitter-based effects	Neuropsychiatric outcomes in TBI	(Zeiler et al., 2019b; Zeiler, et al., 2019a)
CYP19A1	Cytochrome P450 family 19 subfamily A member 1, important for biosynthesis of cholesterol and other lipids	Survival/global functioning	(Kurowski et al., 2017)
DBH	Dopamine beta-hydroxylase, modulator of cognition	Cognitive	(Kurowski et al., 2017)
EDN1	Endothelin 1, exerts endothelial mediated effects	Increased risk of aneurysmal subarachnoid hemorrhage	Gupta et al. (2017), Padmanabhan et al. (2017), Szpecht et al. (2017), Zeiler et al. (2019a)
EDNRA	Endothelin 1 receptor, exerts endothelial mediated effects	Subarachnoid hemorrhage related vasospasm	Hendrix et al. (2017), Zeiler et al. (2019a)
FAAH	Fatty acid amide hydrolase, endocannabinoid-degrading enzyme	Behavioural/ emotional	(Kurowski et al., 2017)
GAD1	Glutamate decarboxylase 1, converts glutamate to GABA	Survival/global functioning; cognitive; medical sequelae	(Kurowski et al., 2017)
GRIN 2A	Glutamate receptor ionotropic, NMDA subtype 2A, having neurotransmitter-based effects	Associated with slow recovery from mTBI	McDevitt et al. (2015), Zeiler et al. (2019a)
mtDNA (Haplotypes J, T, U, K)	Aerobic metabolism, have metabolic effects	Associated with global outcome in TBI	(Bulstrode et al. 2014)
mtDNA (10398A) and (10398G)	Aerobic metabolism, have metabolic effects	Associated with global outcome in TBI	(Conley et al. (2014))
IGF-1	Insulin growth factor, have neuroprotective effects	The minor allele of rs7136446 was found in a large no: of mTBI patients. Patients with SNPs, rs7136446 and rs972936 showed more dizziness and multiple neuropsychiatric symptoms after brain injury.	(Wang, 2019)
IL	Interleukin, has immune functions	Survival/global functioning	(Kurowski et al., 2017)
IL-1B	Interleukin 1 beta, important for cell proliferation and differentiation	Survival/global functioning; medical sequelae	(Kurowski et al., 2017)

Continued

TABLE 10.1
List of Genes and SNPs Associated with TBI.—cont'd

Gene	Description	Association	Reference
IL-RN	Interleukin 1 receptor antagonist	Survival/global functioning	(Kurowski et al., 2017)
MAO-A	Monoamine oxidase type A, having neurotransmitter-based effects	Associated with aggression in persons with TBI	(Pardini et al. (2011); Zeiler et al., 2019b; Zeiler, et al., 2019a)
MME	Neprilysin is a zinc-dependent metalloprotease	The length of the dinucleotide GT repeats in neprilysin was more (>41 repeats) in patients with Aβ accumulation following TBI compared to Aβ nonaccumulators.	(Bennett, 2016)
MTHFR	Methylenetetrahydrofolate reductase (NAD(P)H), mutations are linked to AD	Medical sequelae	(Kurowski et al., 2017)
NADH	NADH: ubiquinone reductase (H$^+$-translocating)	Survival/global functioning; behavioural/emotional	(Kurowski et al., 2017)
NGB	Neuroglobin, have neuroprotective roles	Survival/global functioning; behavioural/emotional	(Kurowski et al., 2017)
NOS3	Endothelial NOS, exerts endothelial mediated effects	Essential hypertension, poor global outcome in TBI, Impaired autoregulation in TBI, Impaired CO2 reactivity in TBI, increased subarachnoid hemorrhage	(Lin et al. 2006, Robertson et al. 2011, Levinsson et al. 2014, Rosalind Lai and Du 2015, Hendrix et al. 2017, Szpecht et al. 2017, Zeiler et al. 2019a)
NT5E	5'-Nucleotidase ecto, have important role in cell adhesion and migration	Medical sequelae	(Kurowski et al., 2017)
PARP-1	Poly(ADP-ribose) polymerase 1, functions in DNA repair and cellular response to stress,	Over activity of PARP-1 adversely affect brain injury by depleting NAD$^+$ resulting in energy failure. An A/T SNP (rs3219119) can predict neurological outcome.	(Sarnaik, 2009)
PPP3CC	encodes a regulatory subunit for calcineurin, a calcium-dependent protein phosphatase.	A/G polymorphism (rs2443504) is a predictor of susceptibility and recovery in TBI,	(Bennett, 2016)
RBMS3	RNA binding motif single stranded interacting protein 3, modulates TGF-β signaling	Medical sequelae	(Kurowski et al., 2017)
SLC6A4	Sodium dependent serotonin transporter, have neurotransmitter-based effects	Associated with post TBI depression	(Failla et al. 2013; Zeiler et al., 2019b; Zeiler, et al., 2019a)
SLC17A7	Vesicular glutamate transporter 1, have neurotransmitter-based effects	Linked to slow recovery from TBI	(Madura et al. 2016; Zeiler et al., 2019b; Zeiler, et al., 2019a)
SNCA	Synuclein alpha, mutations are linked to Parkinson's Disease (PD)	Medical sequelae	(Kurowski et al., 2017)
TNF A	Tumor necrosis factor	Survival/global functioning; medical sequelae	(Kurowski et al., 2017)

TABLE 10.1			
List of Genes and SNPs Associated with TBI.—cont'd			
TP53	Tumor protein P53	Survival/global functioning	(Kurowski et al., 2017)
TRPM4	Trancient receptor potential cation channel subfamily-M, important for cell survival and growth	Associated with cerebral edema and intracranial hypertension in traumatic brain injury. SNPs, rs8104571 (intron-20) and rs150391806 (exon-24) can predict elevated intracranial pressure	(Jha, 2019)

RPL15) in CA3 neurons. Neurotransmitter pathways (CAMK2A, PPP1R1B, GRIN2B, CAV2, PRKCG) are upregulated in CA subtype 1, neurons under conditions of TBI. Cell signaling (PTPRN, PENK, NPY, INHBA, PCDH8), cell migratory (WASL, ARPC3, PIK3CA) and proteasomal pathways (RPN1, PSMC6, PSMA1) are upregulated in DG granule cells, GABAergic interneurons and neuronal subtype, respectively. When neuroplasticity is affected in TBI, the genes SET and BDNF are upregulated, and the genes CHL1, NTF3 and PCP4 are downregulated in DG granule cells (Arneson et al., 2018).

Kurowski and colleagues have compiled a list of genes associated with significant clinical and functional outcome of TBI (Survival/Global Functioning (e.g., Glasgow Outcome Scale), Cognitive (e.g., Neuropsychological/Intelligence Measures), Behavioral/Emotional (e.g., Child Behavioral Checklist, Anxiety, Depression), and Medical Sequelae (e.g., Seizures, Dementia)), from 1269 literatures available from PUBMED and by functional enrichment analysis. Among the 33 genes identified, APOE, adenosine A1 receptor (ADORA1), brain-derived neurotrophic factor (BDNF), and glutamate decarboxylase 1 (GAD1) were shown to have associated with the four outcomes post-TBI (Kurowski et al., 2017).

McDevitt in his article has reported the polymorphisms in the genes APOE (involved in plasticity and repair), GRIN2A (synaptic connectivity), CACNA1E (calcium influx), and SLC17A7 (glutamate metabolism) as potential genetic biomarkers of concussion and recovery post-TBI. NOS3, which modulates vascular response, and NGB, which induces hypoxia response, is recognized as genetic factors in concussion risk and outcome (McDevitt and Krynetskiy, 2017).

Lipponen et al., has tried to identify the altered gene expression after TBI. The group has conducted genomewide RNA sequencing at 3 months post-TBI induction

in rats. The study revealed differential expression of 4964 genes in the perilesional cortex, 1966 in the thalamus, and 1 in the hippocampus. From among the DEG sets, on the basis of statistically significant gene expression, they constructed gene signature for TBI consisting of 874 upregulated genes and 464 downregulated genes. Gene ontology analysis revealed enriched genes in the perilesional cortex (57 genes), thalamus (37 genes), and hypothalamus (1 gene). The similarity was seen in the expression patterns of gene sets in perilesional cortex and thalamus. In both perilesional cortex and thalamus, significant downregulation of genes for ion channel and mitochondrial genes were seen, whereas the gene sets for immunity and inflammation were upregulated. In the hippocampus, the gene for a ligand-gated channel was significantly downregulated (Lipponen et al., 2016).

Age also affects the gene-activity following TBI. The young group has an enhanced gene expression of inflammatory regulatory genes at 48 h and 1 week post trauma. Genes such as basic leucine zipper transcription factor 2 (BACH2), leucine-rich repeat neuronal 3 (LRRN3), and lymphoid enhancer-binding factor 1 (LEF1) were higher compared with the old group. The old group had increased activity in genes belonging to S100 family, such as calcium-binding protein P (S100P) and S100 calcium-binding protein A8 (S100A8), which are linked to poor recovery from TBI. The old group also had reduced activity of the noggin (NOG) gene, which is linked to neurorecovery and neuroregeneration, compared with the young group. These results suggest the lesser likelihood of neuronal recovery in older patients (Cho et al., 2016).

TBI affects alternatively spliced genes in the hippocampus and leukocytes involved in coagulation, blood pressure, inflammation, energy management, and extracellular matrix (ECM) regulation. Splicing regulators such as Rbm47, Sfrs18, Snrnp40, and Uhmk1 are signature genes associated with TBI. TBI affects hippocampal and leukocyte genes associated with the ECM (e.g., -*Fmod*, *Dcn*, *Pcolce*, collagens), transcription factors (e.g., *Tcf7l2*, *Cebpd*), and others (e.g., *Anxa2*, *Ogn*). The ECM genes *Fmod* and *Pcolce* have been identified as regulators of cognitive and metabolic functions; *Tcf7l2* and *Cebpd* are key transcription factors involved in the metabolic processes and *Anxa2* (annexin A2), associated with brain tumor formation (Meng et al., 2017) and are involved in the long-term neurological outcomes of focal embolic stroke (Wang et al., 2014). *Ogn* (osteoglycin) is reduced in the amygdala of animals exposed to stress and can be an important link between TBI and psychiatric-like disorders such as anxiety and depression (Max, 2014; Meng, 2017).

The long-term degenerative effects of TBI cannot be treated, and this is mainly attributed to the minimal understanding of chronic TBI. Microarray and pathway analyses revealed expression changes in the rat hippocampus and cortex at several acute, subchronic, and chronic intervals (24 h, 2 weeks, 1, 2, 3, 6, and 12 months) after injury. Expression of genes belonging to canonical pathways associated with the innate immune response (i.e., NF-κB signaling, NFAT signaling, complement system, acute phase response, toll-like receptor signaling, and neuroinflammatory signaling) were dysregulated. This would compromise multiple pathways essential for brain function, such as cell survival and neuroplasticity (Boone et al., 2019).

10.4 CHALLENGES

Majority of the studies on brain injury–related genetic studies are "candidate gene association studies," which differentiate the allele frequency between a test and a control group. However, such an association does not necessarily mean causation of the parameter being assessed. For instance, an allele that may be of excess frequency in certain disease cases is not because it is causative for a disease, but, it may be due to the association of alternative allele with early mortality from the disease (known as "survivor bias"). Unless the gene is solely expressed in a cell-specific manner, the allele association cannot be inferred. Further, the allele expression is influenced by the unidentified functional variation in the neighboring gene than the gene with recognized variation (Wilson and Montgomery, 2007). Such candidate gene association studies have failed to replicate when conducted in independent cohorts (Farrell et al., 2015; Border et al., 2019; Thompson et al., 2020). Apart from these, a variety of environmental factors influence the disease progression

and outcome. Moreover, the physical extent of the brain injury depends on the strength of blow to the head, location, associated injuries, blood vessel friability, subcutaneous tissue, and skin. Several environmental factors like intracranial pressure, cerebral blood flow, inflammatory response, and tissue oxygen level also influence neural injury. All these factors interact with the genes to produce the outcome of injury (Wilson and Montgomery, 2007).

10.5 RECENT ADVANCES IN GENE MAPPING OF BRAIN DISORDERS

This section discusses the diagnostic techniques of TBI and their limitations and how gene mapping can aid in proper diagnosis and treatment of TBI. Of the several types of TBI discussed so far, 77% comprises mTBI or concussion, and 40% of such cases are mostly ignored. Lack of proper diagnosis is one reason for this. It mainly happens because the symptoms associated with mTBI often resolve within a short period of time, such as days or weeks. However, prolonged problems such as impaired memory, headaches, sleep abnormalities anxiety, and depression are found in a substantial number of cases. These observations point to the fact that there can be several determinants for TBI outcome. For instance, the axonal injury might progress to a year after injury. Such observations pose a challenge to the proper diagnosis of TBI. Generally, TBI scoring and neuroimaging combined with TBI scoring are the methods used to assess the severity of the brain injury. Glasgow Coma Score (GCS) is a gold standard method that assesses the eye movement and verbal and motor responses of the patient. This scoring system fails in patients who use alcohol, sedative medicines, and those under stress and also in polytrauma cases. Abbreviated injury scale (AIS) is another scoring system that scores the injury on a 6-point scale. Cranial CT scoring tool (CCTST) is based on head CT, which predicts on an 8-point scale and is said to be better than the GCS and AIS. CT scans often fail to predict the conditions of mTBI due to its limited sensitivity. MRIs are highly effective but are very expensive. There is still a need for a proper, efficient, and affordable diagnostic tool that could predict the outcomes of TBI. Gene mapping could help in finding the gene targets for diagnosis and therapy.

Gene expression studies are one of the routinely performed analysis to identify and understand the changes occurring under a given disease condition. Such studies have identified several signature molecules that aided in the diagnosis and treatment of several diseases. Genome-wide RNA sequencing is another technique used in a recent study by Lipponen et al. to identify tubulin genes such as *Tubb2a, Tubb3, Tubb4b, Nfe2l2, S100a4, Cd44, and Nfkb2* as specific signatures linked with TBI-associated outcomes and desmethylclomipramine as a promising candidate drug that can reverse such changes. Such studies also have the potential to identify compounds or putative drugs that are protective against long-lasting transcriptomic changes associated with TBI. It is also crucial that studies aimed at understanding the molecular level of TBI should focus on the changes at every phase of TBI to get a complete picture of the rescue mechanisms involved during such events. *In silico* analyses of the transcriptomic data might provide novel treatment targets for TBI (Lipponen et al., 2016).

Studies using repetitive TBI mouse models have shown the important role played by miRNAs on injured neurons. miRNAs are stable molecules and can be detected in small amounts. The study showed the upregulation of miR-124-3p in microglial exosomes, which majorly inhibited the inflammatory responses and also resulted in neurite outgrowth, downregulation of neurodegenerative proteins such as Aβ-peptide and p-Tau, and injury-protective protein such as Rho *in vitro*. mToR signaling is inhibited by miR-124-3p mediated through PDE4B. The study also mentions the manipulation of miRNA expression in microglial exosomes as a therapeutic strategy for TBI and other neurological disorders (Huang et al., 2018). miRNAs are also promising as serum diagnostic markers for mild and moderate TBI (mTBI) and severe TBI (sTBI) in humans. mTBI and sTBI groups show at least 10 upregulated miRNAs in common (Bhomia et al., 2016).

10.6 THE TECHNOLOGY INVOLVED IN GENE MAPPING

Recently, rat mitochondrion-neuron focused microarrays (rMNChip) containing 1500 genes including 37 mitochondrial DNA (mtDNA)-encoded genes, 1098 nuclear DNA (nDNA)-encoded and mitochondria-focused genes, and 365 neuron-related genes were used to identify mitochondrial-targeted genes responsible for altered brain energy metabolism in the injured brain. Out of 1500 genes, 235 genes showed changes in expression pattern, whereas 105 genes were differentially expressed in the hippocampus of mTBI rats compared with the naïve rats. Dysregulated genes in this study included several genes of MAPK signaling pathway (Mapk12, fas, Prkcg, Chp, Atf4, Ntrk1), Ndufs2, Atp5o, Cox5b, Apaf1, Chp genes of apoptotic pathways and oxidative phosphorylation pathways.

This study identified several mitochondria-focused and TBI-responsive genes that may play a role in the pathogenesis of TBI. Some of the dysregulated genes identified are also associated with degenerative disorders (Sharma et al., 2012).

RNA-seq analysis was performed in mTBI model of zebrafish to identify DEGs at two-time points, an acute 3 days postinjury (dpi) and 21 dpi. There were 150 DEGs at three dpi and 400 DEGs at 21 dpi. Forty-three percent of DEGs were upregulated at three dpi and fifty-seven percent of DEGs at 21 dpi. The genes upregulated at 3 dpi were belonging to peak injury response pathways and that of 21 dpi were regenerative pathway genes (Maheras et al., 2018).

Integrated analysis approach was used to classify genes based on their expression using available RNA-sequence data. Based on this, 6513 DEGs were identified (6464 upregulated and 49 downregulated). The major genes identified were FOXO3, DGKZ, and ILK. FOXO3, apolipoprotein (APOE), microtubule-associated protein tau (MAPT), and TREM2 were identified as genes associated with the TBI pathological process. Gene ontology showed that leukocyte transendothelial migration, chemokine signaling pathway, neurotrophin signaling pathway, and longevity-regulating pathway were significantly enriched after TBI (Zhao et al., 2019).

Delaney et al. induced mTBI in mice expressing ChR2 (channel rhodopsin2) and electrophysiologically measured the neural function in response to optogenetic photostimulation. This reduced evoked neuronal response and functional neuronal deficits, in pericontusional tissue and tissue in the contralateral hemisphere. This study, for the first time, showed the optogenetic photostimulation approach for the treatment of TBI (Delaney et al., 2020).

Zhao et al., using optogenetics and nanotechnology, studied the ability of neural stem cells to treat TBI. ChR2-EGFP (enhanced green fluorescence protein) coupled with doublecortin (DCX), a microtubule-associated protein expressed by neuronal progenitor and postmitotic neuronal precursor cells, were used to observe the survival and maturation of nascent neuronal cells after TBI. A lentiviral vector carrying the DCX-ChR2-EGFP gene was injected into the hilus of the DG in C57BL/6 mice. After inducing TBI, the number of EGFP-expressing cells were observed which rose and peaked at 3 and 9 days following TBI. Optical stimulation resulted in an upregulation of several neural marker transcripts, such as microtubule-associated protein 2 (MAP2), neuronal nuclei (NeuN) antigen, neurogenin 2 (Neurog2), neuronal differentiation 1 (NeuroD1), and glutamate receptor subunit 2 (GluR2). This technique promoted neurogenesis following depolarization of stem cells expressing the opsin and thus prove to be used in the treatment of TBI and other neurological disorders (Zhao et al., 2018).

10.7 CONCLUSION

Gene mapping is a robust approach to identify the genes and thereby the biological process underlying various pathological conditions. GWASs of SNPs provide a powerful tool in aiding the identifications of genes that have a role in the progression of pathogenesis in post-TBI. The identification of genes involved in the pathogenesis of post-TBI conditions can help to develop a better strategy for the management of the disease condition. The cell-specific DEGs have the potential to be a reliable biomarker in conjunction with various imaging techniques.

REFERENCES

Arneson, D., et al., 2018. Single cell molecular alterations reveal target cells and pathways of concussive brain injury. Nat. Commun. 9 (1), 3894. https://doi.org/10.1038/s41467-018-06222-0. Epub 2018/09/27 PMID: 30254269.

Bennett, E.R., et al., 2016. Genetic influences in traumatic brain injury. In: Translational research in traumatic brain injury. CRC Press/Taylor and Francis Group. PMID: 26583176.

Bhomia, M., et al., 2016. A panel of serum MiRNA biomarkers for the diagnosis of severe to mild traumatic brain injury in humans. Sci. Rep. 6 (1), 1–12.

Blennow, K., et al., 2016. Traumatic brain injuries. Nat. Rev. Dis. Prim. 2 (1), 1–19.

Boone, D.R., et al., 2019. Traumatic brain injury induces long-lasting changes in immune and regenerative signaling. PLoS One 14 (4), e0214741.

Border, R., et al., 2019. No support for historical candidate gene or candidate gene-by-interaction hypotheses for major depression across multiple large samples. Am. J. Psychiatr. 176 (5), 376–387. https://doi.org/10.1176/appi.ajp.2018.18070881.

Bulstrode, H., et al., 2014. Mitochondrial DNA and traumatic brain injury. Ann. Neurol. 75 (2), 186–195.

CDC, 2019. Surveillance Report of Traumatic Brain Injury-Related Emergency Department Visits, Hospitalizations, and Deaths-United States, 2014. Centers for Disease Control and Prevention, U.S. Department of Health and Human Services. Available at: www.cdc.gov/TraumaticBrainInjury.

Cho, Y.E., et al., 2016. Older age results in differential gene expression after mild traumatic brain injury and is linked to imaging differences at acute follow-up. Front. Aging Neurosci. 8, 168.

Conley, Y.P., et al., 2014. Mitochondrial polymorphisms impact outcomes after severe traumatic brain injury. J. Neurotrauma 31 (1), 34–41.

Cousar, J.L., et al., 2013. Influence of ATP-binding cassette polymorphisms on neurological outcome after traumatic brain injury. Neurocrit. Care 19 (2), 192–198.

Czosnyka, M., et al., 1997. Continuous assessment of the cerebral vasomotor reactivity in head injury. Neurosurgery 41 (1), 11–19.

Dardiotis, E., et al., 2010. Genetic association studies in patients with traumatic brain injury. Neurosurg. Focus 28 (1), E9.

Dardiotis, E., et al., 2014. AQP4 tag single nucleotide polymorphisms in patients with traumatic brain injury. J. Neurotrauma 31 (23), 1920–1926.

Delaney, S.L., et al., 2020. Optogenetic modulation for the treatment of traumatic brain injury. Stem Cell. Dev. 29 (4), 187–197.

Dennis, E.L., et al., 2020. ENIGMA brain injury: framework, challenges, and opportunities. Hum. Brain Mapp. 1–18.

Dewan, M.C., et al., 2019. Estimating the global incidence of traumatic brain injury. J. Neurosurg. 130 (4), 1080–1097.

Diaz-Arrastia, R., et al., 2006. Genetic factors in outcome after traumatic brain injury: what the human genome project can teach us about brain trauma. J. Head Trauma Rehabil. 21 (4), 361–374.

Failla, M.D., et al., 2013. Variants of SLC6A4 in depression risk following severe TBI. Brain Inj. 27 (6), 696–706.

Farrell, M.S., et al., 2015. Evaluating historical candidate genes for schizophrenia. Mol. Psychiatr. 20 (5), 555–562.

Ferrari, M.D., 2009. Early Seizures and Cerebral Edema after Trivial Head Trauma Associated with the CACNA1A S218L Mutation.

Griessenauer, C.J., et al., 2017. Associations of renin-angiotensin system genetic polymorphisms and clinical course after aneurysmal subarachnoid hemorrhage. J. Neurosurg. 126 (5), 1585–1597.

Griessenauer, C.J., et al., 2018. Associations between endothelin polymorphisms and aneurysmal subarachnoid hemorrhage, clinical vasospasm, delayed cerebral ischemia, and functional outcome. J. Neurosurg. 128 (5), 1311–1317.

Gupta, R.M., et al., 2017. A genetic variant associated with five vascular diseases is a distal regulator of endothelin-1 gene expression. Cell 170 (3), 522–533.

Hayes, A., et al., 2018. BDNF genotype is associated with hippocampal volume in mild traumatic brain injury. Genes, Brain Behav. 17 (2), 107–117. https://doi.org/10.1111/gbb.12403.

Hendrix, P., et al., 2017. Endothelial nitric oxide synthase polymorphism is associated with delayed cerebral ischemia following aneurysmal subarachnoid hemorrhage. World Neurosurg. 101, 514–519.

Huang, S., et al., 2018. Increased miR-124-3p in microglial exosomes following traumatic brain injury inhibits neuronal inflammation and contributes to neurite outgrowth via their transfer into neurons. FASEB J. 32 (1), 512–528.

Jha, R.M., et al., 2017. ABCC8 single nucleotide polymorphisms are associated with cerebral edema in severe TBI. Neurocrit. Care 26 (2), 213–224.

Jha, R.M., et al., 2019. Downstream *TRPM4* polymorphisms are associated with intracranial hypertension and statistically interact with *ABCC8* polymorphisms in a prospective cohort of severe traumatic brain injury. J. Neurotrauma 36 (11). PMID: 30484364.

Kayani, S.N., et al., 2020. A neurologic gene map. In: Rosenberg's Molecular and Genetic Basis of Neurological and Psychiatric Disease. Academic Press, pp. 701–791.

Koch, L., 2020. Contact maps and brain disease risk. Nat. Rev. Genet. 21 (2) pp.69–69.

Kurowski, B.G., et al., 2017. Applying systems biology methodology to identify genetic factors possibly associated with recovery after traumatic brain injury. J. Neurotrauma 34 (14), 2280–2290.

Lei, P., et al., 2009. Microarray based analysis of microRNA expression in rat cerebral cortex after traumatic brain injury. Brain Res. 1284, 191–201.

Levinsson, A., et al., 2014. Nitric oxide synthase (NOS) single nucleotide polymorphisms are associated with coronary heart disease and hypertension in the INTERGENE study. Nitric Oxide 39, 1–7.

Lin, C.L., et al., 2006. The effect of an adenosine A1 receptor agonist in the treatment of experimental subarachnoid hemorrhage-induced cerebrovasospasm. Acta Neurochir. 148 (8), 873–879.

Lin, C.L., et al., 2007. Attenuation of experimental subarachnoid hemorrhage–induced cerebral vasospasm by the adenosine A2A receptor agonist CGS 21680. J. Neurosurg. 106 (3), 436–441.

Lipponen, et al., 2016. Analysis of post-traumatic brain injury gene expression signature reveals tubulins, Nfe2l2, Nfkb, Cd44, and S100a4 as treatment targets. Sci. Rep. 6, 31570.

Madura, S.A., et al., 2016. Genetic variation in SLC17A7 promoter associated with response to sport-related concussions. Brain Inj. 30 (7), 908–913.

Maheras, A.L., et al., 2018. Genetic pathways of neuroregeneration in a novel mild traumatic brain injury model in adult zebrafish. Eneuro 5 (1).

Maurano, M.T., et al., 2012. Systematic localization of common disease-associated variation in regulatory DNA. Science 337 (6099), 1190–1195.

McDevitt, J., Krynetskiy, E., 2017. Genetic findings in sport-related concussions: potential for individualized medicine? Concussion 2 (1), CNC26.

Max, J.E., 2014. Neuropsychiatry of pediatric traumatic brain injury. Psyc. Clin. North America 37 (1), 125–140. https://doi.org/10.1016/j.psc.2013.11.003.

McAllister, T.W., et al., 2008. Single nucleotide polymorphisms in ANKK1 and the dopamine D2 receptor gene affect cognitive outcome shortly after traumatic brain injury: A replication and extension study. Brain Injury 22 (9), 705–714. PMID: 18698520.

McDevitt, J., et al., 2015. Association between GRIN2A promoter polymorphism and recovery from concussion. Brain Inj. 29 (13–14), 1674–1681.

Meng, Q., et al., 2017. Traumatic brain injury induces genome-wide transcriptomic, methylomic, and network perturbations in brain and blood predicting neurological disorders. EBioMedicine 16, 184–194.

Nicoll, J.A., et al., 1995. Apolipoprotein E ε4 allele is associated with deposition of amyloid β-protein following head injury. Nat. Med. 1 (2), 135–137.

NIH, 2020. Traumatic Brain Injury: Hope through Research. NIH Publication No. 20-NS-158. Available at: https://www.ninds.nih.gov/Disorders/Patient-Caregiver-Education/Hope-Through-Research/Traumatic-Brain-Injury-Hope-Through (Accessed 17 June 2020).

Nott, A., et al., 2019. Cell Type-specific Enhancer-Promoter Connectivity Maps in the Human Brain and Disease Risk Association. bioRxiv, p. 778183.

Padmanabhan, S., et al., 2017. Genomics of hypertension. Pharmacol. Res. 121, 219–229.

Pardini, M., et al., 2011. Prefrontal cortex lesions and MAO-A modulate aggression in penetrating traumatic brain injury. Neurology 76 (12), 1038–1045.

Raizman, R., et al., 2020. Traumatic brain injury severity in a network perspective: a diffusion MRI based connectome study. Sci. Rep. 10 (1), 1–12.

Robertson, C.S., et al., 2011. Variants of the endothelial nitric oxide gene and cerebral blood flow after severe traumatic brain injury. J. Neurotrauma 28 (5), 727–737.

Rosalind Lai, P.M., Du, R., 2015. Role of genetic polymorphisms in predicting delayed cerebral ischemia and radiographic vasospasm after aneurysmal subarachnoid hemorrhage: a meta-analysis. World Neurosurg. 84 (4), 933–941.

Rusnak, M., 2013. Giving voice to a silent epidemic. Nat. Rev. Neurol. 9 (4), 186–187.

Sarnaik, A.R., et al., 2009. Influence of PARP-1 polymorphisms in patients after traumatic brain injury. J. Neurotrauma 27 (3), 465–471. https://doi.org/10.1089/neu.2009.1171.

Sey, N.Y., et al., 2020. A Computational Tool (H-MAGMA) for Improved Prediction of Brain-Disorder Risk Genes by Incorporating Brain Chromatin Interaction Profiles. Nature Publishing Group, pp. 1–11.

Sharma, P., et al., 2012. Mitochondrial targeted neuron focused genes in hippocampus of rats with traumatic brain injury. Int. J. Crit. Illn. & Inj. Sci. 2 (3), 172.

Szpecht, D., et al., 2017. Role of endothelial nitric oxide synthase and endothelin-1 polymorphism genes with the pathogenesis of intraventricular hemorrhage in preterm infants. Sci. Rep. 7, 42541.

Tang, L., 2020. Predicting brain-disorder risk genes. Nat. Methods 17 (5) pp.459–459.

Thompson, P.M., et al., 2020. ENIGMA and global neuroscience: a decade of large-scale studies of the brain in health and disease across more than 40 countries. Transl. Psychiatry 10 (1), 1–28.

Vaishnavi, S., et al., 2009. Neuropsychiatric problems after traumatic brain injury: unraveling the silent epidemic. Psychosomatics 50 (3), 198–205.

Wang, X., et al., 2014. Effects of tissue plasminogen activator and annexin A2 combination therapy on long-term neurological outcomes of rat focal embolic stroke. Stroke 45 (2), 619–622.

Wang, Y.-J., et al., 2019. The functional roles of IGF-1 variants in the susceptibility and clinical outcomes of mild traumatic brain injury. J. Biomed. Sci. 26 (94). https://doi.org/10.1186/s12929-019-0587-9.

Wilson, M., Montgomery, H., 2007. Impact of genetic factors on outcome from brain injury. Br. J. Anaesth. 99 (1), 43–48.

Zeiler, F.A., et al., 2018. Critical thresholds of intracranial pressure-derived continuous cerebrovascular reactivity indices for outcome prediction in noncraniectomized patients with traumatic brain injury. J. Neurotrauma 35 (10), 1107–1115.

Zeiler, F.A., et al., 2019a. Genetic drivers of cerebral blood flow dysfunction in TBI: a speculative synthesis. Nat. Rev. Neurol. 15 (1), 25–39.

Zeiler, F.A., et al., 2019b. Genetic influences on patient-oriented outcomes in traumatic brain injury: a living systematic review of non-apolipoprotein E single-nucleotide polymorphisms. J. Neurotrauma. https://doi.org/10.1089/neu.2017.5583.

Zhao, M.L., et al., 2018. Optical depolarization of DCX-expressing cells promoted cognitive recovery and maturation of newborn neurons via the Wnt/β-Catenin pathway. J. Alzheim. Dis. 63 (1), 303–318.

Zhao, J., et al., 2019. Identification of target genes in neuroinflammation and neurodegeneration after traumatic brain injury in rats. PeerJ 7, e8324.

Zhou, W., et al., 2008. Meta-analysis of APOE4 allele and outcome after traumatic brain injury. J. Neurotrauma 25 (4), 279–290.

FURTHER READING

Centers for Disease Control and Prevention, 2003. Report to Congress on Mild Traumatic Brain Injury in the United States: Steps to Prevent a Serious Public Health Problem. Centers for Disease Control and Prevention, Atlanta, GA, p. 45.

CHAPTER 11

Immunogenetics in Migraine

GYANESH M. TRIPATHI, MSC, PHD • SWATI TRIPATHI, MSC

11.1 INTRODUCTION

Migraine is a common neurological syndrome; the typical features are moderate to severe, recurrent, unilateral or bilateral, throbbing headache lasting for hours to days, which is usually accompanied by a group of symptoms, that is, nausea, photophobia, phonophobia, and degraded by usual physical exertion (Headache Classification Committee of the International Headache, 2013). Headache is the most noticeable and clinically main symptom of migraine, but its cause is still inexplicable. The severity of pain is varying from mild to moderate, severe, and even very severe pain, which brings impairment in work and social events (Waldie and Poulton, 2002). For many migraine patients, the quality of life and activity is greatly diminished during attacks, and their frequency can interfere with the ability to work or to perform activities of daily living (Society, 2013).

Migraine affects around 14% of the population in which 15–30% are female, and 10–15% are male (Vetvik, 2019). Before puberty, there is generally no sex difference, but the incidence rate generally increases with age, especially in girls, until midpuberty and then falls (Vetvik and MacGregor, 2017). Migraine is now acknowledged as a disorder that affects around one in seven people, two-thirds of whom are women, and is recognized by the World Health Organization as the sixth highest cause worldwide of years lost due to disability.

11.2 PATHOPHYSIOLOGY OF MIGRAINE

From the last few decades, our comprehension of migraine pathophysiology has advanced significantly; yet, the precise migraine mechanisms are not clear. Three main theories have been suggested pertaining to the main mechanisms involved in the migraine. The first theory is the vascular theory, attributes the phenomenon of pain in migraine attack to vasodilatation. The second hypothesis, the neurological theory, deems migraine attacks as a consequence of neuronal events that take place in various brain regions, which are mediated by alterations in neurotransmission systems. This theory emphasizes the phenomenon of cortical spreading depression (CSD), an expanding depolarization of cortical neurons underlying the symptoms of aura (Arulmozhi et al., 2005).

A current unified pathophysiological view considers migraine as a neurovascular disorder (Pietrobon and Moskowitz, 2013). This theory is an effort to merge the vascular alterations with neuronal dysfunction. The initiation of migraine pain seems to be the effect of numerous pathophysiological alterations in meningeal tissues, the trigeminal ganglion, trigeminal brainstem nuclei, and descending inhibitory systems, based on particular characteristics of the trigeminovascular system (Messlinger, 2009). Current research suggests that the trigeminovascular system plays a significant role in migraine due to its critical interaction with the meningeal vasculature and various neurotransmitters, peptides, receptors, and transporters found in this system (Lambert and Zagami, 2009; Messlinger, 2009; Parsons and Strijbos, 2003). This theory considers the release of inflammatory neuropeptides from the trigeminal system, with a consequent dilatation of meningeal vessels as a major pathogenic step of migraine (Pietrobon, 2005; Moskowitz, 1993).

Triggering circumstances like diet, chronic stress, hormonal variability, or CSD-like proceedings can create a state of sterile inflammation in the intracranial meninges leads to the sensitization and instigation of trigeminal meningeal nociceptors. This sterile inflammatory phenotype is depicted by the discharge of neurochemicals such as substance-P (SP), calcitonin gene–related peptide (CGRP), pituitary adenylate cyclase-activating polypeptide (PACAP) from the trigeminal innervation (Ramachandran, 2018b). They are produced by sensory nerves and increased during migraine attacks, inducing mast cells to release vasoactive amines and cytokine release, that enable the infiltration of neutrophils and T-cells (Taracanova et al., 2017). Meningeal nerve fibers also contain neurotransmitters such as glutamate, serotonin (5-HT), hormones such as prostaglandins that can influence the activation

The Molecular Immunology of Neurological Diseases. https://doi.org/10.1016/B978-0-12-821974-4.00006-6

and release of neuropeptides driving neuro-inflammation interaction.

11.3 ROLE OF INFLAMMATION IN MIGRAINE

It has been suggested that the migraine pathophysiology comprises the neuroimmuno interaction at meninges, and it is an important mechanism for the sensitization of nociceptors (Pinho-Ribeiro et al., 2017; Yuan and Silberstein, 2018; Aich et al., 2015). Headache triggered by the augmentation of neurotransmitters, which provokes mast cell degranulation, interacts with blood vessels, neurons and immune cells, initiating inflammation by inducing vasodilation, plasma extravasation secondary to capillary leakage, edema, glial cell activation, and chemotaxis of inflammatory cells (with the production of proinflammatory family members) (Pinho-Ribeiro et al., 2017; Markowitz et al., 1988) that may change the local chemical environment and activate the nerve terminal (Fig. 11.1) (Zhang et al., 2011; Theoharides et al., 2010).

11.3.1 Neurogenic Inflammation

The role of neurogenic inflammation has been considered in the origin and upkeeping of migraine pain (Malhotra, 2016). Although certain features of neurogenic inflammation have been overlooked in the manifestation of migraine pain, the targeted study of factors has provided us the links between the neurons and immune cells in driving such a sterile neuroinflammatory state in migraine pathophysiology.

Most of the studies that are related to neurogenic inflammation are mainly conducted in rodents (Pietrobon and Moskowitz, 2013), and their results have suggested the role of neurogenic inflammation in migraine pathology. However, these outcomes have not been supported by clinical data obtained during acute migraine attacks (Peroutka, 2005) as illustrated by the lack of effect of SP receptor antagonists in this perspective (Diener, 2003). Therefore, it can be assumed that neurogenic inflammation might occur in conjunction with a migraine attack but is improbable to be the commencing cause of migraine pathology. Additionally, only a few shreds of evidence exist for the role of

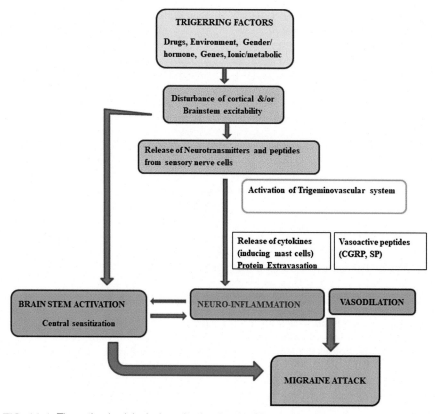

FIG. 11.1 The pathophysiological mechanism involved in migraine (neuroimmune interaction).

the central nerve system (CNS) neuroinflammation in migraine. CNS inflammation may not be involved in initiating migraine attacks, as blood-brain-barrier alterations, glial cell activation, and leukocyte infiltration have not been observed in migraineurs (Edvinsson et al., 2019a). Furthermore, sensitization might be caused by a phenomenon, whereby activation of trigeminal sensory pathways leads to an organized inflammatory response involving immune cells, vascular cells, and neurons (Xanthos and Sandkuhler, 2014).

11.3.2 Neurogenic Neuroinflammation Mechanism

Another emphasis of migraine research has been the plausible role of inflammation in chronification through the neurogenic neuroinflammation mechanism. It can be described as inflammatory reactions in central and peripheral regions of the trigeminovascular system in reaction to neuronal activity (Ramachandran, 2018a) (Bernstein and Burstein, 2012). This phrase is mainly concerned with CNS structures but can also be applied to peripheral nerve system (PNS) components like the trigeminal ganglion. It comprises the vascular and glial cell components of brain tissue and can recruit immune cells from the bloodstream (Xanthos and Sandkuhler, 2014; Edvinsson et al., 2019b).

Chronification and sensitization, PNS, in particular, the trigeminal ganglion, could be the site of inflammation and peripheral sensitization may play an important factor in migraine chronification, as opposed to migraine triggering (Edvinsson et al., 2019b). It has been hypothesized that continued stimulation of C-fibers during repeated migraine attacks and the ensuing activation of Aδ-fibers and satellite glial cells (SGCs), leads to neurogenic neuroinflammation in the trigeminovascular system, thereby promoting the chronification process. One study has supported this premise as the increase of (interleukin-1) IL-1 receptor expression in trigeminal ganglion neurons and that inflammation potentiated the excitability of Aδ and C-fibers (Takeda et al., 2007). This increased Aδ-fiber activity caused by inflammation could also be blocked by an IL-1 receptor type-1 antagonist (Takeda et al., 2008).

11.4 FACTORS AFFECTING INFLAMMATION IN MIGRAINE

The inflammatory reaction comprises numerous factors, including cytokines and chemokines that are released in the endothelium and cerebral vessels and play a crucial role in the CNS (Hanisch, 2002). The blood vessels that convey immune cells pass through the meninges and alter immunocyte trafficking in the brain (Kivisakk et al., 2003). The effect of mast cells resident in the meninges can play a critical role in the development of inflammation in many diseases, including migraine (Levy, 2009), and several genetic studies also endorse in the evolution of our understanding of migraine, which is now primarily considered as a neurogenic disorder. It is suggested that both genetic and environmental factors may modify individual vulnerability by slashing the CSD threshold (Van Harreveld, 1978). This segment will suggest some factors, which will influence the possible involvement of neuroinflammation.

11.4.1 Calcitonin Gene–Related Peptide

It can be postulated that prolonged activation of the trigeminovascular system leads to sustained release of CGRP and possibly other neuropeptides. This continued stimulation could result in neurogenic neuroinflammation in the trigeminovascular system.

During a migraine attack, the level of CGRP measured in plasma from the jugular vein is increased in patients (Goadsby et al., 1990). In whole-cell recordings on isolated trigeminal neurons, CGRP was shown to be excitatory, either by lowering the threshold for action potentials or by causing spontaneous firing (Meng et al., 2009). It has been observed in animal models that continuous CGRP release can induce peripheral sensitization (Nakamura-Craig and Gill, 1991), probably through the release of inflammatory mediators such as bradykinin and prostaglandins from nerve endings and cells of the immune system (Wang et al., 2006; Schaible and Schmidt, 1988).

The stimulation of CGRP in trigeminal neurons leads to modification in the action of intracellular signaling molecules that are relevant to pain, which in turn leads to an increase in expression of inflammatory cytokines in the dura mater and probably also support in rises in neuronal cell bodies in the trigeminal ganglion. This mechanism of CGRP-induced inflammation was recently described in an animal model where in addition to causing heat hyperalgesia, intraganglionic administration of CGRP increased the expression of mRNA encoding proinflammatory cytokines such as IL-1β, possibly localized to the SGCs (Afroz et al., 2019). Furthermore, transgenic mice sensitized to CGRP through elevated expression of a CGRP receptor subunit have demonstrated the symptoms that are consistent with migraine-like photophobia and mechanical allodynia, which confirms the hypothesis that CGRP through its receptors induces neurogenic neuroinflammation (Russo et al., 2009).

It has been hypothesized that in migraineurs, CGRP might evoke continuous activation of Aδ-fibers or SGCs the cells that express receptors for CGRP in the trigeminal ganglia (Haanes and Edvinsson, 2019). The past few years have seen the development of an exciting new class of medications for migraine prevention monoclonal antibodies that target CGRP or the CGRP receptor (Edvinsson et al., 2018) also provides the evidence of immunogenetics in the pathomechanism and treatment of migraine.

11.4.2 Mast Cell

Mast cells are present in a great number within the meninges, and their products cause inflammation and may contribute to migraine (Williamson and Hargreaves, 2001). Earlier the influence of mast cells in the migraine attack was introduced by Sicuteri et al. in the 1950s; they demonstrated that slow carotid artery infusion of compound 48/80 (a polymer used to promote mast cell degranulation) in human and dog induced migraine-like symptoms (Sicuteri et al., 1957). This was supported by one more study where patients with certain food allergies linked to migraine showed no symptoms following oral mast cell stabilizing agent cromolyn (Monro et al., 1984).

Mast cell degranulation is a common phenomenon observed in conditions like mastocytosis, rhinitis, and asthma with headache being one of the major symptoms associated with these conditions and therefore underlines the connection of mast cells in engendering headaches (Smith et al., 2011). Numerous reports have also linked migraine with allergic diseases including eczema, hay fever, and asthma (Gurkan et al., 2000). However, there is no indication that individuals with migraine are more susceptible to allergies than others; also there is no direct relationship between headache, and the mast cell allergy mediated by IgE (Yuan and Silberstein, 2018).

Degranulation of mast cells releases molecules such as histamine, serotonin, tumor necrosis factor (TNF), kinins, and proteases that are collected in secretory granules (Theoharides et al., 2012). Mast cells can selectively release proinflammatory cytokines synthesized de novo like TNF-α, IL-1, and IL-6 and also produce lipid-derived mediators, that is, leukotrienes and prostanoids without undertaking degranulation (Aich et al., 2015). IL-17 supplying mast cells can also promote inflammatory response-mediated migraine (Conti et al., 2018). It should be perceived that almost all the studies that have been reported migraine in relation to mast cells concern in vitro and animal model experiments.

11.4.3 Histamine

It is observed from the studies that exogenous SP and CGRP in vitro provoked histamine discharge from mast cells as both neurokinin-1 receptor and CGRP receptor antagonists prevent this effect (Ottosson and Edvinsson, 1997). It should be remarked that a confounding element that comes out from the preclinical studies is that mast cell degranulation brought by CGRP appears to be rodent-specific, as human mast cells do not show the receptor components that are desired for a CGRP response (Eftekhari et al., 2013). This outcome explains human mast cells do not release histamine in response to CGRP (Ottosson and Edvinsson, 1997).

11.4.4 Hormones

Hormones have also been proposed as recruiters and intermediaries of dural mast cell degranulation. Ovarian hormonal level showed variations in the dural mast cell density throughout the estrus cycle in female rats with estrogen promoting an increase in mast cell population (Boes and Levy, 2012). Estrogen has been shown to degranulate mast cells and release its mediators in mast cell lines and primary cultures of bone marrow–derived mast cells via its action on estrogen receptor, although it is not known whether estrogen can activate human dural mast cells.

11.5 INFLAMMATORY MARKERS AND CYTOKINES

The association of inflammation with the alternation in levels of inflammatory markers associated with the trigeminovascular system has been confirmed by a few studies (Noseda and Burstein, 2013; Ashina et al., 2019; Samsam et al., 2007). Cytokines are considered to be the pain mediators in neurovascular inflammation (Lipnik-Stangelj, 2013). Furthermore, cytokines may be considered as a cause of migraine pain as the high level of chemokines could stimulate the activation of trigeminal nerves, the release of vasoactive peptides or other biochemical mediators, such as nitric oxide, and then to cause inflammation (Harriott et al., 2019). Cytokines are important mediators of the immune and inflammatory pathways, and their receptors are also widely expressed in the CNS by all cell types, including neurons, indicating that they can act on neuronal receptors (Bruno et al., 2007).

Some major cytokines including TNF, IL-1β, and IL-6 have been associated with migraine pathophysiology, as their levels have been changed in migraine patients (Johnson and Bolay, 2006). There is an increase in levels of IL-1β, IL-6, and TNF in serum during a migraine attack have also been described and Sarchielli et al. found an increase in IL-6 and TNF levels in the jugular venous blood during a migraine attack, which promotes pain and thrombosis in migraine (Yücel et al., 2016; Sarchielli et al., 2006). Even, plasma and CSF levels of afferent peptides CGRP, histamine, proteases, and proinflammatory cytokines such as TNFα and IL-1β are raised during the migraine attacks reflect activation of neuroinflammatory cascades in migraine pathogenesis (Rozen and Swidan, 2007; Sarchielli et al., 2006; Perini et al., 2005).

11.5.1 Tumor Necrosis Factor

TNF, an effective mast cell–derived proinflammatory cytokine, mediates sensitization of the meningeal nociceptors (Moretti et al., 2014) and therefore plays an important role in the pathophysiology of migraine (Levy, 2012). TNF influences the reactivity of signal nociceptors to the brain and increases its blood levels during headache; therefore, it plays an important role in the initiation of migraine. Studies have reported changes in TNF levels in plasma, serum, and/or urine levels of patients with migraine during headache and headache-free duration, recommending the presence of inflammation in support a pathogenic role for TNF in migraine (Tanure et al., 2010). The increased level of TNF serum in humans, even the interictal period, also supports a pathogenic role for TNF in migraine (Covelli et al., 1991; Perini et al., 2005).

The action of TNF is conducted through the binding to its receptor subtype TNF receptor-1 (TNFR-1) expressed on the cells of the somatic and vascular nociceptors (Zhang et al., 2011) leads to the generation and release of inflammatory mediators produced by mast cells. These cells, following activation via FcεRI (high-affinity Immunoglobulin E receptor), can rapidly produce and release preformed TNF, physically associated with the secretory granules, or later produce TNF resulting from the induction of the corresponding mRNA (Conti et al., 2018). Franceschini and coworkers found that mRNA expression of TNF was upregulated in the trigeminal ganglia in a knock-in animal model of familial hemiplegic migraine (Franceschini et al., 2013).

11.5.2 Toll-like Receptor

The toll-like receptor (TLR) domain participates in the adaptation of innate immunity and neuroinflammation. The activation of TLRs results in the upregulation of NF-kB (Nuclear Factor kappa-light-chain-enhancer of activated B-cells, a small family of inducible transcription factors) with an increase of gene transcriptions

encoding IL-1 family cytokines and TNF. In addition, TLRs are crucial for signaling via the MyD88 pathway (The MyD88 gene delivers information for generating a protein concerned in signaling within immune cells), which can be inhibited by IL-37 (Downer et al., 2013). Inhibition of MyD88, and therefore of NF-kB, in astrocytes can be beneficial in CNS diseases, including migraine, with the reduction of ganglion cell death after injury, axonal regeneration, and reduced inflammation (Zhan et al., 2017).

Furthermore, TLR2 is a membrane receptor that is found in many areas, including neurons, where it induces the generation of inflammatory cytokines, activating NF-kB, with consequent pain (Barua et al., 2015). In migraine and other primary headache, there is an increase in the expression of TLR2 has been noticed (Marshall et al., 2003).

TLR4 was first associated in mediating immune–neuraxial interactions in several inflammatory pain states, and more recently, it became of interest in migraine (Ramachandran et al., 2019). TLR4s is expressed on primary afferent neurons, microglia, and astrocytes (Qi et al., 2011; Lin et al., 2015). Several endogenous TRL4 ligands are of particular interest in regard to migraine. Of particular relevance, high mobility group box 1 (HMGB1) is released in the cortex following CSD (Karatas et al., 2013). Increased levels of heat-shock protein 70 (HSP70), another ligand, has been demonstrated in trigeminal ganglion following neuroinflammation (Ohara et al., 2013). Similarly, other endogenous TLR4 ligands such as fibronectin and tenascin-C are of interest as they are expressed by fibroblasts that are the primary resident cell type in the meninges and have been suggested to play a potential role in headache pathophysiology (Wei et al., 2014). These studies suggest that endogenous processes such as CSD and inflammation that are linked to migraine might influence TLR4 activation.

An interesting covariate of the role of TLR4 signaling in the migraine attacks is that the naloxone effect was reported to be effective in treating acute migraine attacks (Nicolodi and Sicuteri, 1992; Centonze et al., 1983). A study has shown that both the opioid-active naloxone and the nonopioid-active isomer naloxone can block the TLR receptor function (Lewis et al., 2012). Thus, the naloxone effect, which was elucidated as being applicable to the role of endogenous opioids, may now, in light of the proposed role of TLR4 in migraine manifestation, as it is well interpreted as reflecting a TLR4 mechanism.

Additionally, under the influence of chemokines and adhesion molecules, opioid peptide-containing

TABLE 11.1
Summary of Studies Related to Immunogenetics in the Animal Models of Migraine.

Title	Author	Design of Study	Detail of the Study	Conclusion
1. CGRP induces differential regulation of cytokines from satellite glial cells in trigeminal ganglia and orofacial nociception	Afroz et al. (2019)	Quantitative PCR	To evaluate the mRNA expression of IL-1β, IL-6, TNF-α, IL-1 receptor antagonist (IL-1RA), sodium channel 1.7 (NaV 1.7, for assessment of neuronal activation) and glial fibrillary acidic protein (GFAP, a marker of glial activation)	An increase in the mRNA expression of pro-inflammatory cytokines IL-1β and anti-inflammatory cytokine IL-1RA.
2. A potential preclinical migraine model: CGRP-sensitized mice	Russo et al. (2009)	Transgenic mice and antagonist study	Clinical studies suggested that migraineurs are more sensitive to CGRP than people who do not suffer from migraine. However, a major challenge for studying CGRP actions is the lack of animal models for migraine. The transgenic mouse that is sensitized to CGRP (nestin/hRAMP1 mice) have elevated expression of a subunit of the CGRP receptor, human receptor activity–modifying protein 1 (hRAMP1).	CGRP acts as a neuromodulator to increase sensory responses and that regulation of a single gene, hRAMP1, could potentially contribute to migraine susceptibility.
3. Enhanced excitability of nociceptive trigeminal ganglion (TRG) neurons by satellite glial cytokine following peripheral inflammation.	Takeda et al. (2007)	Immunohistochemistry	Activation of satellite glial cells modulates the excitability of TRG neurons via IL-1beta following inflammation. Fluorogold (FG) labeling identified the site of inflammation.	The activation of satellite glial cells modulates the excitability of small-diameter TRG neurons via IL-1beta following inflammation, and that the upregulation of IL-1RI in the soma may contribute to the mechanism underlying inflammatory hyperalgesia.
4. Contribution of activated interleukin receptors in TRG neurons to hyperalgesia via satellite glial interleukin-1β paracrine mechanism	Takeda et al. (2008)	Extracellular electrophysiological recording with multibarrel electrodes.	The present study investigated whether under in vivo conditions, inflammation alters the excitability of nociceptive Aδ- TRG neurons innervating the facial skin via a cytokine paracrine mechanism.	Inflammation modulates the excitability of nociceptive Aδ-TRG neurons innervating the facial skin via IL-1β paracrine action within TRGs. Such an IL-1β release could be important in determining trigeminal inflammatory hyperalgesia

TABLE 11.1
Summary of Studies Related to Immunogenetics in the Animal Models of Migraine.—cont'd

Title	Author	Design of Study	Detail of the Study	Conclusion
5. TNF alpha levels and macrophages expression reflect an inflammatory the potential of TRGs in a mouse model of familial hemiplegic migraine.	Franceschini et al. (2013)	Confocal immunohistochemistry.	Whether cytokines such as TNFα contribute to a local inflammatory phenotype, in trigeminal sensory ganglia of a transgenic knock-in mouse model of familial hemiplegic migraine type-1 (FHM-1) was investigated.	R192Q KI (R192Q mutant CaV2.1 Ca^{2+} channel, a genetic model of FHM-1) TRGs constitutively expressed higher mRNA levels of IL1β, IL6, IL10, and TNFα cytokines and the MCP-1 chemokine. TNFα expression and macrophage occurrence were significantly higher in R192Q KI ganglia with respect to wild-type ganglia.
6. Toll-like receptor 4 signaling in neurons of TRG contributes to nociception induced by acute pulpitis in rats.	Lin et al. (2015)	Immunohistochemistry	To study the underlying mechanism of pain in the animal model and its relation with neuroinflammation.	A consistent up-regulation of toll-like receptor 4 (TLR4) in the trigeminal ganglion (TG) ipsilateral to the injured pulp was found; and downstream signaling components of TLR4, including MyD88, TRIF and NF-κB and cytokines such as TNF-α and IL-1β, were also increased.
7. α7 Nicotinic acetylcholine receptor-mediated the anti-inflammatory effect in a chronic migraine rat model via the attenuation of glial cell activation.	Liu et al. (2018)	The expression levels of α- 7nAChR, tumor necrosis factor-alpha, and interleukin-1 beta was analyzed by Western blot and real-time fluorescence quantitative PCR.	To investigate the role of α 7nAChR in chronic migraine and provide a new therapeutic target. The expression of α 7nAChR was reduced after repeated inflammatory soup administration.	The activation of α 7nAChR increased the mechanical threshold and alleviated pain in the CM rat model. α7nAChR activation also decreased the up-regulation of astrocytes and microglia.
8. Basal astrocyte and microglia activation in the central nervous system of familial hemiplegic migraine type I mice (FHM1)	Magni et al., 2019	The signs of reactive glia in brains from naive FHM1 mutant mice were evaluated in assessment with wild-type animals by immunohistochemistry and Western blotting.	To evaluate the role of reactive astrocyte and microglial activation in the central nervous system of FHM1 mutant mice.	Indications of reactive astrogliosis and microglial activation in the naive FHM1 mutant mouse brain were reported, and these data reinforce the involvement of glial cells in migraine.

TABLE 11.2
Summary of Studies Related to Immunogenetics in Migraine Patients.

Title	Author	Design of Study	Detail of the Study	Conclusion
1. Upregulation of inflammatory gene transcripts in periosteum of chronic migraineurs: Implications for the extracranial origin of headache	Perry et al. (2016)	RT PCR	The expression of mRNA, which encodes proteins that are important in immune reactions, was measured in concerned (i.e., where the head hurts) calvarial periosteum of (1) patients who are associated with muscle tenderness and (2) patients with no history of headache.	Expression of proinflammatory genes (e.g., *CCL8*, *TLR2*) in the calvarial periosteum significantly upregulated in chronic migraine patients attesting to muscle tenderness; however, the expression of genes that subdue inflammation and immune cell differentiation (e.g.,; *IL10RA*, *CSF1R*) declined.
2. The synergistic effects of ω-3 fatty acids and nano-curcumin supple-mentation on tumor necrosis factor (TNF)-α gene expression and serum level in migraine patients	Abdolahi et al. (2017)	The gene expression of TNF-α and level of TNF-α in serum were estimated by real-time PCR and ELISA methods.	The intention of the study was to evaluate the synergistic impacts of ω-3 fatty acids and nanocurcumin on TNF-α gene expression and serum levels in migraineurs.	The combination of ω-3 fatty acids and nanocurcumin downregulated TNF-α messenger RNA (mRNA) significantly in a synergistic manner ($P < .05$). As relative to gene expression, a significantly greater reduction in the level of TNF-α was observed in the combination group. However, supplementation of ω-3 fatty acids or nanocurcumin alone did not exhibit significant association with mRNA or serum levels of TNF-α.
3. Association between interleukin-4 (IL-4), gene polymorphisms (C-589T, +2979G, and C-33T) and migraine susceptibility in Iranian population: a case-control study	Ramroodi et al. (2017)	PCR-RFLP.	This study was aimed to investigate the possible associations between IL-4 Single nucleotide polymorphisms (SNPs) and susceptibility to migraine patients.	In IL-4 rs2243250 and rs2227284 genotypes and allele frequencies have a role in susceptibility to migraine in our population. Therefore, it is suggested that in addition to other factors, IL-4 genetic variations also play a pivotal role in the progress of migraine.

TABLE 11.2
Summary of Studies Related to Immunogenetics in Migraine Patients.—cont'd

Title	Author	Design of Study	Detail of the Study	Conclusion
				There was no statistically significant relationship between these SNPs and different subclasses (common, classic and complicated) of migraine.
4. Association of the long pentraxin PTX3 gene polymorphism (rs3816527) with migraine in an Iranian population	Zandifar et al. (2015)	RFLP	Migraine could be a form of sterile neurogenic inflammation. The association of rs3816527 polymorphism of the PTX3 gene and migraine in an Iranian population was investigated.	The association between the PTX3 rs3816527 gene polymorphism with susceptibility to migraine only in male patients was observed. Total HIT-6 scores as a scale for assessment of the severity were also linked to the PTX3 gene polymorphism. However, no association was found between this particular polymorphism and headache frequency.
5. HLA antigens in migraine	O'Neill et al. (1979)	Microlymphocytotoxicity assay	Patients with migraine attacks were assessed for the role of 10 HLA-A and 14 HLA-B antigens. Comparisons were made between different types/groups of migraine (i.e., classical vs. common and patients with/without a family history of migraine).	All groups of patients showed an association with B8 and B27. This study, therefore, recommends that the postulated multifactorial inheritance for migraine probably does not entail the contribution of the HLA-A or B locus of the major histocompatibility complex.

immune cells extravasate and accumulate in the injured tissues (Sehgal et al., 2011). Reports from animal and human clinical studies support the involvement of endogenous opioids and its receptors in analgesia, particularly in the presence of inflammation where both opioid receptor expression and efficacy are increased, and their role has also been mentioned in the migraine attack (Bach et al., 1992).

Moreover, in the inflammatory model, treatment of mice with IL-37 downregulates the proinflammatory cytokines of the IL-1 family and an increase in anti-inflammatory IL-10 (Liu et al., 2017; Barbagallo et al., 2017) thus demonstrating its therapeutic efficacy against inflammatory processes, and it emerges as an inhibitor and protector of neuroinflammatory diseases including migraine (Barbagallo et al., 2017; Dinarello et al., 2016).

Tables 11.1 and 11.2 briefly summarize the studies related to immunogenetics, which have been in animal and migraine patients till date. These studies have confirmed the role of inflammatory markers, their related factors, and receptors in migraine.

11.6 OTHER FACTORS

Few more components that have been initially studied and more elaborated work for any confirmatory role of these factors in migraine are required, that is:

i) Microglial activation is regulated by the chemokine CX3CR1 receptor genes and represents a distress signal leading to headache (Ransohoff and Brown, 2012), and its role may be involved in migraine.

ii) Various stimuli can directly activate microglia through the NALP3 inflammasome pathway and therefore the generation of IL-1 mediates neuroinflammation and migraine (Shi et al., 2012).

iii) After activation, the (T helper) Th1, Th2, and Th17 effector cells carry a series of cytokines that act on the innate immune cells to fight infections and may cause migraine (Cseh et al., 2013).

11.7 CONCLUSION

Few pieces of evidence are available to recommend the usual neuronal inflammatory response in combination with acute migraine attacks. In view of the lack of standard markers of CNS inflammation, it is believed that CNS inflammation is not concerned with the initiation of a migraine attack. However, the role of inflammation in the sensitization process that leads to enhanced responsiveness of target tissues, both in the PNS and the CNS, has been reported. The idea of neurogenic neuroinflammation in the trigeminal ganglion could explain the findings of inflammatory markers in migraine and lead the way toward elucidating and treating migraine chronification.

New approaches toward genetics, transgenic animals, immunomolecular studies and recent advances in ligand-receptor studies, imaging studies of inflammatory markers and their receptors have now suggested more links between neuro and inflammation interaction in migraine pathogenesis. Genetic confirmations of newly approached receptors and their legends that are related to signaling and regulatory pathway could also be beneficial for therapeutic aspects of migraine.

REFERENCES

Abdolahi, M., Tafakhori, A., Togha, M., Okhovat, A.A., Siassi, F., Eshraghian, M.R., Sedighiyan, M., Djalali, M., Honarvar, N.M., Djalali, M., 2017. The synergistic effects of ω-3 fatty acids and nano-curcumin supplementation on tumor necrosis factor (TNF)-α gene expression and serum level in migraine patients. Immunogenetics 69, 371–378.

Afroz, S., Arakaki, R., Iwasa, T., Oshima, M., Hosoki, M., Inoue, M., Baba, O., Okayama, Y., Matsuka, Y., 2019. CGRP induces differential regulation of cytokines from satellite glial cells in trigeminal ganglia and orofacial nociception. Int. J. Mol. Sci. 20, 711.

Aich, A., Afrin, L.B., Gupta, K., 2015. Mast cell-mediated mechanisms of nociception. Int. J. Mol. Sci. 16, 29069–29092.

Arulmozhi, D.K., Veeranjaneyulu, A., Bodhankar, S.L., 2005. Migraine: current concepts and emerging therapies. Vasc. Pharmacol. 43, 176–187.

Ashina, M., Hansen, J.M., Do, T.P., Melo-Carrillo, A., Burstein, R., Moskowitz, M.A., 2019. Migraine and the trigeminovascular system—40 years and counting. Lancet Neurol. 18, 795–804.

Bach, F.W., Langemark, M., Secher, N.H., Olesen, J., 1992. Plasma and cerebrospinal fluid beta-endorphin in chronic tension-type headache. Pain 51, 163–168.

Barbagallo, M., Vitaliti, G., Greco, F., Pavone, P., Matin, N., Panta, G., Lubrano, R., Falsaperla, R., 2017. Idiopathic intracranial hypertension in a paediatric population: a retrospective observational study on epidemiology, symptoms and treatment. J. Biol. Regul. Homeost. Agents 31, 195–200.

Barua, R.S., Sharma, M., Dileepan, K.N., 2015. Cigarette smoke amplifies inflammatory response and atherosclerosis progression through activation of the H1R-TLR2/4-COX2 axis. Front. Immunol. 6, 572.

Bernstein, C., Burstein, R., 2012. Sensitization of the trigeminovascular pathway: perspective and implications to migraine pathophysiology. J. Clin. Neurol. 8, 89–99.

Boes, T., Levy, D., 2012. Influence of sex, estrous cycle, and estrogen on intracranial dural mast cells. Cephalalgia 32, 924–931.

Bruno, P.P., Carpino, F., Carpino, G., Zicari, A., 2007. An overview on immune system and migraine. Eur. Rev. Med. Pharmacol. Sci. 11, 245–248.

Centonze, V., Brucoli, C., Macinagrossa, G., Attolini, E., Campanozzi, F., Albano, O., 1983. Non-familial hemiplegic migraine responsive to naloxone. Cephalalgia 3, 125–127.

Conti, P., D'ovidio, C., Conti, C., Gallenga, C.E., Lauritano, D., Caraffa, A., Kritas, S.K., Ronconi, G., 2018. Progression in migraine: role of mast cells and pro-inflammatory and anti-inflammatory cytokines. Eur. J. Pharmacol. 844, 87–94.

Covelli, V., Munno, I., Pellegrino, N., Altamura, M., Decandia, P., Marcuccio, C.A., Di, A.V., Jirillo, E., 1991. Are TNF-alpha and IL-1 beta relevant in the pathogenesis of migraine without aura? Acta Neurol. 13, 205–211.

Cseh, A., Farkas, K.M., Derzbach, L., Muller, K., Vasarhelyi, B., Szalay, B., Treszl, A., Farkas, V., 2013. Lymphocyte subsets in pediatric migraine. Neurol. Sci. 34, 1151–1155.

Diener, H., 2003. RPR100893, a substance-P antagonist, is not effective in the treatment of migraine attacks. Cephalalgia 23, 183–185.

Dinarello, C.A., Nold-Petry, C., Nold, M., Fujita, M., Li, S., Kim, S., Bufler, P., 2016. Suppression of innate inflammation and immunity by interleukin-37. Eur. J. Immunol. 46, 1067–1081.

Downer, E.J., Johnston, D.G., Lynch, M.A., 2013. Differential role of Dok1 and Dok2 in TLR2-induced inflammatory signaling in glia. Mol. Cell. Neurosci. 56, 148–158.

Edvinsson, L., Haanes, K.A., Warfvinge, K., 2019a. Does inflammation have a role in migraine? Nat. Rev. Neurol. 15, 483–490.

Edvinsson, L., Haanes, K.A., Warfvinge, K., 2019b. Does inflammation have a role in migraine? Nat. Rev. Neurol. 15, 483–490.

Edvinsson, L., Haanes, K.A., Warfvinge, K., Krause, D.N., 2018. CGRP as the target of new migraine therapies—successful translation from bench to clinic. Nat. Rev. Neurol. 14, 338–350.

Eftekhari, S., Warfvinge, K., Blixt, F.W., Edvinsson, L., 2013. Differentiation of nerve fibers storing CGRP and CGRP receptors in the peripheral trigeminovascular system. J. Pain 14, 1289–1303.

Franceschini, A., Vilotti, S., Ferrari, M.D., Van Den Maagdenberg, A.M., Nistri, A., Fabbretti, E., 2013. TNFα levels and macrophages expression reflect an inflammatory potential of trigeminal ganglia in a mouse model of familial hemiplegic migraine. PLoS One 8, e52394.

Goadsby, P., Edvinsson, L., Ekman, R., 1990. Vasoactive peptide release in the extracerebral circulation of humans during migraine headache. Ann. Neurol. 28, 183–187.

Gurkan, F., Ece, A., Haspolat, K., Dikici, B., 2000. Parental history of migraine and bronchial asthma in children. Allergol. Immunopathol. 28, 15–17.

Haanes, K.A., Edvinsson, L., 2019. Pathophysiological mechanisms in migraine and the identification of new therapeutic targets. CNS Drugs 33, 525–537.

Hanisch, U.K., 2002. Microglia as a source and target of cytokines. Glia 40, 140–155.

Harriott, A.M., Strother, L.C., Vila-Pueyo, M., Holland, P.R., 2019. Animal models of migraine and experimental techniques used to examine trigeminal sensory processing. J. Headache Pain 20, 91.

Headache Classification Committee of the International Headache Society, 2013. The international classification of headache disorders, 3rd edition (beta version). Cephalalgia 33, 629–808.

Johnson, K.W., Bolay, H., 2006. Neurogenic Inflammatory Mechanisms. Lippincott, Williams & Wilkins, PA, USA.

Karatas, H., Erdener, S.E., Gursoy-Ozdemir, Y., Lule, S., Eren-Koçak, E., Sen, Z.D., Dalkara, T., 2013. Spreading depression triggers headache by activating neuronal Panx1 channels. Science 339, 1092–1095.

Kivisakk, P., Mahad, D.J., Callahan, M.K., Trebst, C., Tucky, B., Wei, T., Wu, L., Baekkevold, E.S., Lassmann, H., Staugaitis, S.M., Campbell, J.J., Ransohoff, R.M., 2003. Human cerebrospinal fluid central memory CD4+ T cells: evidence for trafficking through choroid plexus and meninges via P-selectin. Proc. Natl. Acad. Sci. USA 100, 8389–8394.

Lambert, G.A., Zagami, A.S., 2009. The mode of action of migraine triggers: a hypothesis. Headache 49, 253–275.

Levy, D., 2009. Migraine pain, meningeal inflammation, and mast cells. Curr. Pain Headache Rep. 13, 237–240.

Levy, D., 2012. Endogenous mechanisms underlying the activation and sensitization of meningeal nociceptors: the role of immuno-vascular interactions and cortical spreading depression. Curr. Pain Headache Rep. 16, 270–277.

Lewis, S.S., Loram, L.C., Hutchinson, M.R., Li, C.-M., Zhang, Y., Maier, S.F., Huang, Y., Rice, K.C., Watkins, L.R., 2012. (+)-naloxone, an opioid-inactive toll-like receptor 4 signaling inhibitor, reverses multiple models of chronic neuropathic pain in rats. J. Pain 13, 498–506.

Lin, J.-J., Du, Y., Cai, W.-K., Kuang, R., Chang, T., Zhang, Z., Yang, Y.-X., Sun, C., Li, Z.-Y., Kuang, F., 2015. Toll-like receptor 4 signaling in neurons of trigeminal ganglion contributes to nociception induced by acute pulpitis in rats. Sci. Rep. 5, 12549.

Lipnik-Stangelj, M., 2013. Mediators of inflammation as targets for chronic pain treatment. Mediat. Inflamm. 2013, 783235.

Liu, L., Li, X., Chang, G., Wang, Z., Zhang, S., Ju, X., 2017. Sibelium in combination with dibazole in the treatment of angioneurotic headache. J. Biol. Regul. Homeost. Agents 31, 653–657.

Liu, Q., Liu, C., Jiang, L., Li, M., Long, T., He, W., Qin, G., Chen, L., Zhou, J., 2018. α7 nicotinic acetylcholine receptor-mediated anti-inflammatory effect in a chronic migraine rat model via the attenuation of glial cell activation. J. Pain Res. 11, 1129.

Magni, G., Boccazzi, M., Bodini, A., Abbracchio, M.P., Van Den Maagdenberg, A.M., Ceruti, S., 2019. Basal astrocyte and microglia activation in the central nervous system of familial hemiplegic migraine type I mice. Cephalalgia 39, 1809–1817.

Malhotra, R., 2016. Understanding migraine: potential role of neurogenic inflammation. Ann. Indian Acad. Neurol. 19, 175–182.

Markowitz, S., Saito, K., Moskowitz, M.A., 1988. Neurogenically mediated plasma extravasation in dura mater: effect of ergot alkaloids: a possible mechanism of action in vascular headache. Cephalalgia 8, 83–91.

Marshall, J.S., Mccurdy, J.D., Olynych, T., 2003. Toll-like receptor-mediated activation of mast cells: implications for allergic disease? Int. Arch. Allergy Immunol. 132, 87–97.

Meng, J., Ovsepian, S.V., Wang, J., Pickering, M., Sasse, A., Aoki, K.R., Lawrence, G.W., Dolly, J.O., 2009. Activation of TRPV1 mediates calcitonin gene-related peptide release, which excites trigeminal sensory neurons and is attenuated by a retargeted botulinum toxin with anti-nociceptive potential. J. Neurosci. 29, 4981–4992.

Messlinger, K., 2009. Migraine: where and how does the pain originate? Exp. Brain Res. 196, 179–193.

Monro, J., Carini, C., Brostoff, J., 1984. Migraine is a food-allergic disease. Lancet 324, 719–721.

Moretti, S., Bozza, S., Oikonomou, V., Renga, G., Casagrande, A., Iannitti, R.G., Puccetti, M., Garlanda, C., Kim, S., Li, S., 2014. IL-37 inhibits inflammasome activation and disease severity in murine aspergillosis. PLoS Pathog. 10, e1004462.

Moskowitz, M.A., 1993. Neurogenic inflammation in the pathophysiology and treatment of migraine. Neurology 43, S16–S20.

Nakamura-Craig, M., Gill, B.K., 1991. Effect of neurokinin A, substance P and calcitonin gene related peptide in peripheral hyperalgesia in the rat paw. Neurosci. Lett. 124, 49–51.

Nicolodi, M., Sicuteri, F., 1992. Chronic naloxone administration, a potential treatment for migraine, enhances morphine-induced miosis. Headache J. Head Face Pain 32, 348–352.

Noseda, R., Burstein, R., 2013. Migraine pathophysiology: anatomy of the trigeminovascular pathway and associated neurological symptoms, CSD, sensitization and modulation of pain. Pain 154 (Suppl. 1).

O'neill, B.P., Kapur, J.J., Good, A.E., 1979. HLA antigens in migraine. Headache J. Head Face Pain 19, 071–073.

Ohara, K., Shimizu, K., Matsuura, S., Ogiso, B., Omagari, D., Asano, M., Tsuboi, Y., Shinoda, M., Iwata, K., 2013. Toll-like receptor 4 signaling in trigeminal ganglion neurons contributes tongue-referred pain associated with tooth pulp inflammation. J. Neuroinflammation 10, 139.

Ottosson, A., Edvinsson, L., 1997. Release of histamine from dural mast cells by substance P and calcitonin gene-related peptide. Cephalalgia 17, 166–174.

Parsons, A.A., Strijbos, P.J., 2003. The neuronal versus vascular hypothesis of migraine and cortical spreading depression. Curr. Opin. Pharmacol. 3, 73–77.

Perini, F., D'andrea, G., Galloni, E., Pignatelli, F., Billo, G., Alba, S., Bussone, G., Toso, V., 2005. Plasma cytokine levels in migraineurs and controls. Headache J. Head Face Pain 45, 926–931.

Peroutka, S.J., 2005. Neurogenic inflammation and migraine: implications for the therapeutics. Mol. Interv. 5, 304.

Perry, C.J., Blake, P., Buettner, C., Papavassiliou, E., Schain, A.J., Bhasin, M.K., Burstein, R., 2016. Upregulation of inflammatory gene transcripts in periosteum of chronic migraineurs: implications for extracranial origin of headache. Ann. Neurol. 79, 1000–1013.

Pietrobon, D., 2005. Migraine: new molecular mechanisms. Neuroscientist 11, 373–386.

Pietrobon, D., Moskowitz, M.A., 2013. Pathophysiology of migraine. Annu. Rev. Physiol. 75, 365–391.

Pinho-Ribeiro, F.A., Verri Jr., W.A., Chiu, I.M., 2017. Nociceptor sensory neuron–immune interactions in pain and inflammation. Trends Immunol. 38, 5–19.

Qi, J., Buzas, K., Fan, H., Cohen, J.I., Wang, K., Mont, E., Klinman, D., Oppenheim, J.J., Howard, O.Z., 2011. Painful pathways induced by TLR stimulation of dorsal root ganglion neurons. J. Immunol. 186, 6417–6426.

Ramachandran, R., 2018a. Neurogenic inflammation and its role in migraine. Semin. Immunopathol. 40, 301–314.

Ramachandran, R., 2018b. Neurogenic inflammation and its role in migraine. Semin. Immunopathol. 301–314.

Ramachandran, R., Wang, Z., Saavedra, C., Dinardo, A., Corr, M., Powell, S.B., Yaksh, T.L., 2019. Role of Toll-like receptor 4 signaling in mast cell-mediated migraine pain pathway. Mol. Pain 15, 1744806919867842.

Ramroodi, N., Javan, M.R., Sanadgol, N., Jahantigh, M., Khodakheir, T.N., Ranjbar, N., 2017. Association between interleukin-4 (IL-4), gene polymorphisms (C-589t, T+ 2979G, and C-33T) and migraine susceptibility in Iranian population: a case–control study. Egypt. J. Med. Hum. Genet. 18, 29–34.

Ransohoff, R.M., Brown, M.A., 2012. Innate immunity in the central nervous system. J. Clin. Invest. 122, 1164–1171.

Rozen, T., Swidan, S.Z., 2007. Elevation of CSF tumor necrosis factor α levels in new daily persistent headache and treatment refractory chronic migraine. Headache J. Head Face Pain 47, 1050–1055.

Russo, A.F., Kuburas, A., Kaiser, E.A., Raddant, A.C., Recober, A., 2009. A potential preclinical migraine model: CGRP-sensitized mice. Mol. Cell. Pharmacol. 1, 264.

Samsam, M., Coveñas, R., Ahangari, R., Yajeya, J., Narváez, J., 2007. Role of neuropeptides in migraine: where do they stand in the latest expert recommendations in migraine treatment? Drug Dev. Res. 68, 294–314.

Sarchielli, P., Alberti, A., Baldi, A., Coppola, F., Rossi, C., Pierguidi, L., Floridi, A., Calabresi, P., 2006. Proinflammatory cytokines, adhesion molecules, and lymphocyte integrin expression in the internal jugular blood of migraine patients without aura assessed ictally. Headache J. Head Face Pain 46, 200–207.

Schaible, H., Schmidt, R.F., 1988. Excitation and sensitization of fine articular afferents from cat's knee joint by prostaglandin E2. J. Physiol. 403, 91–104.

Sehgal, N., Smith, H.S., Manchikanti, L., 2011. Peripherally acting opioids and clinical implications for pain control. Pain Physician 14, 249–258.

Shi, F., Yang, L., Kouadir, M., Yang, Y., Wang, J., Zhou, X., Yin, X., Zhao, D., 2012. The NALP3 inflammasome is involved in neurotoxic prion peptide-induced microglial activation. J. Neuroinflammation 9, 73.

Sicuteri, F., Ricci, M., Monfardini, R., Ficini, M., 1957. Experimental headache with endogenous histamine: first results obtained by 48/80, a histawine-liberator drug, in the cephalic and peripheral circulatory systems of man. Allergy 11, 188–192.

Smith, J.H., Butterfield, J.H., Cutrer, F.M., 2011. Primary headache syndromes in systemic mastocytosis. Cephalalgia 31, 1522–1531.

Society, H.C.C.O.T.I.H., 2013. The international classification of headache disorders, (beta version). Cephalalgia 33, 629–808.

Takeda, M., Takahashi, M., Matsumoto, S., 2008. Contribution of activated interleukin receptors in trigeminal ganglion neurons to hyperalgesia via satellite glial interleukin-1β paracrine mechanism. Brain Behav. Immun. 22, 1016–1023.

Takeda, M., Tanimoto, T., Kadoi, J., Nasu, M., Takahashi, M., Kitagawa, J., Matsumoto, S., 2007. Enhanced excitability of nociceptive trigeminal ganglion neurons by satellite glial cytokine following peripheral inflammation. Pain 129, 155–166.

Tanure, M.T.A., Gomez, R.S., Hurtado, R.C.L., Teixeira, A.L., Domingues, R.B., 2010. Increased serum levels of brain-derived neurotropic factor during migraine attacks: a pilot study. J. Headache Pain 11, 427.

Taracanova, A., Alevizos, M., Karagkouni, A., Weng, Z., Norwitz, E., Conti, P., Leeman, S.E., Theoharides, T.C.,

2017. SP and IL-33 together markedly enhance TNF synthesis and secretion from human mast cells mediated by the interaction of their receptors. Proc. Natl. Acad. Sci. USA 114, E4002–E4009.

Theoharides, T.C., Alysandratos, K.-D., Angelidou, A., Delivanis, D.-A., Sismanopoulos, N., Zhang, B., Asadi, S., Vasiadi, M., Weng, Z., Miniati, A., 2012. Mast cells and inflammation. Biochim. Biophys. Acta Mol. Basis Dis. 1822, 21–33.

Theoharides, T.C., Zhang, B., Kempuraj, D., Tagen, M., Vasiadi, M., Angelidou, A., Alysandratos, K.-D., Kalogeromitros, D., Asadi, S., Stavrianeas, N., 2010. IL-33 augments substance P–induced VEGF secretion from human mast cells and is increased in psoriatic skin. Proc. Natl. Acad. Sci. USA 107, 4448–4453.

Van Harreveld, A., 1978. Two mechanisms for spreading depression in the chicken retina. J. Neurobiol. 9, 419–431.

Vetvik, K.G., 2019. Epidemiology of migraine in men and women. In: Maassen Van Den Brink, A., Macgregor, E.A. (Eds.), Gender and Migraine. Springer International Publishing, Cham.

Vetvik, K.G., Macgregor, E.A., 2017. Sex differences in the epidemiology, clinical features, and pathophysiology of migraine. Lancet Neurol. 16, 76–87.

Waldie, K.E., Poulton, R., 2002. The burden of illness associated with headache disorders among young adults in a representative cohort study. Headache 42, 612–619.

Wang, H., Ehnert, C., Brenner, G.J., Woolf, C.J., 2006. Bradykinin and peripheral sensitization. Biol. Chem. 387, 11–14.

Wei, X., Melemedjian, O.K., Ahn, D.D.-U., Weinstein, N., Dussor, G., 2014. Dural fibroblasts play a potential role in headache pathophysiology. Pain 155, 1238–1244.

Williamson, D.J., Hargreaves, R.J., 2001. Neurogenic inflammation in the context of migraine. Microsc. Res. Tech. 53, 167–178.

Xanthos, D.N., Sandkuhler, J., 2014. Neurogenic neuroinflammation: inflammatory CNS reactions in response to neuronal activity. Nat. Rev. Neurosci. 15, 43–53.

Yuan, H., Silberstein, S.D., 2018. Histamine and migraine. Headache J. Head Face Pain 58, 184–193.

Yücel, M., Kotan, D., Gurol Çiftçi, G., Çiftçi, I., Cikriklar, H., 2016. Serum levels of endocan, claudin-5 and cytokines in migraine. Eur. Rev. Med. Pharmacol. Sci. 20, 930–936.

Zandifar, A., Iraji, N., Taheriun, M., Tajaddini, M., Javanmard, S.H., 2015. Association of the long pentraxin PTX3 gene polymorphism (rs3816527) with migraine in an Iranian population. J. Neurol. Sci. 349, 185–189.

Zhan, Q., Zeng, Q., Song, R., Zhai, Y., Xu, D., Fullerton, D.A., Dinarello, C.A., Meng, X., 2017. IL-37 suppresses MyD88-mediated inflammatory responses in human aortic valve interstitial cells. Mol. Med. 23, 83–91.

Zhang, X.-C., Kainz, V., Burstein, R., Levy, D., 2011. Tumor necrosis factor-α induces sensitization of meningeal nociceptors mediated via local COX and p38 MAP kinase actions. Pain 152, 140–149.

CHAPTER 12

Immunogenetics of Neuropathy Disease

KUMARI SWATI, MSC • VIJAY KUMAR, PHD

12.1 PERIPHERAL NEUROPATHY

The peripheral nervous system (PNS) is the part of the nervous system that is outside the brain and spinal cord. The primary function of the PNS is to connect the central nervous system (CNS) to the rest of the body and the external environment. It perceives the environment and mobilizes the body to respond. Peripheral nerves have motor, sensory, and autonomic fibers, and they set up communication between the CNS and body part. PPNs are diseases that disrupt these information pathways and hamper normal function (Deer et al., 2020; Jameson, 2018; Knierim et al., 2018).

Sensory, motor, or autonomic functions were gotten impaired in PPNs, either singly or in combination. PPN is linked with various health conditions that include diabetes, autoimmune diseases, infections, vitamin deficiency, and other diseases. This chapter discusses the neuropathies in the background of autoimmune disease.

12.2 GENES ASSOCIATED WITH AUTOIMMUNE DISEASES—ASSOCIATED NEUROPATHY

Several predisposing genetic factors are linked with autoimmune diseases—associated neuropathy (ADAN) susceptibility, severity, and outcome. Human leukocyte antigen (HLA), T cell, Fc receptors, and cytokine genes are associated with ADAN (Hewagama and Richardson, 2009; Vandenbroeck, 2012). In the following, we study candidate genes that are associated with ADAN.

12.2.1 Human Leukocyte Antigen

HLA regions play a leading role in autoimmunity. The HLA system has a significant role in the antigen presentation of extracellular and intracellular peptides and the regulation of adaptive and innate immune responses. Associations between different HLA SNPs with GBS, SLE, and RA have been reported. The major histocompatibility complex (MHC) has many genes related to the immune system functions, including genes encoding HLA molecules and cell surface antigen-presenting proteins. The primary functions of HLA molecules are antigen presentation to T cells. HLA genes are the strongest predisposing genetic factors in autoimmune disease (Bodis et al., 2018; Gough and Simmonds, 2007; Matzaraki et al., 2017; Rose and Bona, 1993).

12.2.2 Cluster of Differentiation 1

Cluster of differentiation 1 (CD1) is a family of transmembrane glycoproteins expressed on the surface of antigen-presenting cells. It is crucial for capturing and presenting a variety of both microbial and self-glycolipids to diverse T cells and natural killer T (NKT) cells. They are related to the class I MHC molecules and recognize lipid antigens in association with T cell (Barral and Brenner, 2007; De Libero and Mori, 2005; Shahine et al., 2019).

12.3 CYTOKINE

12.3.1 Interleukin-10

Interleukin (IL)-10 is a pleiotropic immunomodulatory cytokine that plays a significant role in the regulation of cellular and humoral immune responses. It is secreted by both macrophages and T lymphocytes and involved in the immune response of both autoimmune and infectious diseases. It has proinflammatory, antiinflammatory, and stimulatory activities (Iyer and Cheng, 2012; Ng et al., 2013).

12.3.2 Tumor Necrosis Factor

TNF is produced by T lymphocytes, natural killer cells, activated macrophages, and monocytes. TNF is a proinflammatory cytokine and functions as an endogenous pyrogen and has a significant role in the early phase of immunity and inflammation. TNF mediates an extensive biological process including immunostimulatory, cell proliferation, cell differentiation, apoptosis, lipid metabolism, coagulation, and antiviral responses (Croft, 2009; Turner et al., 2014).

The Molecular Immunology of Neurological Diseases. https://doi.org/10.1016/B978-0-12-821974-4.00011-X

12.4 AUTOIMMUNE DISEASES—ASSOCIATED NEUROPATHY

Autoimmune disease is a condition arising when our immune system attacks its body part. Our immune system fails to differentiate between foreign cells and its own cells. In ADAN conditions, the immune system can directly attack nerves and myelin sheath, which are often triggered by recent infections. Damage to the motor fibers causes muscle weakness and muscle shrinking. ADAN includes rheumatoid arthritis (RA), systemic lupus erythematosus (SLE), Guillain–Barré syndrome (GBS), chronic inflammatory demyelinating polyneuropathy (CIDP), Sjogren's syndrome (SJS), and vasculitis (Katz et al., 2011; Smith and Germolec, 1999; Theofilopoulos et al., 2017; Thorsby and Lie, 2005). Genetic variations in the HLA, cytokine, and other immune-related genes are associated with the susceptibility for ADAN (Inshaw et al., 2018; Smith and Germolec, 1999; Theofilopoulos et al., 2017).

12.5 GUILLAIN–BARRÉ SYNDROME

GBS is a rare autoimmune disorder of PNS in which the body's own immune system mistakenly damages the nerves. GBS is characterized by acute onset ascending sensorimotor neuropathy, demyelination, muscle weakness, loss of tendon reflexes, numbness, and sometimes paralysis. Infiltration of autoreactive leukocyte into the PNS leads to demyelination, axonal degeneration, and neuroinflammation. In GBS, PNS is attacked by the immune system, which results in impaired signaling from the brain to muscles, resulting in irregular nerve conduction velocity, such as slow signal conduction, abnormal sensations, and symmetric limb weakness (Leonhard et al., 2019; van den Berg et al., 2014; Willison et al., 2016).

GBS susceptibility is associated with HLA-DRB1, HLA-DRB4, HLA-DR3, and HLA-DR7 (Fekih-Mrissa et al., 2014; Gorodezky et al., 1983; Hasan et al., 2014). HLA-DQ-β epitopes are associated with acute inflammatory demyelinating polyneuropathy (AIDP) form of GBS (Magira et al., 2003). HLA-DR5 has a slight negative association with GBS (McCombe et al., 2006).

Humans with CD1E*01/01 genotype have a higher risk to develop GBS, whereas humans with CD1E*01/02 or CD1A*01/02 have a lower relative risk (Caporale et al., 2006). A study of polymorphisms of TNFα C(-863)A SNP in GBS patients found that this SNP is associated with the degree of weakness and poor outcome (Geleijns et al., 2007). TNFα-308G/A polymorphisms were associated with the acute motor axonal neuropathy (AMAN) and acute motor-sensory axonal neuropathy (AMSAN) forms of GBS. GBS patients have a higher frequency of TNFα 308AA genotype (Jiao et al., 2012; Zhang et al., 2007). A metaanalysis study shows that TNFα 308 G/A polymorphisms are moderately associated with the risk of GBS (Wu et al., 2012). IL-23 is associated with the immune-mediated demyelination of the peripheral nerve (Hu et al., 2006). Polymorphisms in the IL-10 promoter region are associated with GBS susceptibility but do not influence the disease severity (Myhr et al., 2003).

MMP9, TNF-α, and IL-1β are expressed in the peripheral nerve of GBS and are upregulated in peripheral blood in the progressive phase (Nyati et al., 2010). Geleijns et al. found MMP9 C(-1562)T SNP in GBS patients who are associated with the severity of disease and poor outcome (Geleijns et al., 2007). Polymorphisms in genes encoding FCγRIIIA and FcγRIIA are associated with GBS (Sinha et al., 2010). The FcγRIIIa gene variants are also associated with the severity of GBS (van Sorge et al., 2005). Dekker et al. found that polymorphisms of the glucocorticoid receptor are not linked to susceptibility but are related to outcome (Dekker et al., 2009). MBL2 gene (encoding mannose-binding protein C) polymorphism is also associated with and the severity and outcome of GBS (Geleijns et al., 2006).

12.6 CHRONIC INFLAMMATORY DEMYELINATING POLYNEUROPATHY

Chronic inflammatory demyelinating polyneuropathy is a heterogeneous autoimmune disease affecting PNS. HLA typing study found that CIDP patients have a stronger association with HLA-Cw7 and a slight association with HLA-B8 (Vaughan et al., 1990). HLA typing of CIDP patients showed increased frequencies of HLA-DR2, HLA-A3, and HLA-B7 as well as concomitantly decreased frequencies of HLA-DR7 and HLA-44. The strongest associations are seen with HLA-DR7, HLA-DR2, and HLA-B44 in CIDP patients (Feeney et al., 1990). DRB1*15 alleles are a strong risk factor associated with CIDP (Martinez-Martinez et al., 2017). Piccinelli et al. found that HLA-DR3 and HLA-DR3/DQ2 are more frequent in CIDP patients and associated with disease severities (Cotti Piccinelli et al., 2019).

12.7 SYSTEMIC LUPUS ERYTHEMATOSUS

Systemic lupus erythematosus (SLE) is a chronic inflammatory disorder; it mainly affects the joints, skin, kidneys, blood cells, and nervous system. SLE is the most common type of lupus in which the immune system attacks its tissues (Kaul et al., 2016; Tamirou et al., 2018). The prevalence of SLE-associated peripheral neuropathies has been reported to occur in 1.5%–32% in SLE patients (Florica et al., 2011; Omdal et al., 1991; Oomatia et al., 2014; Xianbin et al., 2015).

A genetic involvement to SLE is well proven, and many genes are associated with SLE manifestations. *PTPN22, TNFSF4, PDCD1, IL16, IL10, BCL6, TYK2, ITGAM, MECP2/IRAK1, PRL,* and *STAT4 genes coding for* component of the complement system, Fcγ receptors, and HLA region genes are involved in the SLE manifestations (Jonsen et al., 2004; Karassa et al., 2004; Moser et al., 2009; Sestak et al., 2011; Tsokos, 2011). A metaanalysis study found that interferon regulatory factor 5 (IRF5) gene polymorphism is associated with SLE in different populations. IRF5 rs2004640 is significantly associated with SLE in populations of Asian, European, and Latin American people. IRF5 rs10954213 is associated with SLE in patients of European origin but not in Asian origin patients (Hu and Ren, 2011).

IL-10/TNFα interactions (Suarez et al., 2005) and the IL10 promoter haplotype (Rosado et al., 2008), components of complement pathway (C1Q, C2, C3, and C4), and TNFα and TNFβ gene polymorphisms are associated with SLE susceptibility (Hu and Ren, 2011; Miyagawa et al., 2008; Sullivan et al., 2007; Tarassi et al., 1998; Tsokos, 2011). Polymorphisms in C4A, C4B, and C2 are in linkage disequilibrium with *HLA-DR* and *HLA-B* alleles (Speirs et al., 1989) and may cause SLE. Also, Castano-Rodriguez et al. found a significant association of the *HLA-DRB1* gene with SLE in Latin American populations (Castano-Rodriguez et al., 2008). The *CDKN1A* gene is located on chromosome 6p21.2, encoding a cell cycle inhibitor, p21(WAF1/CIP1). An SNP at position -899 in *CDKN1A* causes a decreased level of p21 and associated with SLE susceptibility. *PTPN22 R620W* polymorphism is associated with a higher risk of SLE (Kyogoku et al., 2004).

An intronic programmed cell death 1 gene (*PDCD1*) SNP has a strong association with SLE (Prokunina et al., 2002). Genetic variability in *CTLA4* has been associated with SLE (Barreto et al., 2004; Graham et al., 2006). The SNP rs7601754, rs7582694, and rs7574865 in *STAT4* are associated with SLE disease severity (Sigurdsson et al., 2008; Taylor et al., 2008; Yang et al., 2009).

12.8 RHEUMATOID ARTHRITIS
Rheumatoid arthritis (RA) is a chronic inflammatory autoimmune disorder and can lead to severe joint damage and disability. RA is characterized by inflammation of the cartilage and bone, leading to pain, swelling, and stiffness of joints (Aletaha and Smolen, 2018; Guo et al., 2018; Smolen et al., 2016).

HLA region genes are predominantly responsible for the genetic susceptibility of RA. *PTPN22, TRAF1-C5, PADI4,* and *STAT4* are other significant genes associated with RA susceptibilities. Polymorphisms of genes encoding TNF, IL-1, IL-4, IL-5, IL-6, and PRF1 are the association between radiographic damage (Ceccarelli et al., 2017).

The prevalence of clinical neuropathy in RA patients varies from 0.5% to 85% and may be vasculitis in nature (Agarwal et al., 2008; Kaeley et al., 2019; Nadkar et al., 2001; Sim et al., 2014). Polymorphisms in *DRB1, NOD2,* and *CSF1* gene are associated with IL-6 levels, and polymorphisms in *PTPN2, DRB1,* and *NOD2* gene are associated with TNFα levels (Solus et al., 2015).

HLA-DRB1 allele variants show the strongest linkage to RA, and their predisposition to RA is clear (Orozco et al., 2006). The *PTPN21858T* variant and rs7574865 SNP of STAT4 is found to be associated with both SLE and RA (Kobayashi et al., 2008; Orozco et al., 2008). However, the inconsistent association of CTLAþ49/G SNP with RA has been obtained in the different ethnic group (Zhernakova et al., 2005).

12.9 SJÖGREN'S SYNDROME
Sjögren's syndrome (SJS) is a chronic, systemic immune-mediated inflammatory disease. It is characterized by infiltration and proliferation of lymphocytes into affected glands (mainly the salivary and lacrimal glands), which lead to progressive exocrinopathy. It can occur at any age but is most common in older women (Brito-Zeron et al., 2016; Tarn et al., 2019). A wide variety of neurological complications are characteristic features of SJS, of which PPN is a major neurological manifestation (Koike and Sobue, 2013). Many patients develop SJS as a complication of another autoimmune disease, such as RA or lupus (Kivity et al., 2014). Studies on the polymorphisms of HLA-DR and HLA-DQ gene regions in SJS patients show differential susceptibility to the syndrome as the result of distinct types of the resulting autoantibody production (Bolstad et al., 2001; Cruz-Tapias et al., 2012). Polymorphisms in the *HCP5, IL10,* and *STAT4* genes are associated with SJS susceptibility. *STAT4* (rs7574865) and *HCP5* (rs3099844) genes are significantly more prevalent in SJS patients than in controls (Colafrancesco et al., 2019; Korman et al., 2008). IL1RN*2 is a marker of more severe forms of SJS (Perrier et al., 1998).

12.10 VASCULITIS

Vasculitic neuropathies (VNs) are a heterogeneous group of PNS disorders associated with vasculitis. In VN, blood vessel walls are infiltrated and damaged by inflammatory cells, leading to thrombosis and ischemia. In VN, there is damage to the vessels that supply blood to the nerves. It can occur as a primary or secondary phenomenon related to several autoimmune diseases. VN may present in several patterns, more commonly in a mononeuropathy multiplex or asymmetrical polyneuropathy with painful sensory symptoms or sensorimotor complaints. Electrodiagnostic studies show an axonal loss in sensory and motor fibers in a patchy distribution (Burns et al., 2007; Collins et al., 2019; Collins and Hadden, 2017; Graf and Imboden, 2019; Said and Lacroix, 2005; Schaublin et al., 2005).

Ombrello et al. found an association between *HLA-B51* and Behçet's disease (Hughes et al., 2013; Ombrello et al., 2014). Several genetic studies have identified an association between *ERAP1*, *CCR1-CCR3*, *STAT4*, *KLRC4*, *GIMAP4*, and *TNFAIP3* genes and Behçet's disease (Kim et al., 2013; Kirino et al, 2013a,b; Wallace, 2014). *NMNAT2*, *HCP5*, *BLK*, and *CD40* genes are associated with Kawasaki disease (Burgner et al., 2009; Khor et al., 2011; Kim et al., 2017; Onouchi et al., 2012; Tsai et al., 2011). *CD226*, *CTLA-4*, *FCGR2A*, *HLA-B*, *HLA-DP*, *HLA-DQ*, *HLA-DR*, *HSD17B8*, *IRF5*, *PTPN22*, *RING1/RXRB*, *RXRB*, *STAT4*, *SERPINA1*, and *TLR9* genes are associated with antineutrophil cytoplasmic antibodies—associated vasculitis (Bonatti et al., 2014; Lee et al., 2019; Lyons et al., 2019; Rahmattulla et al., 2016; Slot et al., 2008). Takayasu arteritis—associated vasculitis is associated with the *HLA-B*52*, *FCGR2A/FCGR3A*, *IL12B*, *IL6*, and *RPS9/LILRB3* gene variations (Renauer and Sawalha, 2017; Terao, 2016; Terao et al., 2018).

12.11 CONCLUSION

In conclusion, genetic studies toward autoimmune diseases and genetic factors that affect neuropathies will help in a better understanding of the immunogenetics of neuropathic disorder. These new pieces of knowledge about and HLA and other genes will help us to better understand the pathogenic mechanisms of autoimmune-associated neuropathy. Genes and other factors that cause autoimmunity are needed to explore. Immune-mediated neuropathies are complex, polygenic disorders sharing common predisposing genetic factors with related autoimmune diseases, such as GBS, SLE, and RA.

12.12 ABBREVIATIONS

ADAN	Autoimmune diseases—associated neuropathy
CIDP	Chronic inflammatory demyelinating polyneuropathy
HLA	Human leukocyte antigen
PNS	Peripheral nervous system
GBS	Guillain—Barré syndrome
RA	Rheumatoid arthritis
SLE	Systemic lupus erythematosus
VN	Vasculitic neuropathies
SJS	Sjögren's syndrome

REFERENCES

Agarwal, V., Singh, R., Wiclaf, Chauhan, S., Tahlan, A., Ahuja, C.K., Goel, D., Pal, L., 2008. A clinical, electrophysiological, and pathological study of neuropathy in rheumatoid arthritis. Clin. Rheumatol. 27 (7), 841—844.

Aletaha, D., Smolen, J.S., 2018. Diagnosis and management of rheumatoid arthritis: a review. J. Am. Med. Assoc. 320 (13), 1360—1372.

Barral, D.C., Brenner, M.B., 2007. CD1 antigen presentation: how it works. Nat. Rev. Immunol. 7 (12), 929—941.

Barreto, M., Santos, E., Ferreira, R., Fesel, C., Fontes, M.F., Pereira, C., Martins, B., Andreia, R., Viana, J.F., Crespo, F., Vasconcelos, C., Ferreira, C., Vicente, A.M., 2004. Evidence for CTLA4 as a susceptibility gene for systemic lupus erythematosus. Eur. J. Hum. Genet. 12 (8), 620—626.

Bodis, G., Toth, V., Schwarting, A., 2018. Role of human leukocyte antigens (HLA) in autoimmune diseases. Rheumatol. Ther. 5 (1), 5—20.

Bolstad, A.I., Wassmuth, R., Haga, H.J., Jonsson, R., 2001. HLA markers and clinical characteristics in Caucasians with primary Sjogren's syndrome. J. Rheumatol. 28 (7), 1554—1562.

Bonatti, F., Reina, M., Neri, T.M., Martorana, D., 2014. Genetic susceptibility to ANCA-associated vasculitis: state of the art. Front. Immunol. 5 (577), 577.

Brito-Zeron, P., Baldini, C., Bootsma, H., Bowman, S.J., Jonsson, R., Mariette, X., Sivils, K., Theander, E., Tzioufas, A., Ramos-Casals, M., 2016. Sjogren syndrome. Nat. Rev. Dis. Prim. 2 (1), 16047.

Burgner, D., Davila, S., Breunis, W.B., Ng, S.B., Li, Y., Bonnard, C., Ling, L., Wright, V.J., Thalamuthu, A., Odam, M., Shimizu, C., Burns, J.C., Levin, M., Kuijpers, T.W., Hibberd, M.L., International Kawasaki Disease Genetics Consortium, 2009. A genome-wide association study identifies novel and functionally related susceptibility Loci for Kawasaki disease. PLoS Genet. 5 (1), e1000319.

Burns, T.M., Schaublin, G.A., Dyck, P.J., 2007. Vasculitic neuropathies. Neurol. Clin. 25 (1), 89—113.

Caporale, C.M., Papola, F., Fioroni, M.A., Aureli, A., Giovannini, A., Notturno, F., Adorno, D., Caporale, V.,

Uncini, A., 2006. Susceptibility to Guillain-Barre syndrome is associated to polymorphisms of CD1 genes. J. Neuroimmunol. 177 (1−2), 112−118.

Castano-Rodriguez, N., Diaz-Gallo, L.M., Pineda-Tamayo, R., Rojas-Villarraga, A., Anaya, J.M., 2008. Meta-analysis of HLA-DRB1 and HLA-DQB1 polymorphisms in Latin American patients with systemic lupus erythematosus. Autoimmun. Rev. 7 (4), 322−330.

Ceccarelli, F., Agmon-Levin, N., Perricone, C., 2017. Genetic factors of autoimmune diseases 2017. J. Immunol. Res. 2017, 2789242.

Colafrancesco, S., Ciccacci, C., Priori, R., Latini, A., Picarelli, G., Arienzo, F., Novelli, G., Valesini, G., Perricone, C., Borgiani, P., 2019. STAT4, TRAF3IP2, IL10, and HCP5 polymorphisms in Sjogren's syndrome: association with disease susceptibility and clinical aspects. J. Immunol. Res. 2019, 7682827.

Collins, M.P., Dyck, P.J.B., Hadden, R.D.M., 2019. Update on classification, epidemiology, clinical phenotype and imaging of the nonsystemic vasculitic neuropathies. Curr. Opin. Neurol. 32 (5), 684−695.

Collins, M.P., Hadden, R.D., 2017. The nonsystemic vasculitic neuropathies. Nat. Rev. Neurol. 13 (5), 302−316.

Cotti Piccinelli, S., Carella, G., Frassi, M., Caria, F., Gallo Cassarino, S., Baldelli, E., Marini, M., Tincani, A., Padovani, A., Filosto, M., 2019. Human leukocyte antigens class II in CIDP spectrum neuropathies. J. Neurol. Sci. 407, 116533.

Croft, M., 2009. The role of TNF superfamily members in T-cell function and diseases. Nat. Rev. Immunol. 9 (4), 271−285.

Cruz-Tapias, P., Rojas-Villarraga, A., Maier-Moore, S., Anaya, J.M., 2012. HLA and Sjogren's syndrome susceptibility. A meta-analysis of worldwide studies. Autoimmun. Rev. 11 (4), 281−287.

De Libero, G., Mori, L., 2005. Recognition of lipid antigens by T cells. Nat. Rev. Immunol. 5 (6), 485−496.

Deer, T.R., Naidu, R., Strand, N., Sparks, D., Abd-Elsayed, A., Kalia, H., Hah, J.M., Mehta, P., Sayed, D., Gulati, A., 2020. A review of the bioelectronic implications of stimulation of the peripheral nervous system for chronic pain conditions. Bioelectron. Med. 6, 9.

Dekker, M.J., van den Akker, E.L., Koper, J.W., Manenschijn, L., Geleijns, K., Ruts, L., van Rijs, W., Tio-Gillen, A.P., van Doorn, P.A., Lamberts, S.W., Jacobs, B.C., 2009. Effect of glucocorticoid receptor gene polymorphisms in Guillain-Barre syndrome. J. Peripher. Nerv. Syst. 14 (2), 75−83.

Feeney, D.J., Pollard, J.D., McLeod, J.G., Stewart, G.J., Doran, T.J., 1990. HLA antigens in chronic inflammatory demyelinating polyneuropathy. J. Neurol. Neurosurg. Psychiatry 53 (2), 170−172.

Fekih-Mrissa, N., Mrad, M., Riahi, A., Sayeh, A., Zaouali, J., Gritli, N., Mrissa, R., 2014. Association of HLA-DR/DQ polymorphisms with Guillain-Barre syndrome in Tunisian patients. Clin. Neurol. Neurosurg. 121, 19−22.

Florica, B., Aghdassi, E., Su, J., Gladman, D.D., Urowitz, M.B., Fortin, P.R., 2011. Peripheral neuropathy in patients with systemic lupus erythematosus. Semin. Arthritis Rheum. 41 (2), 203−211.

Geleijns, K., Emonts, M., Laman, J.D., van Rijs, W., van Doorn, P.A., Hermans, P.W., Jacobs, B.C., 2007. Genetic polymorphisms of macrophage-mediators in Guillain-Barre syndrome. J. Neuroimmunol. 190 (1−2), 127−130.

Geleijns, K., Roos, A., Houwing-Duistermaat, J.J., van Rijs, W., Tio-Gillen, A.P., Laman, J.D., van Doorn, P.A., Jacobs, B.C., 2006. Mannose-binding lectin contributes to the severity of Guillain-Barre syndrome. J. Immunol. 177 (6), 4211−4217.

Gorodezky, C., Varela, B., Castro-Escobar, L.E., Chavez-Negrete, A., Escobar-Gutierrez, A., Martinez-Mata, J., 1983. HLA-DR antigens in Mexican patients with Guillain-Barre syndrome. J. Neuroimmunol. 4 (1), 1−7.

Gough, S.C., Simmonds, M.J., 2007. The HLA region and autoimmune disease: associations and mechanisms of action. Curr. Genom. 8 (7), 453−465.

Graf, J., Imboden, J., 2019. Vasculitis and peripheral neuropathy. Curr. Opin. Rheumatol. 31 (1), 40−45.

Graham, D.S.C., Wong, A.K., McHugh, N.J., Whittaker, J.C., Vyse, T.J., 2006. Evidence for unique association signals in SLE at the CD28−CTLA4−ICOS locus in a family-based study. Hum. Mol. Genet. 15 (21), 3195−3205.

Guo, Q., Wang, Y., Xu, D., Nossent, J., Pavlos, N.J., Xu, J., 2018. Rheumatoid arthritis: pathological mechanisms and modern pharmacologic therapies. Bone Res. 6 (1), 15.

Hasan, Z.N., Zalzala, H.H., Mohammedsalih, H.R., Mahdi, B.M., Abid, L.A., Shakir, Z.N., Fadhel, M.J., 2014. Association between human leukocyte antigen-DR and demylinating Guillain-Barre syndrome. Neurosciences 19 (4), 301−305.

Hewagama, A., Richardson, B., 2009. The genetics and epigenetics of autoimmune diseases. J. Autoimmun. 33 (1), 3−11.

Hu, W., Dehmel, T., Pirhonen, J., Hartung, H.P., Kieseier, B.C., 2006. Interleukin 23 in acute inflammatory demyelination of the peripheral nerve. Arch. Neurol. 63 (6), 858−864.

Hu, W., Ren, H., 2011. A meta-analysis of the association of IRF5 polymorphism with systemic lupus erythematosus. Int. J. Immunogenet. 38 (5), 411−417.

Hughes, T., Coit, P., Adler, A., Yilmaz, V., Aksu, K., Duzgun, N., Keser, G., Cefle, A., Yazici, A., Ergen, A., Alpsoy, E., Salvarani, C., Casali, B., Kotter, I., Gutierrez-Achury, J., Wijmenga, C., Direskeneli, H., Saruhan-Direskeneli, G., Sawalha, A.H., 2013. Identification of multiple independent susceptibility loci in the HLA region in Behcet's disease. Nat. Genet. 45 (3), 319−324.

Inshaw, J.R.J., Cutler, A.J., Burren, O.S., Stefana, M.I., Todd, J.A., 2018. Approaches and advances in the genetic causes of autoimmune disease and their implications. Nat. Immunol. 19 (7), 674−684.

Iyer, S.S., Cheng, G., 2012. Role of interleukin 10 transcriptional regulation in inflammation and autoimmune disease. Crit. Rev. Immunol. 32 (1), 23−63.

Jameson, J.L., 2018. Harrison's Principles of Internal Medicine. McGraw-Hill Education.

Jiao, H., Wang, W., Wang, H., Wu, Y., Wang, L., 2012. Tumor necrosis factor alpha 308 G/A polymorphism and Guillain-Barre syndrome risk. Mol. Biol. Rep. 39 (2), 1537−1540.

Jonsen, A., Bengtsson, A.A., Sturfelt, G., Truedsson, L., 2004. Analysis of HLA DR, HLA DQ, C4A, FcgammaRIIa, FcgammaRIIIa, MBL, and IL-1Ra allelic variants in Caucasian systemic lupus erythematosus patients suggests an effect of the combined FcgammaRIIa R/R and IL-1Ra 2/2 genotypes on disease susceptibility. Arthritis Res. Ther. 6 (6), R557–R562.

Kaeley, N., Ahmad, S., Pathania, M., Kakkar, R., 2019. Prevalence and patterns of peripheral neuropathy in patients of rheumatoid arthritis. J. Fam. Med. Prim. Care 8 (1), 22–26.

Karassa, F.B., Trikalinos, T.A., Ioannidis, J.P., 2004. The role of FcgammaRIIA and IIIA polymorphisms in autoimmune diseases. Biomed. Pharmacother. 58 (5), 286–291.

Katz, U., Shoenfeld, Y., Zandman-Goddard, G., 2011. Update on intravenous immunoglobulins (IVIg) mechanisms of action and off- label use in autoimmune diseases. Curr. Pharmaceut. Des. 17 (29), 3166–3175.

Kaul, A., Gordon, C., Crow, M.K., Touma, Z., Urowitz, M.B., van Vollenhoven, R., Ruiz-Irastorza, G., Hughes, G., 2016. Systemic lupus erythematosus. Nat. Rev. Dis. Prim. 2 (1), 16039.

Khor, C.C., Davila, S., Breunis, W.B., Lee, Y.C., Shimizu, C., Wright, V.J., Yeung, R.S., Tan, D.E., Sim, K.S., Wang, J.J., Wong, T.Y., Pang, J., Mitchell, P., Cimaz, R., Dahdah, N., Cheung, Y.F., Huang, G.Y., Yang, W., Park, I.S., Lee, J.K., Wu, J.Y., Levin, M., Burns, J.C., Burgner, D., Kuijpers, T.W., Hibberd, M.L., Hong Kong-Shanghai Kawasaki Disease Genetics, C., Korean Kawasaki Disease Genetics, C., Taiwan Kawasaki Disease Genetics, C., International Kawasaki Disease Genetics, C., Consortium, U.S.K.D.G., Blue Mountains Eye, S., 2011. Genome-wide association study identifies FCGR2A as a susceptibility locus for Kawasaki disease. Nat. Genet. 43 (12), 1241–1246.

Kim, J.J., Yun, S.W., Yu, J.J., Yoon, K.L., Lee, K.Y., Kil, H.R., Kim, G.B., Han, M.K., Song, M.S., Lee, H.D., Ha, K.S., Sohn, S., Johnson, T.A., Takahashi, A., Kubo, M., Tsunoda, T., Ito, K., Onouchi, Y., Hong, Y.M., Jang, G.Y., Lee, J.K., Korean Kawasaki Disease Genetics Consortium, 2017. A genome-wide association analysis identifies NMNAT2 and HCP5 as susceptibility loci for Kawasaki disease. J. Hum. Genet. 62 (12), 1023–1029.

Kim, S.J., Lee, S., Park, C., Seo, J.S., Kim, J.I., Yu, H.G., 2013. Targeted resequencing of candidate genes reveals novel variants associated with severe Behcet's uveitis. Exp. Mol. Med. 45 (10), e49.

Kirino, Y., Bertsias, G., Ishigatsubo, Y., Mizuki, N., Tugal-Tutkun, I., Seyahi, E., Ozyazgan, Y., Sacli, F.S., Erer, B., Inoko, H., Emrence, Z., Cakar, A., Abaci, N., Ustek, D., Satorius, C., Ueda, A., Takeno, M., Kim, Y., Wood, G.M., Ombrello, M.J., Meguro, A., Gul, A., Remmers, E.F., Kastner, D.L., 2013a. Genome-wide association analysis identifies new susceptibility loci for Behcet's disease and epistasis between HLA-B*51 and ERAP1. Nat. Genet. 45 (2), 202–207.

Kirino, Y., Zhou, Q., Ishigatsubo, Y., Mizuki, N., Tugal-Tutkun, I., Seyahi, E., Ozyazgan, Y., Ugurlu, S., Erer, B., Abaci, N., Ustek, D., Meguro, A., Ueda, A., Takeno, M.,

Inoko, H., Ombrello, M.J., Satorius, C.L., Maskeri, B., Mullikin, J.C., Sun, H.W., Gutierrez-Cruz, G., Kim, Y., Wilson, A.F., Kastner, D.L., Gul, A., Remmers, E.F., 2013b. Targeted resequencing implicates the familial Mediterranean fever gene MEFV and the toll-like receptor 4 gene TLR4 in Behcet disease. Proc. Natl. Acad. Sci. U.S.A. 110 (20), 8134–8139.

Kivity, S., Arango, M.T., Ehrenfeld, M., Tehori, O., Shoenfeld, Y., Anaya, J.M., Agmon-Levin, N., 2014. Infection and autoimmunity in Sjogren's syndrome: a clinical study and comprehensive review. J. Autoimmun. 51, 17–22.

Knierim, M.T., Rissler, R., Dorner, V., Maedche, A., Weinhardt, C., 2018. The Psychophysiology of Flow: A Systematic Review of Peripheral Nervous System Features. Springer International Publishing, Cham.

Kobayashi, S., Ikari, K., Kaneko, H., Kochi, Y., Yamamoto, K., Shimane, K., Nakamura, Y., Toyama, Y., Mochizuki, T., Tsukahara, S., Kawaguchi, Y., Terai, C., Hara, M., Tomatsu, T., Yamanaka, H., Horiuchi, T., Tao, K., Yasutomo, K., Hamada, D., Yasui, N., Inoue, H., Itakura, M., Okamoto, H., Kamatani, N., Momohara, S., 2008. Association of STAT4 with susceptibility to rheumatoid arthritis and systemic lupus erythematosus in the Japanese population, 58 (7), 1940–1946.

Koike, H., Sobue, G., 2013. Sjogren's syndrome-associated neuropathy. Brain Nerve 65 (11), 1333–1342.

Korman, B.D., Alba, M.I., Le, J.M., Alevizos, I., Smith, J.A., Nikolov, N.P., Kastner, D.L., Remmers, E.F., Illei, G.G., 2008. Variant form of STAT4 is associated with primary Sjögren's syndrome. Gene Immun. 9 (3), 267–270.

Kyogoku, C., Langefeld, C.D., Ortmann, W.A., Lee, A., Selby, S., Carlton, V.E., Chang, M., Ramos, P., Baechler, E.C., Batliwalla, F.M., Novitzke, J., Williams, A.H., Gillett, C., Rodine, P., Graham, R.R., Ardlie, K.G., Gaffney, P.M., Moser, K.L., Petri, M., Begovich, A.B., Gregersen, P.K., Behrens, T.W., 2004. Genetic association of the R620W polymorphism of protein tyrosine phosphatase PTPN22 with human SLE. Am. J. Hum. Genet. 75 (3), 504–507.

Lee, K.S., Kronbichler, A., Pereira Vasconcelos, D.F., Pereira da Silva, F.R., Ko, Y., Oh, Y.S., Eisenhut, M., Merkel, P.A., Jayne, D., Amos, C.I., Siminovitch, K.A., Rahmattulla, C., Lee, K.H., Shin, J.I., 2019. Genetic variants in antineutrophil cytoplasmic antibody-associated vasculitis: a bayesian approach and systematic review. J. Clin. Med. 8 (2), 266.

Leonhard, S.E., Mandarakas, M.R., Gondim, F.A.A., Bateman, K., Ferreira, M.L.B., Cornblath, D.R., van Doorn, P.A., Dourado, M.E., Hughes, R.A.C., Islam, B., Kusunoki, S., Pardo, C.A., Reisin, R., Sejvar, J.J., Shahrizaila, N., Soares, C., Umapathi, T., Wang, Y., Yiu, E.M., Willison, H.J., Jacobs, B.C., 2019. Diagnosis and management of Guillain-Barre syndrome in ten steps. Nat. Rev. Neurol. 15 (11), 671–683.

Lyons, P.A., Peters, J.E., Alberici, F., Liley, J., Coulson, R.M.R., Astle, W., Baldini, C., Bonatti, F., Cid, M.C., Elding, H., Emmi, G., Epplen, J., Guillevin, L., Jayne, D.R.W., Jiang, T., Gunnarsson, I., Lamprecht, P., Leslie, S., Little, M.A., Martorana, D., Moosig, F., Neumann, T.,

Ohlsson, S., Quickert, S., Ramirez, G.A., Rewerska, B., Schett, G., Sinico, R.A., Szczeklik, W., Tesar, V., Vukcevic, D., European Vasculitis Genetics, C., Terrier, B., Watts, R.A., Vaglio, A., Holle, J.U., Wallace, C., Smith, K.G.C., 2019. Genome-wide association study of eosinophilic granulomatosis with polyangiitis reveals genomic loci stratified by ANCA status. Nat. Commun. 10 (1), 5120.

Magira, E.E., Papaioakim, M., Nachamkin, I., Asbury, A.K., Li, C.Y., Ho, T.W., Griffin, J.W., McKhann, G.M., Monos, D.S., 2003. Differential distribution of HLA-DQ beta/DR beta epitopes in the two forms of Guillain-Barre syndrome, acute motor axonal neuropathy and acute inflammatory demyelinating polyneuropathy (AIDP): identification of DQ beta epitopes associated with susceptibility to and protection from AIDP. J. Immunol. 170 (6), 3074−3080.

Martinez-Martinez, L., Lleixa, M.C., Boera-Carnicero, G., Cortese, A., Devaux, J., Siles, A., Rajabally, Y., Martinez-Pineiro, A., Carvajal, A., Pardo, J., Delmont, E., Attarian, S., Diaz-Manera, J., Callegari, I., Marchioni, E., Franciotta, D., Benedetti, L., Lauria, G., de la Calle Martin, O., Juarez, C., Illa, I., Querol, L., 2017. Anti-NF155 chronic inflammatory demyelinating polyradiculoneuropathy strongly associates to HLA-DRB15. J. Neuroinflammation 14 (1), 224.

Matzaraki, V., Kumar, V., Wijmenga, C., Zhernakova, A., 2017. The MHC locus and genetic susceptibility to autoimmune and infectious diseases. Genome Biol. 18 (1), 76.

McCombe, P.A., Csurhes, P.A., Greer, J.M., 2006. Studies of HLA associations in male and female patients with Guillain-Barre syndrome (GBS) and chronic inflammatory demyelinating polyradiculoneuropathy (CIDP). J. Neuroimmunol. 180 (1−2), 172−177.

Miyagawa, H., Yamai, M., Sakaguchi, D., Kiyohara, C., Tsukamoto, H., Kimoto, Y., Nakamura, T., Lee, J.H., Tsai, C.Y., Chiang, B.L., Shimoda, T., Harada, M., Tahira, T., Hayashi, K., Horiuchi, T., 2008. Association of polymorphisms in complement component C3 gene with susceptibility to systemic lupus erythematosus. Rheumatology 47 (2), 158−164.

Moser, K.L., Kelly, J.A., Lessard, C.J., Harley, J.B., 2009. Recent insights into the genetic basis of systemic lupus erythematosus. Gene Immun. 10 (5), 373−379.

Myhr, K.M., Vagnes, K.S., Maroy, T.H., Aarseth, J.H., Nyland, H.I., Vedeler, C.A., 2003. Interleukin-10 promoter polymorphisms in patients with Guillain-Barre syndrome. J. Neuroimmunol. 139 (1−2), 81−83.

Nadkar, M.Y., Agarwal, R., Samant, R.S., Chhugani, S.J., Idgunji, S.S., Iyer, S., Borges, N.E., 2001. Neuropathy in rheumatoid arthritis. J. Assoc. Phys. India 49, 217−220.

Ng, T.H., Britton, G.J., Hill, E.V., Verhagen, J., Burton, B.R., Wraith, D.C., 2013. Regulation of adaptive immunity; the role of interleukin-10. Front. Immunol. 4, 129.

Nyati, K.K., Prasad, K.N., Verma, A., Paliwal, V.K., 2010. Correlation of matrix metalloproteinases-2 and -9 with proinflammatory cytokines in Guillain-Barre syndrome. J. Neurosci. Res. 88 (16), 3540−3546.

Ombrello, M.J., Kirino, Y., de Bakker, P.I., Gul, A., Kastner, D.L., Remmers, E.F., 2014. Behcet disease-associated MHC class I residues implicate antigen binding and regulation of cell-mediated cytotoxicity. Proc. Natl. Acad. Sci. U.S.A. 111 (24), 8867−8872.

Omdal, R., Henriksen, O.A., Mellgren, S.I., Husby, G., 1991. Peripheral neuropathy in systemic lupus erythematosus. Neurology 41 (6), 808−811.

Onouchi, Y., Ozaki, K., Burns, J.C., Shimizu, C., Terai, M., Hamada, H., Honda, T., Suzuki, H., Suenaga, T., Takeuchi, T., Yoshikawa, N., Suzuki, Y., Yasukawa, K., Ebata, R., Higashi, K., Saji, T., Kemmotsu, Y., Takatsuki, S., Ouchi, K., Kishi, F., Yoshikawa, T., Nagai, T., Hamamoto, K., Sato, Y., Honda, A., Kobayashi, H., Sato, J., Shibuta, S., Miyawaki, M., Oishi, K., Yamaga, H., Aoyagi, N., Iwahashi, S., Miyashita, R., Murata, Y., Sasago, K., Takahashi, A., Kamatani, N., Kubo, M., Tsunoda, T., Hata, A., Nakamura, Y., Tanaka, T., Japan Kawasaki Disease Genome Consortium, US Kawasaki Disease Genetics Consortium, 2012. A genome-wide association study identifies three new risk loci for Kawasaki disease. Nat. Genet. 44 (5), 517−521.

Oomatia, A., Fang, H., Petri, M., Birnbaum, J., 2014. Peripheral neuropathies in systemic lupus erythematosus: clinical features, disease associations, and immunologic characteristics evaluated over a twenty-five-year study period. Arthritis Rheum. 66 (4), 1000−1009.

Orozco, G., Pascual-Salcedo, D., Lopez-Nevot, M.A., Cobo, T., Cabezon, A., Martin-Mola, E., Balsa, A., Martin, J., 2008. Auto-antibodies, HLA and PTPN22: susceptibility markers for rheumatoid arthritis. Rheumatology 47 (2), 138−141.

Orozco, G., Rueda, B., Martin, J., 2006. Genetic basis of rheumatoid arthritis. Biomed. Pharmacother. 60 (10), 656−662.

Perrier, S., Coussediere, C., Dubost, J.J., Albuisson, E., Sauvezie, B., 1998. IL-1 receptor antagonist (IL-1RA) gene polymorphism in Sjogren's syndrome and rheumatoid arthritis. Clin. Immunol. Immunopathol. 87 (3), 309−313.

Prokunina, L., Castillejo-Lopez, C., Oberg, F., Gunnarsson, I., Berg, L., Magnusson, V., Brookes, A.J., Tentler, D., Kristjansdottir, H., Grondal, G., Bolstad, A.I., Svenungsson, E., Lundberg, I., Sturfelt, G., Jonssen, A., Truedsson, L., Lima, G., Alcocer-Varela, J., Jonsson, R., Gyllensten, U.B., Harley, J.B., Alarcon-Segovia, D., Steinsson, K., Alarcon-Riquelme, M.E., 2002. A regulatory polymorphism in PDCD1 is associated with susceptibility to systemic lupus erythematosus in humans. Nat. Genet. 32 (4), 666−669.

Rahmattulla, C., Mooyaart, A.L., van Hooven, D., Schoones, J.W., Bruijn, J.A., Dekkers, O.M., European Vasculitis Genetics Consortium, Bajema, I.M., 2016. Genetic variants in ANCA-associated vasculitis: a meta-analysis. Ann. Rheum. Dis. 75 (9), 1687−1692.

Renauer, P., Sawalha, A.H., 2017. The genetics of Takayasu arteritis. Presse Med. 46 (7−8 Pt 2), e179−e187.

Rosado, S., Rua-Figueroa, I., Vargas, J.A., Garcia-Laorden, M.I., Losada-Fernandez, I., Martin-Donaire, T., Perez-Chacon, G., Rodriguez-Gallego, C., Naranjo-

Hernandez, A., Ojeda-Bruno, S., Citores, M.J., Perez-Aciego, P., 2008. Interleukin-10 promoter polymorphisms in patients with systemic lupus erythematosus from the Canary Islands. Int. J. Immunogenet. 35 (3), 235–242.

Rose, N.R., Bona, C., 1993. Defining criteria for autoimmune diseases (Witebsky's postulates revisited). Immunol. Today 14 (9), 426–430.

Said, G., Lacroix, C., 2005. Primary and secondary vasculitic neuropathy. J. Neurol. 252 (6), 633–641.

Schaublin, G.A., Michet Jr., C.J., Dyck, P.J., Burns, T.M., 2005. An update on the classification and treatment of vasculitic neuropathy. Lancet Neurol. 4 (12), 853–865.

Sestak, A.L., Furnrohr, B.G., Harley, J.B., Merrill, J.T., Namjou, B., 2011. The genetics of systemic lupus erythematosus and implications for targeted therapy. Ann. Rheum. Dis. 70 (Suppl. 1), i37–43.

Shahine, A., Reinink, P., Reijneveld, J.F., Gras, S., Holzheimer, M., Cheng, T.Y., Minnaard, A.J., Altman, J.D., Lenz, S., Prandi, J., Kubler-Kielb, J., Moody, D.B., Rossjohn, J., Van Rhijn, I., 2019. A T-cell receptor escape channel allows broad T-cell response to CD1b and membrane phospholipids. Nat. Commun. 10 (1), 56.

Sigurdsson, S., Nordmark, G., Garnier, S., Grundberg, E., Kwan, T., Nilsson, O., Eloranta, M.L., Gunnarsson, I., Svenungsson, E., Sturfelt, G., Bengtsson, A.A., Jonsen, A., Truedsson, L., Rantapaa-Dahlqvist, S., Eriksson, C., Alm, G., Goring, H.H., Pastinen, T., Syvanen, A.C., Ronnblom, L., 2008. A risk haplotype of STAT4 for systemic lupus erythematosus is over-expressed, correlates with anti-dsDNA and shows additive effects with two risk alleles of IRF5. Hum. Mol. Genet. 17 (18), 2868–2876.

Sim, M.K., Kim, D.Y., Yoon, J., Park, D.H., Kim, Y.G., 2014. Assessment of peripheral neuropathy in patients with rheumatoid arthritis who complain of neurologic symptoms. Ann. Rehabil. Med. 38 (2), 249–255.

Sinha, S., Prasad, K.N., Jain, D., Nyati, K.K., Pradhan, S., Agrawal, S., 2010. Immunoglobulin IgG Fc-receptor polymorphisms and HLA class II molecules in Guillain-Barre syndrome. Acta Neurol. Scand. 122 (1), 21–26.

Slot, M.C., Sokolowska, M.G., Savelkouls, K.G., Janssen, R.G., Damoiseaux, J.G., Tervaert, J.W., 2008. Immunoregulatory gene polymorphisms are associated with ANCA-related vasculitis. Clin. Immunol. 128 (1), 39–45.

Smith, D.A., Germolec, D.R., 1999. Introduction to immunology and autoimmunity. Environ. Health Perspect. 107 (Suppl. 5), 661–665.

Smolen, J.S., Aletaha, D., McInnes, I.B., 2016. Rheumatoid arthritis. Lancet 388 (10055), 2023–2038.

Solus, J.F., Chung, C.P., Oeser, A., Li, C., Rho, Y.H., Bradley, K.M., Kawai, V.K., Smith, J.R., Stein, C.M., 2015. Genetics of serum concentration of IL-6 and TNFalpha in systemic lupus erythematosus and rheumatoid arthritis: a candidate gene analysis. Clin. Rheumatol. 34 (8), 1375–1382.

Speirs, C., Fielder, A.H., Chapel, H., Davey, N.J., Batchelor, J.R., 1989. Complement system protein C4 and susceptibility to hydralazine-induced systemic lupus erythematosus. Lancet 1 (8644), 922–924.

Suarez, A., Lopez, P., Mozo, L., Gutierrez, C., 2005. Differential effect of IL10 and TNF{alpha} genotypes on determining susceptibility to discoid and systemic lupus erythematosus. Ann. Rheum. Dis. 64 (11), 1605–1610.

Sullivan, K.E., Suriano, A., Dietzmann, K., Lin, J., Goldman, D., Petri, M.A., 2007. The TNFalpha locus is altered in monocytes from patients with systemic lupus erythematosus. Clin. Immunol. 123 (1), 74–81.

Tamirou, F., Arnaud, L., Talarico, R., Scire, C.A., Alexander, T., Amoura, Z., Avcin, T., Bortoluzzi, A., Cervera, R., Conti, F., Cornet, A., Devilliers, H., Doria, A., Frassi, M., Fredi, M., Govoni, M., Houssiau, F., Llado, A., Macieira, C., Martin, T., Massaro, L., Moraes-Fontes, M.F., Pamfil, C., Paolino, S., Tani, C., Tas, S.W., Tektonidou, M., Tincani, A., Van Vollenhoven, R.F., Bombardieri, S., Burmester, G., Eurico, F.J., Galetti, I., Hachulla, E., Mueller-Ladner, U., Schneider, M., Smith, V., Cutolo, M., Mosca, M., Costedoat-Chalumeau, N., 2018. Systemic lupus erythematosus: state of the art on clinical practice guidelines. RMD Open 4 (2), e000793.

Tarassi, K., Carthy, D., Papasteriades, C., Boki, K., Nikolopoulou, N., Carcassi, C., Ollier, W.E., Hajeer, A.H., 1998. HLA-TNF haplotype heterogeneity in Greek SLE patients. Clin. Exp. Rheumatol. 16 (1), 66–68.

Tarn, J.R., Howard-Tripp, N., Lendrem, D.W., Mariette, X., Saraux, A., Devauchelle-Pensec, V., Seror, R., Skelton, A.J., James, K., McMeekin, P., Al-Ali, S., Hackett, K.L., Lendrem, B.C., Hargreaves, B., Casement, J., Mitchell, S., Bowman, S.J., Price, E., Pease, C.T., Emery, P., Lanyon, P., Hunter, J., Gupta, M., Bombardieri, M., Sutcliffe, N., Pitzalis, C., McLaren, J., Cooper, A., Regan, M., Giles, I., Isenberg, D., Saravanan, V., Coady, D., Dasgupta, B., McHugh, N., Young-Min, S., Moots, R., Gendi, N., Akil, M., Griffiths, B., Johnsen, S.J.A., Norheim, K.B., Omdal, R., Stocken, D., Everett, C., Fernandez, C., Isaacs, J.D., Gottenberg, J.-E., Ng, W.-F., Devauchelle-Pensec, V., Dieude, P., Dubost, J.J., Fauchais, A.-L., Goeb, V., Hachulla, E., Larroche, C., Le Guern, V., Morel, J., Perdriger, A., Puéchal, X., Rist, S., Sen, D., Sibilia, J., Vittecoq, O., Benessiano, J., Tubiana, S., Inamo, K., Gaete, S., Batouche, D., Molinari, D., Randrianandrasana, M., Pane, I., Abbe, A., Baron, G., Ravaud, P., Gottenberg, J.-E., Ravaud, P., Puéchal, X., Le Guern, V., Sibilia, J., Larroche, C., Saraux, A., Devauchelle-Pensec, V., Morel, J., Hayem, G., Hatron, P., Perdriger, A., Sene, D., Zarnitsky, C., Batouche, D., Furlan, V., Benessiano, J., Perrodeau, E., Seror, R., Mariette, X., Brown, S., Navarro, N.C., Pitzalis, C., Emery, P., Pavitt, S., et al. 2019. Symptom-based stratification of patients with primary Sjögren's syndrome: multi-dimensional characterisation of international observational cohorts and reanalyses of randomised clinical trials. Lancet Rheumatol. 1 (2), e85–e94.

Taylor, K.E., Remmers, E.F., Lee, A.T., Ortmann, W.A., Plenge, R.M., Tian, C., Chung, S.A., Nititham, J., Hom, G., Kao, A.H., Demirci, F.Y., Kamboh, M.I., Petri, M., Manzi, S., Kastner, D.L., Seldin, M.F., Gregersen, P.K., Behrens, T.W., Criswell, L.A., 2008. Specificity of the

STAT4 genetic association for severe disease manifestations of systemic lupus erythematosus. PLoS Genet. 4 (5), e1000084.

Terao, C., 2016. Revisited HLA and non-HLA genetics of Takayasu arteritis–where are we? J. Hum. Genet. 61 (1), 27−32.

Terao, C., Yoshifuji, H., Matsumura, T., Naruse, T.K., Ishii, T., Nakaoka, Y., Kirino, Y., Matsuo, K., Origuchi, T., Shimizu, M., Maejima, Y., Amiya, E., Tamura, N., Kawaguchi, T., Takahashi, M., Setoh, K., Ohmura, K., Watanabe, R., Horita, T., Atsumi, T., Matsukura, M., Miyata, T., Kochi, Y., Suda, T., Tanemoto, K., Meguro, A., Okada, Y., Ogimoto, A., Yamamoto, M., Takahashi, H., Nakayamada, S., Saito, K., Kuwana, M., Mizuki, N., Tabara, Y., Ueda, A., Komuro, I., Kimura, A., Isobe, M., Mimori, T., Matsuda, F., 2018. Genetic determinants and an epistasis of LILRA3 and HLA-B*52 in Takayasu arteritis, 115 (51), 13045−13050.

Theofilopoulos, A.N., Kono, D.H., Baccala, R., 2017. The multiple pathways to autoimmunity. Nat. Immunol. 18 (7), 716−724.

Thorsby, E., Lie, B.A., 2005. HLA associated genetic predisposition to autoimmune diseases: genes involved and possible mechanisms. Transpl. Immunol. 14 (3−4), 175−182.

Tsai, F.J., Lee, Y.C., Chang, J.S., Huang, L.M., Huang, F.Y., Chiu, N.C., Chen, M.R., Chi, H., Lee, Y.J., Chang, L.C., Liu, Y.M., Wang, H.H., Chen, C.H., Chen, Y.T., Wu, J.Y., 2011. Identification of novel susceptibility Loci for kawasaki disease in a Han Chinese population by a genome-wide association study. PLoS One 6 (2), e16853.

Tsokos, G.C., 2011. Systemic lupus erythematosus. N. Engl. J. Med. 365 (22), 2110−2121.

Turner, M.D., Nedjai, B., Hurst, T., Pennington, D.J., 2014. Cytokines and chemokines: at the crossroads of cell signalling and inflammatory disease. Biochim. Biophys. Acta 1843 (11), 2563−2582.

van den Berg, B., Walgaard, C., Drenthen, J., Fokke, C., Jacobs, B.C., van Doorn, P.A., 2014. Guillain-Barre syndrome: pathogenesis, diagnosis, treatment and prognosis. Nat. Rev. Neurol. 10 (8), 469−482.

van Sorge, N.M., van der Pol, W.L., Jansen, M.D., Geleijns, K.P., Kalmijn, S., Hughes, R.A., Rees, J.H., Pritchard, J., Vedeler, C.A., Myhr, K.M., Shaw, C., van Schaik, I.N.,

Wokke, J.H., van Doorn, P.A., Jacobs, B.C., van de Winkel, J.G., van den Berg, L.H., 2005. Severity of Guillain-Barre syndrome is associated with Fc gamma Receptor III polymorphisms. J. Neuroimmunol. 162 (1−2), 157−164.

Vandenbroeck, K., 2012. Cytokine gene polymorphisms and human autoimmune disease in the era of genome-wide association studies. J. Interferon Cytokine Res. 32 (4), 139−151.

Vaughan, R.W., Adam, A.M., Gray, I.A., Hughes, R.A., Sanders, E.A., van Dam, M., Welsh, K.I., 1990. Major histocompatibility complex class I and class II polymorphism in chronic idiopathic demyelinating polyradiculoneuropathy. J. Neuroimmunol. 27 (2−3), 149−153.

Wallace, G.R., 2014. HLA-B*51 the primary risk in Behcet disease. Proc. Natl. Acad. Sci. U.S.A. 111 (24), 8706−8707.

Willison, H.J., Jacobs, B.C., van Doorn, P.A., 2016. Guillain-Barré syndrome. Lancet 388 (10045), 717−727.

Wu, L.Y., Zhou, Y., Qin, C., Hu, B.L., 2012. The effect of TNF-alpha, FcγR and CD1 polymorphisms on Guillain-Barré syndrome risk: evidences from a meta-analysis. J Neuroimmunol 243 (1−2), 18−24. https://doi.org/10.1016/j.jneuroim.2011.12.003.

Xianbin, W., Mingyu, W., Dong, X., Huiying, L., Yan, X., Fengchun, Z., Xiaofeng, Z., 2015. Peripheral neuropathies due to systemic lupus erythematosus in China. Medicine (Baltim.) 94 (11), e625.

Yang, W., Ng, P., Zhao, M., Hirankarn, N., Lau, C.S., Mok, C.C., Chan, T.M., Wong, R.W., Lee, K.W., Mok, M.Y., Wong, S.N., Avihingsanon, Y., Lee, T.L., Ho, M.H., Lee, P.P., Wong, W.H., Lau, Y.L., 2009. Population differences in SLE susceptibility genes: STAT4 and BLK, but not PXK, are associated with systemic lupus erythematosus in Hong Kong Chinese. Gene Immun. 10 (3), 219−226.

Zhang, J., Dong, H., Li, B., Li, C.Y., Guo, L., 2007. Association of tumor necrosis factor polymorphisms with Guillain-Barre syndrome. Eur. Neurol. 58 (1), 21−25.

Zhernakova, A., Eerligh, P., Barrera, P., Wesoly, J.Z., Huizinga, T.W., Roep, B.O., Wijmenga, C., Koeleman, B.P., 2005. CTLA4 is differentially associated with autoimmune diseases in the Dutch population. Hum. Genet. 118 (1), 58−66.

Index

Note: Page numbers followed by "f" indicate figures and "t" indicate tables.

Printed in the United States
By Bookmasters